Benchmark Papers
in Acoustics

Series Editor: R. Bruce Lindsay
Brown University

Published Volumes and Volumes in Preparation

UNDERWATER SOUND
   Vernon M. Albers
ACOUSTICS: Historical and Philosophical Development
   R. Bruce Lindsay
SPEECH SYNTHESIS
   James L. Flanagan and L. R. Rabiner
PHYSICAL ACOUSTICS
   . R. Bruce Lindsay
ARCHITECTURAL ACOUSTICS
   Thomas D. Northwood
MUSICAL ACOUSTICS: The Violin Family
   Carleen Hutchins
PSYCHOLOGICAL ACOUSTICS
   Arnold M. Small, Jr.
PHYSIOLOGICAL ACOUSTICS
   Arnold M. Small, Jr., and Joel S. Wernick
LIGHT AND SOUND INTERACTION
   Osman K. Mawardi
VIBRATION PROBLEMS
   Arturs Kalnins
ACOUSTICAL INSTRUMENTATION
   Benjamin B. Bauer
NOISE AND NOISE CONTROL
   Lewis S. Goodfriend

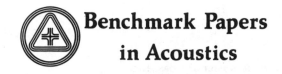

Benchmark Papers
in Acoustics

———— A *BENCHMARK* TM Books Series ————

ACOUSTICS:

Historical and
Philosophical Development

Edited by
R. BRUCE LINDSAY
Brown University

Dowden, Hutchinson
& Ross, Inc.
Stroudsburg, Pennsylvania

Copyright © by **Dowden, Hutchinson & Ross, Inc.**
Library of Congress Catalog Card Number: 72–90974
ISBN: 0–87933–015–5

Library of Congress Cataloging in Publication Data

Lindsay, Robert Bruce, 1900–   comp.
  Acoustics: historical and philosophical development.

  (Benchmark papers in acoustics)
  Thirty-nine articles, some of which have been
translated into English for the first time from the
French, German or Latin.
  1. Sound—History—Sources. 2. Sound—Early works
to 1800. I.   Title.
QC221.7.L56   534'.09   72–90974
ISBN 0–87933–015–5

Manufactured in the United States of America.

Exclusive distributor outside the United States and Canada:
John Wiley & Sons, Inc.

Acknowledgment
and Permissions

ACKNOWLEDGEMENT

Dover Publications, Inc.—*Collected Papers on Acoustics*
   "Reverberation"

Dover Publications, Inc.—*On the Sensation of Tone*
   Selections from "On the Sensation of Tone"

Dover Publications, Inc.—*Scientific Papers of Lord Rayleigh*
   "On the Theory of Resonance"
   "Our Perception of the Direction of a Source of Sound"
   "On the Application of the Principle of Reciprocity to Acoustics"
   "On the Physics of Media That are Composed of Free and Perfectly Elastic Molecules in a State
      of Motion"
   "On the Cooling of Air by Radiation and Conduction, and on the Propagation of Sound"

Dover Publications, Inc.—*The Ten Books on Architecture*
   "Acoustics of the Theater"

PERMISSIONS

The following papers have been reprinted with the permission of the authors and copyright owners.

American Institute of Physics—*Journal of the Acoustical Society of America*
   "The Story of Acoustics"

Cambridge University Press—*De Anima*
   Selection from *De Anima*

The Clarendon Press—*The Oxford Translation of Aristotle*
   Selection from *De Audibilibus*

New York Academy of Medicine—*Treatise on the Diseases of the Chest*
   Selection from "Treatise on the Diseases of the Chest"

Series Editor's Preface

The "Benchmark Papers in Acoustics" constitute a series of volumes that make available to the reader in carefully organized form important papers in all branches of acoustics. The literature of acoustics is vast in extent and much of it, particularly the earlier part, is inaccessible to the average acoustical scientist and engineer. These volumes aim to provide a practical introduction to this literature, since each volume offers an expert's selection of the seminal papers in a given branch of the subject, that is, those papers which have significantly influenced the development of that branch in a certain direction and introduced concepts and methods that possess basic utility in modern acoustics as a whole. Each volume provides a convenient and economical summary of results as well as a foundation for further study for both the person familiar with the field and the person who wishes to become acquainted with it.

Each volume has been organized and edited by an authority in the area to which it pertains. In each volume there is provided an editorial introduction summarizing the technical significance of the field being covered. Each article is accompanied by editorial commentary, with necessary explanatory notes, and an adequate index is provided for ready reference. Articles in languages other than English are either translated or abstracted in English. It is the hope of the publisher and editor that these volumes will constitute a working library of the most important technical literature in acoustics of value to students and research workers.

The present volume, *Acoustics: Historical and Philosophical Development*, has been edited by the series editor. It is intended to serve as an introduction to the series as a whole, in the sense that it emphasizes through its 39 articles the historical and philosophical growth of the whole subject from very early times up to approximately 1900. The nature of the book is discussed in greater detail in the Introductory Essay.

R. Bruce Lindsay

Editor's Acknowledgment

I am deeply indebted to Patricia Galkowski of the Sciences Library of Brown University for help in connection with the location of source material. Acknowledgment is also made of the assistance rendered by the Photo Laboratory of Brown University and the Library of the Massachusetts Institute of Technology in connection with the microfilming of material. I am very grateful to Susan Desilets Proto of the Department of Physics of Brown University for the typing of translated material.

My greatest debt is to those great workers of the past who by their labors have created the science of acoustics.

R. B. L.

Contents

Contents by Author

Introduction
Acoustics: Science, Technology, and Art

Acoustics is the name given to that branch of science which deals with the phenomena of sound. Appropriately enough in a certain sense the word acoustics comes from the Greek meaning hearing, though audible sound now forms only a small part of its field of application. What is sound? When a person opens his mouth and speaks, he is said to utter sound and another person in the vicinity, with so-called normal hearing, is said to hear him. The key idea is motion, motion which is first produced in the air near the mouth of the speaker and is later reproduced near the ear of the listener through the agency of wave propagation.

We live immersed in a world of sound, produced not only by ourselves and other living things but also by inanimate nature on all sides. It is not surprising then that acoustics has a long history and that as a subject it appears in many guises, impinging as it does on such a wide variety of aspects of our total human experience.

As a way of looking at and manipulating human experience, acoustics functions as a science, a technology, and an art. What does this mean? Acoustics is a science in the sense that it strives to describe, create, and understand a portion of human experience. It describes by seeking to establish order and regularity in the world of sound and ultimately talking about this by means of acoustical laws, i.e., statements of patterns of experience like the relation connecting the reverberation time in a room with its volume and the amount of absorbing material in it. It creates by establishing through experiment new and previously undetected sound phenomena, such as, for example, the dependence of the range of an underwater sound signal on the frequency and the mode of variation of sound velocity with depth. Finally it seeks to understand by developing theories of sound propagation, as for example, the one relating propagation to the molecular constitution of the medium. In these respects, acoustics is like all science, and plays its role in the unified structure of science as a whole.

But this does not exhaust the impact of acoustics on human thought and activity.

1

For acoustics is also a branch of technology, which is deliberate activity on the part. of man to modify his environment so as to make living more comfortable and interesting. The scientific principles of acoustics have been applied, for example, to the recording and reproduction of sound, thus creating a vast new industry which has served to bring the pleasures of music to millions who normally would be able to hear it but rarely.

Finally, acoustics is not only a science and a technology but also an art. This is shown by its concern for music, out of which indeed the science of acoustics developed and whose association with technical acoustics is becoming ever closer.

It is clear that acoustics is a wide-ranging discipline. Its many ramifications are well displayed in the accompanying circular chart, which will repay attention. The

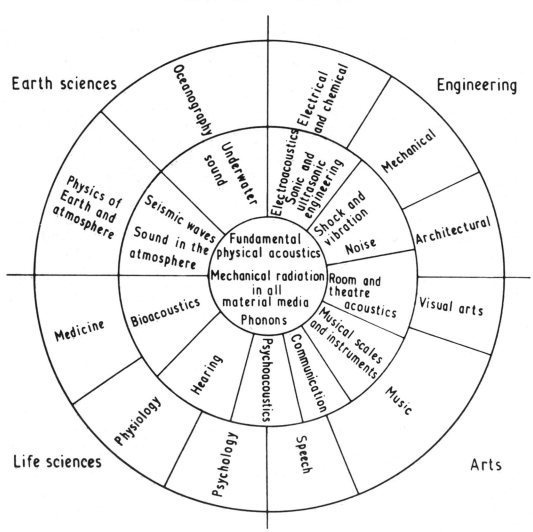

center circle represents fundamental physical acoustics, the basis of all aspects of sound. Sound is essentially mechanical radiation in material media. By mechanical radiation we mean the propagation of disturbances in matter which are connected with relative motions of the points of the medium. This is to be distinguished from electromagnetic radiation in which the propagated disturbances are changes in an electromagnetic field. In the term mechanical radiation we also include the origin of the disturbance, that is, the source of the sound, as well as its reception by suitable artificial devices in addition to the ear.

Surrounding the inner circle are two annular rings, containing wedge-shaped sections. The annular ring region is further divided into four quadrants corresponding to the engineering, earth sciences, life sciences, and arts aspects of the subject. The segments in the first annular ring refer to the various branches of acoustics into which acoustical research and the associated professional literature are divided. Indeed, the various captions here correspond rather closely to the index classification for an acoustical journal such as the *Journal of the Acoustical Society of America*.

The outer annular ring contains captions relating to the various technical and artistic fields to which acoustics in its many branches has been and is being applied. Even a casual glance at the chart serves to emphasize the enormous range of application of acoustics. There is scarcely a phase of human experience into which acoustics does not enter in some significant measure. Its interdisciplinary character is well brought out when we note that it not only enters into oceanography through underwater sound, but also into speech and music through psychoacoustics and general communication principles. It not only impinges on mechanical engineering via its emphasis on vibration and shock phenomena, but it makes great contributions to medicine through the agency of bioacoustics. At the same time it is clear that no chart can do complete justice to the interdisciplinary ramifications of acoustics. Thus the chart puts ultrasonics in the upper right quadrant to emphasize its important connection with engineering problems. But ultrasonics could equally well have been placed also in the lower left quadrant, since its relation to medicine through its use as a diagnostic tool as well as in therapy forms a large part of its use.

The diversified character of acoustics exhibited in the chart implies that the subject must have had a somewhat elaborate historical development. An overall understanding of the science of acoustics can best be gained by reference to this development. It is the purpose of this volume to provide a picture of the growth of ideas of acoustics in terms of some of the important papers which acoustical scientists and engineers have published over the years.

The first article is a general review of the history of acoustics. It summarizes briefly the principal steps in the development of acoustics from the earliest times in terms of the three fundamental aspects of the production, propagation, and reception of sound. The principal figures who have contributed to the evolution of our ideas on sound are named, along with their dates, and brief indications of their important contributions are provided. Emphasis is laid on the unique role of Lord Rayleigh, whose great work "The Theory of Sound," first published in 1877–1878, set the stage for what may be called the modern age of acoustics.

This summary history is followed by an anthology of 39 articles representing landmarks in the evolution of acoustics as a science. These are arranged in chronological order, beginning with the ideas of the early Greek and Roman philosophers.

The choice of order has been deliberate; no attempt is made to segregate the material with respect to the various branches of the subject. In this way it is possible to trace the kinds of acoustical problems which were thought worthy of study at any particular epoch. This arrangement brings out clearly the increase in elaborateness and sophistication of the questions investigated as time went on. No work subsequent to 1900 has been included. Classic papers in 20th century acoustics will be found in other volumes of the series in their respective fields.

It is hoped that the reader will glimpse the interesting way in which the desire to understand certain sound phenomena (e.g., the behavior of musical instruments) led to the attempt to apply fundamental mechanical principles to their elucidation (e.g., the invention of the theory of mechanical vibrations). The desire to solve such problems as the sound from a vibrating string and a blown organ pipe provided an impetus for the development of more powerful mathematical methods; this will be clearly evident from a study of the articles in this volume. It is fascinating to note how even the greatest minds working on the problem of sound propagation through a medium like air were puzzled for over a century by the failure of Newton's theoretically established value of the velocity of sound in air to agree with that experimentally measured. One can hardly refrain from a feeling of wonder in noting how the great Laplace, primarily a mathematician and cosmologist and not at all what we would call an acoustical scientist, solved the problem in a few pages (actually it was not quite so simple as that!).

Most of the items in the anthology are articles which originally appeared in journals of learned societies, but frequently extracts are taken from books, as providing a simpler source. Occasionally more informal material is introduced, such as Euler's famous 1759 letter to Lagrange, said to constitute a turning point in the whole development of the theory of sound propagation, and Colladon's entertaining account of how he measured the velocity of sound in the water of Lake Geneva, taken from his very readable autobiography.

All material is presented in English, the earlier works in Latin, French, and German having been translated by the editor of this volume, save where otherwise indicated. Each entry is prefaced by a brief biographical note as an indication of the significance of the extract. Frequent editorial notes help to clear up obscure points in the text, particularly with reference to older, unfamiliar terminology.

In such an anthology the choice of entries is obviously arbitrary to a considerable extent, reflecting as it does the views of the editor as to what is really significant. Nevertheless, the editor believes that those articles chosen for inclusion will be recognized by authorities in the field as landmarks in the early development of acoustics. The reader will observe that most of the names appearing in the bibliography of the historical review article "The Story of Acoustics" are represented somewhere in this anthology.

In the translation and editing of such a large body of materials errors are bound to occur. The editor realizes that he has doubtless sinned in several places. He will be grateful to all who will bring such errors to his attention.

Reprinted from THE JOURNAL OF ACOUSTICAL SOCIETY OF AMERICA, Vol. 39, No 4, 629–644, April 1966

1

The Story of Acoustics

R. BRUCE LINDSAY

Brown University, Providence, Rhode Island 02912

The historical progress of the science of acoustics is surveyed from the earliest recorded phenomena and theories to the present status of the subject. Considerable attention is paid to the development of both mathematical and experimental tools for studying the production, propagation, and reception of sound, particularly in the 18th and 19th centuries. The impact of Rayleigh's work on modern acoustics is estimated. Contemporary developments are treated only briefly. The endeavor has been made to refer in almost all cases to original sources.

INTRODUCTION

ALTHOUGH the history of science was long neglected by professional historians, this unhappy situation is now being rectified through an awareness of the significant influence the growth of science has had on the development of civilization. No apology is therefor necessary for a concern with this field of scholarship. The question has indeed been raised whether knowledge of the history of science has any value for the practicing scientist, and arguments have been presented on both sides. It appears to the present writer that the weight of the historical evidence itself is in favor of the affirmative view in this matter. A knowledge of the evolution of the concepts basic to a given branch of science can often suggest useful ways of approaching current experience, and actually has done so in numerous instances. It is with this premise in mind that the following brief survey of the history of acoustics is presented.

We begin by observing that acoustics occupies a somewhat anomalous position in the hierarchy of the sciences. Though hearing is obviously one of the most important of human sensations, and it has been a common observation from the earliest times that we live in a world of sound, the intensive historical study of the development of human ideas on this subject has until comparatively recent times been greatly neglected. Why is this? It was suggested a good many years ago [79] that the accepted root ideas on the origin, propagation and reception of sound were proposed at a very early stage in the development of human thought. It seems clear that the ancient Greek philosophers were convinced that the origin of sound is to be sought in motion of the parts of bodies, that it is transmitted through the air by means of some undefined motions of the latter, and that this motion in the neighborhood of the ear produces the

sensation of hearing. These ideas were vague enough, but they were much closer to what came to be the accepted theory of sound than the ancient notions of the motion of large scale objects, to say nothing of the primitive theories of light and heat. The latter branches of physics suffered many vicissitudes of treatment in which theory succeeded theory until the present point of view was attained. But in acoustics all that was really needed was the elaboration and refinement of the basic idea by the necessary mathematical analysis and its application to new phenomena as they were discovered. On its theoretical side in particular, the history of acoustics thus tends to be merged in the general evolution of mathematical mechanics as a whole.

This indeed seems at first to be a plausible point of view, closely connected with the somewhat perverse attitude of many modern physicists that the essential physics of sound was worked out so long ago that it is no longer a physical subject but rather a branch of electrical engineering or possibly also of physiology. But this opinion is in fact a distorted one and has no more justification than the associated claim that the subject has no history worth mentioning, since the fundamental notions were laid down early and have not suffered serious change in the passage of time. In this review, we refute this point of view by surveying in some detail the fascinating history of those parts of mechanics and other branches of physics that have a definite bearing on acoustical theory as well as acoustical practice.

The problems of acoustics are most conveniently divided into three main groups: viz, (1) the production of sound, (2) the propagation of sound, and (3) the reception of sound. The following historical outline is organized accordingly.

the journal of the Acoustical Society of America 629

I. PRODUCTION OF SOUND

The fact that when a solid body is struck a sound is produced must have been observed from the very earliest times. The additional observation that under certain circumstances the sounds so produced are particularly agreeable to the ear furnished the basis for the creation of music, which must have originated long before the beginning of recorded history, and which was, of course, also closely associated with the pleasant sounds that, again under favorable circumstances, may be emitted from the mouths of human beings, either directly to the ambient air or by means of a tube of appropriate shape. But music was an art for millenia before, so far as we know from the available record, its nature began to be examined in a scientific manner. It is usually assumed that the first Greek philosopher to study the origin of musical sounds was Pythagoras, who established his school in Crotone in southern Italy in the 6th century B.C. He is supposed to have been impressed by the fact that, of two stretched strings fastened at the ends, the note of higher pitch is emitted by the shorter one, and that indeed, if one has twice the length of the other, the shorter will emit a note an octave above the other. The story is probably legendary [10]. It is usually cited to provide a basis for the obsession that Pythagoras and his followers appeared to have for integral numbers as fundamental for the understanding of experience. It seems clear that the germ of the idea that pitch depends somehow on the frequency of vibration of the sound-producing object was in the minds of Greek philosophers of the Pythagorean school, such as Archytas of Tarentum in southern Italy, who flourished around 375 B.C. A fairly clear presentation of this point of view is to be found in the writings on music of the Roman philosopher, Boethius, in the 6th century A.D. For the modern scientific basis of this relation, it has been customary to look to Galileo Galilei (1564–1642). At the very end of the "First Day" of the great Italian's *Dialogues concerning Two New Sciences*, first published in 1638, there is a remarkable discussion of the vibration of bodies [27]. Beginning with the well-known observations on the isochronism of the simple pendulum (in which Galileo does indeed make the mistake, perhaps excusable in his day, of concluding that the period of the pendulum is independent of the amplitude no matter how large the latter is) and the dependence of the frequency of vibration on the length of the suspension, the author goes on to describe the phenomenon of sympathetic vibration or resonance, by which the vibration of one body can produce similar vibration in another distant body. He reviews the common notions about the relation of the pitch of a vibrating string to its length and expresses the opinion that the physical meaning of the relation is to be found in the number of vibrations per unit time, i.e., what we now call the frequency. He says that his view of this was confirmed by two observations. The first was that of a glass goblet placed with its base fixed to the bottom of a large vessel and filled with water almost up to the brim of the goblet. By rubbing the edge of the goblet with the finger, the goblet can be made to vibrate and emit a sound. At the same time, ripples are observed to run across the surface of the water. And when, as occasionally happens, the note from the goblet rises an octave in pitch, the ripples in the water are "divided in two," i.e., what we should now call the wavelength is halved. The second observation was the result of an accident in which he happened to scrape a brass plate with an iron chisel in order to remove some spots from it. Once in a while, the scraping would be accompanied by a sharp whistling sound of definite musical character. In this case, he always observed a long row of parallel fine streaks on the surface of the brass, equidistant from each other. He further noticed that the pitch of the whistling note could be increased by increasing the speed of scraping and, in this case, the separation of the streaks decreased.

Galileo states that he was able to tune spinet strings with the aid of these chisel-scraping tones and found that, when the musical interval between two such strings was judged by ear to be a fifth, the average spacings between the lines on the brass plate for the corresponding scraping tones were in the ratio of three to two. It seems clear from a careful reading of Galileo's writings that he had a clear understanding of the dependence of the frequency of a stretched string on the length, tension, and density, though much of his knowledge was undoubtedly learned from predecessors. He made an interesting comparison between the vibrations of strings and those of pendulums in the endeavor to make clear why sounds of certain frequencies—i.e., those whose frequencies are in the ratio of two small integers—appear to the ear to combine pleasantly whereas others not possessing this property sound discordant. Galileo observed that a set of pendulums of different lengths, set oscillating about a common axis and viewed in the original plane, presents to the eye (at least it did to his eye) a pleasing pattern if their frequencies are simply commensurable, whereas it forms a complicated jumble otherwise. One must admit that this was a kinematic observation of great ingenuity and formed the basis of a suggestive analogy.

It is well to keep in mind that the history of science, like all history, depends to a considerable extent on the historian. It is not surprising, therefore, that Galileo's achievements in acoustics have been questioned. Clifford Truesdell [70] in his elaborate history of the mechanics of elasticity expresses the opinion that the importance of Galileo's contribution to the mechanics of vibration has been exaggerated. He points out that, though much of the material on general mechanics in Galileo's famous book dates from the early 17th century, when Galileo evidently first thought it out consistently, most of his results on vibrations appeared in the dialogues for the

6

first time. In the meantime, several investigators had apparently come up with the fundamental ideas that Galileo so engagingly expresses. The Frenchman, Isaac Beeckman (1588–1637), had evidently thought a good deal about the vibration of strings and printed some of his speculations* as early as 1618. Here, he manifested his confidence in the relation between pitch and frequency and gave arguments in its favor. He is usually credited with the initiation of René Descartes into the study of physics. Even earlier than Beeckman was the Italian, Giovanni Battista Benedetti (1530–1590), who in a work on musical intervals published in Turin in 1585 stated clearly his belief in the equality between the ratio of pitches and the ratio of the frequencies of the vibrating motions corresponding to the production of the sounds. More elaborate were the studies of the French Minim priest, Marin Mersenne (1588–1648). In 1625, Mersenne published some results that he had obtained from the experimental observation of the vibrations of a stretched string. Here, he recognized that, other things being equal, the frequency of the vibration is inversely proportional to the length of the string, while it is directly proportional to the square root of the cross-sectional area. Truesdell considers that Mersenne definitely anticpated Galileo in these important conclusions about vibrating strings [49,50].

Later experimenters, like Robert Hooke (1635–1703), whose law of elasticity is one of the most widely known in physics, tried to connect frequency of vibration with pitch by allowing a cog wheel to run against the edge of a piece of cardboard, a common lecture demonstration to this day [30].†

The man who made the most thorough-going pioneer studies of frequency in relation to pitch was undoubtedly the Frenchman, Joseph Sauveur (1653–1716). He can claim the distinction of first suggesting that the name *acoustics* be applied to the science of sound [64]. It is well known that this term comes from the Greek word meaning hearing, and is, therefore, to a certain extent appropriate, though modern acoustics far transcends the sounds that we can actually hear. Sauveur was aware of the significance of the beats that are observed when two organ pipes (or similar sound sources) of slightly different pitches are sounded together, and actually used them to calculate the fundamental frequencies of two such pipes that were adjusted by the ear to be a semitone apart—i.e., having frequencies in the ratio $\frac{15}{16}$. By experiment, he found that when sounded together the pipes gave six beats a second. By treating this number as the difference between the frequencies of the pipes, Sauveur reached the conclusion that these latter numbers were, respectively, 90 and 96 cps. Sauveur also experimented with strings and, in 1700, calculated by a somewhat dubious method the frequency of a given stretched string from the measured sag of the central point.

* See Ref. [70], p. 24.
† See Ref. [30], p. 57.

It was reserved to the English mathematician, Brook Taylor (1685–1731), the well-known author of Taylor's theorem on infinite series [68] to be the first to provide a strictly dynamical solution of the vibrating string. This was published in 1713 and was based on an assumed curve for the shape of the string when vibrating in what we now call its fundamental mode (i.e., when all parts of the string are simultaneously on the same side of the equilibrium horizontal position). This curve was taken to be of such a character that every point would reach the horizontal position at the same time. From the equation of this curve and the Newtonian equation of motion, he was able to derive a formula for the frequency of the fundamental vibration that agreed with the experimental law of Mersenne and Galileo. It is of particular interest to note that, as Truesdell has pointed out, this seems to be the first time that the Newtonian equation of motion $F = ma$ was applied to an *element* of a *continuous* medium.

Though Taylor treated only a special case and was clearly unable to progress to the treatment of the general string with all its modes because of his lack of the calculus of partial derivatives, he did pave the way for the more elaborate mathematical techniques of the Swiss, Daniel Bernoulli [2] (1700–1782) (one of a family that produced eight scientists who achieved distinction in the 18th century), the Frenchman, Jean Le Rond d'Alembert [16] (1717–1783), and the Swiss, Leonhard Euler‡ (1707–1783). These gentlemen managed to set up the partial differential equation of motion of the vibrating string and to solve it in what is essentially the modern fashion. It is interesting to reflect here on how the lack of adequate mathematical tools retarded the progress of the science of sound as it held back similarly the advance of mechanics in general. Unfortunately, neither the fluxions of Sir Isaac Newton (1642–1727) nor the differentials of Baron Gottfried Wilhelm von Leibniz (1646–1716) were quite adequate for the handling of the motions of continuous media.

To come back to the physical side of the problem of the vibrating string as a source of sound, it had already been observed, notably by John Wallis [77] (1616–1703) in England as well as by Sauveur [64] in France, that a stretched string can vibrate in parts so that at certain intermediate points, which Sauveur called *nodes*, no motion ever takes place, whereas very violent motion takes place at intermediate points called *loops*. It was soon realized that such vibrations correspond to higher frequencies than that associated with the simple vibration of the string as a whole without nodes, and indeed that these frequencies are integral multiples of the frequency of the simple vibration. The associated emitted sounds were called by Sauveur the *harmonic* tones, while the sound corresponding to the simple vibration was

‡ The most complete study of the contributions of Euler to acoustics and the relation of his work to that of his contemporaries is that of Clifford Truesdell in his introductions to the new edition of Euler's writings. In particular, see Ref. [70].

named the *fundamental*. The notation thus introduced (around 1700) has survived to the present day. Sauveur noted the additional important fact that a vibrating string could produce the sounds corresponding to several of its harmonics at the same time. The dynamical explanation of this was given by Daniel Bernoulli in his famous memoir for the Royal Academy of Berlin [2]. In this, he showed that it is possible for a string to vibrate in such a way that a multitude of simple-harmonic oscillations are present at the same time and that each contributes independently to the resultant vibration, the displacement at any point of the string at any instant being the algebraic sum of the displacements associated with the various simple-harmonic modes. He thus propounded the famous principle of the coexistence of small oscillations, also referred to as the principle of superposition. Bernoulli tried to give a proof of the principle but did not succeed. His grasp of mathematics was not so great as his understanding of physical ideas. The real significance of the superposition principle was pointed out almost immediately by Euler* [25]: namely, that the partial differential equation that governs the motion of the ideal frictionless string is *linear*. With this understanding, the superposition principle can be proved as a theorem.

The whole history of the vibrating string in the middle of the 18th century and even up to 1785 consists of a series of controversies in which ingenious investigators like Bernoulli, Euler, and d'Alembert argued vehemently with each other in their journal memoirs. They took their work very seriously and unfortunately did not hesitate to cast aspersions on each other in rather harsh language. This was the period in which the mathematics needed for the description of the motion of continuous media, so fundamental for the progress of acoustics as an exact science, was being born and the travail was not easy. The standard texts tend to gloss over the unpleasant things that even the great scientists of the time said about each other in their correspondence and their articles, as well as the serious errors that they often made. The possibility of expressing any arbitrary function—e.g., the initial shape of a vibrating string in terms of an infinite series of sines and cosines, implied by the superposition theorem—was hard to accept in terms of mid-18th century mathematics. It was only in 1822 that J. B. J. Fourier [26] (1768–1830), in his analytical theory of heat, based his celebrated theorem on this type of expansion with consequences of the greatest value for the advancement of acoustics.

Among the 18th-century mathematicians who tackled the problem of the vibrating string was J. L. Lagrange (1736–1813), the Italian from Turin, who spent most of his active career in France. He was the author of the treatise, *Mécanique analytique* [41], in which mechanics was reduced to a branch of mathematical analysis, and in the preface of which the author boasted that

---

* This memoir, Ref. [25], was received just after that by Bernoulli, mentioned in Ref. [2].

he had included no figures, for such were unnecessary. The reader who has studied theoretical physics will recall generalized coordinates and Lagrange's equations. In an extensive memoir [42] presented to the Turin Academy in 1759, Lagrange decided to adopt what he claimed was a different and novel approach to the string problem. He assumed the string to be composed of a finite number of equally spaced identical-mass particles tied together by equal segments of stretched weightless string. He then solved the problem of the motion of this system as a dynamical system with many degrees of freedom, and established the existence of a number of independent frequencies equal to the number of the particles. When he passed to the limit and allowed the number of particles to become infinitely great and the mass of each correspondingly small (so that the product equalled the finite mass of the string), these frequencies were found to be precisely the harmonic frequencies of the stretched continuous string. Lagrange felt that his device avoided the analytical difficulties associated with the motion of the continuous string and that he had made a decisive advance. A few observations are in order here. In the first place, Euler [21] had already in 1744 solved the mechanical problem of the motion of n particles on a string, where n is any integer, though he had not been successful in the passage to the limit. In the second place, it has been pointed out by Truesdell that Lagrange's passage to the limit was mathematically faulty and, to be made rigorous, demanded essentially the same kind of mathematical assumptions that he objected to in the analysis of his contemporaries, Bernoulli, Euler, and d'Alembert, for the continuous string. Be this as it may, Lagrange's method was adopted by J. W. Strutt Lord Rayleigh (1842–1919) in his *Theory of Sound* [67] and has found its way into most modern texts in mechanics and acoustics. It is not indeed the most direct way in which to handle the vibrating string, and undoubtedly Lagrange exaggerated the significance of his accomplishment. But his method has the merit of variety and this is important in science; the more ways in which we can look at the same phenomenon, the better is our grasp of it. Truesdell, indeed, considers that Lagrange's contributions to acoustics and to mechanics in general have been overrated.

d'Alembert usually gets the credit for having been the first to develop, in 1747, the partial differential equation of the vibrating string in the form that we now refer to as the wave equation [16]. He also found its general solution in the form of waves traveling in both directions along the string. From this point of view, the vibrations of the string are due to a combination of such traveling waves to form so-called standing or stationary waves. As we have pointed out above, there was a lot of controversy about the meaning and validity of all this, just as there exists controversy today over the development of modern theories like those of relativity and quantum mechanics.

8

It must not be thought that the vibrating string, important as it was, took up all the attention of the 18th-century savants. They were interested in other sound-producing motions as well. In the Lagrange 1759 memoir, for example, there is a treatment of the sounds produced by organ pipes and musical wind instruments in general. The basic experimental facts were already known and Lagrange was able to predict theoretically the approximate harmonic frequencies of closed and open pipes. The boundary conditions gave some trouble, as indeed they still do. In any case, problems of this kind impinge rather closely on the propagation of sound, treated a little further on. We pause only to note that Euler made great contributions to this field also; it is only recently that their magnitude has been appreciated. In a little treatise [23], "Physical Dissertation on Sound," written in Basel in 1727 when Euler was only about twenty years old, the essential features of the overtones of pipes are set forth in essentially modern form. This was, of course, far earlier than the work of Lagrange, just noted. Euler was at that time particularly interested in musical instruments like the flute. Around 1759, there was considerable activity by both Euler and Lagrange on the subject of sound oscillations in tubes and much correspondence between them. Around 1766, Euler produced an elaborate treatise on fluid mechanics, the fourth section [24] of which was entirely devoted to sound waves in tubes. It is a bit hard for us today to appreciate the tremendous zeal with which such problems were tackled and solved by the great mechanically minded mathematicians of this era. It is scarcely an exaggeration to call this the golden age of mathematical physics.

The mathematical scientists of the 18th century, of course, realized that other solid bodies besides strings emit sound when disturbed. For example, they were, of course, familiar with bells, and there must have accumulated by that time a vast amount of empirical knowledge about sound sources of this kind. But the successful application of the mathematical methods described in the preceding paragraphs to the vibrations of metal bars, plates, and shells naturally demanded a knowledge of the relation between the deformation of the solid body and the impressed deforming force. Fortunately, this problem had already been tackled and solved in its simplest form by Robert Hooke [32], who in 1660 discovered and in 1675 announced in the form of the anagram CEIIINOSSSTTUV the law, which in the Latin form he expressed as "ut tensio sic vis," connecting the stress and strain for bodies undergoing *elastic* deformations. It says that within the so-called elastic limit the strain of an elastic body (i.e., the fractional increase in length for a linear rod or bar) is directly proportional to the stress (i.e., the force per unit area of cross section of the rod or bar in the direction of the stretch). This law forms the basis for the whole mathematical theory of elasticity, including elastic vibrations giving rise to sound. Its application to the vibrations of bars sup-

ported and clamped in various ways appears to have been made as early as 1734–1735 by Euler and Daniel Bernoulli.* The mathematical methods used were later systematized and extended by Lord Rayleigh in his *Theory of Sound* [67]. The fundamental idea was to begin with the expression for the energy of a deformed bar and to use the so-called variational technique, which leads to the well-known equation of the fourth order in space derivatives.

The corresponding analytical solution for the vibrations of a solid elastic plate proved much more difficult and came much later, though much useful experimental information was obtained in the latter part of the 18th century by the German scientist, E. F. F. Chladni (1756–1824). In 1787, he published his celebrated treatise *Entdeckungen über die Theorie des Klanges* [8], in which he described his method of using sand sprinkled on vibrating plates to show the nodal lines, i.e., lines of zero displacement. These Chladni figures have long been recognized as things of great beauty. In a general way, they could be accounted for by considerations similar to those explaining the existence of nodes in a vibrating string. The exact forms, however, defied analysis for many years, even after the publication of Chladni's classic work [7], *Die Akustik*, in 1802. The Emperor Napoleon provided for the Institute of France a prize of 3000 francs to be awarded for a satisfactory mathematical theory of the vibrations of plates. This prize was awarded in 1815 to the celebrated woman mathematician Mlle. Sophie Germain, who produced the correct fourth-order equation. Her choice of boundary conditions proved, however, to be incorrect. It was not until 1850 that G. R. Kirchhoff [36] (1824–1887) gave a more accurate theory. Modern technology with its concern for the vibrations of such things as airplane fuselages, etc., still supports active research on the vibrations of plates and solid shells of various shapes.

The first solution of the analogous problem of the vibrations of a flexible membrane, important for the understanding of the sounds emitted by drums, is usually attributed to the French mathematician, S. D. Poisson [58] (1781–1840), though he failed to complete the case of the circular membrane ,which was handled by R. F. A. Clebsch [9] (1833–1872) in 1862. It has been pointed out by Truesdell [70] that this attribution to Poisson neglects a very important work by Euler seventy years earlier [22]. Euler derived the appropriate partial differential equation for the vibrating membrane and expressed it properly for both rectangular and circular shapes, but he made a curious slip in assigning boundary conditions and hence failed to get the right answer for the normal modes (by a factor of two). Poisson apparently never was familiar with this paper by Euler. He got the correct normal modes for the rectangular membrane. It is significant that much of the theoretical work on vibration problems during

---

* See Ref. [70], p. 165.

the 19th century was carried out by persons who called themselves mathematicians. This was natural, since much of the application of the mathematics of that period concerned physical problems of this kind, but it was also rather unfortunate, for the choice of boundary conditions did not always reflect actually experimentally realizable situations. The ability to excite vibrations in media of arbitrary nature, size, and shape and with arbitrary frequency over a wide range had to await the development of electroacoustics, largely a product of 20th-century research [33]. It is true that electrical oscillations and oscillating electric circuits were discovered and invented in the middle of the 19th century, but methods of coupling them to mechanical systems so as to make them produce mechanical oscillations did not arise in a practical fashion until after 1900. While tuning forks remained the only practical frequency standards for sound sources, no great progress in applying the mathematical theory of acoustics to practical cases could be expected. The basic physical principles for this purpose were indeed well known in the 19th century. Thus, the force on a current-carrying conductor in a magnetic field had been discovered in the 1820's. The piezoelectric effect emerged from the experiments of the Curie brothers [14,15] in 1880. This is the property, displayed by some crystals (notably quartz), of having electric charges appear on the faces when subjected to mechanical stresses of various kinds, and conversely of changing dimensions (i.e., exhibiting strain) when placed in an electric field. It was early recognized that herein lay the possibility of developing both controlled sources and receivers of sound waves, but the actual exploitation of the effect for this purpose did not take place until the second decade of the 20th century. The same situation prevailed with the magnetostriction effect, i.e., the tendency of magnetizable materials to change dimensions when placed in a magnetic field. This was discovered by J. P. Joule [60] (1818–1892) in 1842. The advent of the vacuum-tube oscillator and amplifier made the employment of these effects possible in the production and reception of sound at all frequencies and intensities on a precision basis.

The rest of the story of the production of sound, which is largely that of the evolution of electroacoustics, may profitably be left to the end of this sketch, which discusses briefly the contributions of Lord Rayleigh and his successors. However, we ought not to overlook here one of the most important methods for the production of sound: namely, the vocal chords leading to speech in human beings and the noises emitted by lower animals. It is a curious fact that, though these examples of sound production are in may ways the most obvious of all, little attention was paid to them during the historical evolution of acoustics just surveyed. Perhaps it would be more correct to say that human speech did not engage the attention of the mathematicians and physicists, who

were principally concerned with the production of sound. The very fact of the obviousness of speech may well have led to lack of interest in it on the part of those who felt themselves concerned with the physically objective aspects of sound. Speech, after all, seemed closer to language and therefore in the province of the philologists and etymologists.

At the same time, the basic mechanism of human speech—i.e., the combination of vibrating vocal cords and mouth cavities—was considered largely a matter for anatomists and physiologists. At any rate, it is of interest to note that as early as 1629 the Englishman, W. Babington, observed the motions of the vocal chords by means of light reflected from mirrors in the mouth. This was the beginning of the development of what came to be called the laryngoscope, finally perfected by the Czech physiologist, J. N. Czermac, in 1857. Some eighty years later, movies of the vocal chords were made by D. W. Farnsworth at the Bell Telephone Laboratories.

The nature of the vowel sounds of speech and singing was first thoroughly investigated by Hermann von Helmholtz (1821–1894) in 1860, with the use of the famous resonators that bear his name. His results were set forth in the treatise [75] *Die Lehre von den Tonempfindungen als physiologische Grundlage für die Theorie der Musik*, published in 1862. This ranks as one of the great masterpieces of acoustics. It is mentioned again later in connection with the reception of sound by the ear. It is true that earlier investigations (in 1837) by the Englishman, Sir Charles Wheatstone (1802–1875) had led to the development of a harmonic theory for the production of vowel sounds. In accordance with this, the vocal chords vibrate so as to produce both a fundamental frequency and numerous harmonics. It was assumed that these vibrations when communicated to the air are reinforced by resonance in the mouth cavities. Another theory apparently first proposed by W. T. Willis, in England in 1829, assumed that the origin of the vowel sound was not in the continuous vibration of the vocal chords, but rather in puffs of air emitted by them. These transient puffs set the air in the mouth cavities in vibration and the resonance there gives the emitted sound its characteristic quality. von Helmholtz later pointed out that both ideas have elements of correctness, and modern research has confirmed this opinion. Some early workers held that the whole oral cavity acts as a single resonator, while others, notably Alexander Graham Bell, the inventor of the telephone, and von Helmholtz himself [75,55] believed that vowel sounds depend on two characteristic resonances, corresponding to the action of the mouth as a double resonator, i.e., two resonators coupled together. von Helmholtz points out in his famous treatise that the knowledge that the oral cavities can be tuned to different frequencies goes back to the early 17th century, though it is clear there was at that time no highly developed instrumental study of the phenomenon.

II. PROPAGATION OF SOUND

From the earliest recorded observations, there has been general agreement that sound is conveyed from one point in space to another through some activity of the air. Aristotle indeed emphasizes that there is actual motion of air involved, and it is possible to read into the description in his treatise, "De Anima," and to the treatise "De Audibilibus" often ascribed to him the notion that sound is due to compressional waves in air* The usual difficulty of interpretation that forever plagues the history of science occurs again here in annoying fashion, since it is difficult to be sure that Aristotle and his contemporaries had really grasped the idea that in the propagation of sound the air does not move as a whole—i.e., as a stream—in the direction of propagation. It seems clear that the famous Roman architectural engineer, Marcus Vitruvius Pollio,† of the first century B.C. had an adequate grasp of the wave theory of sound in the analogies that he drew with surface waves on water. However, since in the transmission of sound the air certainly does not appear to move, it is not surprising that other and much later philosophers denied these views of Aristotle and Vitruvius. Even during the Galilean period, the Frenchman, Pierre Gassendi [48] (1592–1655), in his revival of the atomic theory, attributed the propagation of sound to the emission of a stream of very small, invisible particles from the sounding body, which, after moving through the air, are able somehow to affect the ear. Otto von Guericke [74] (1602–1686) expressed great doubt that sound is conveyed by a motion of the air, observing that it is transmitted better when the air is still than when there is a wind. Moreover, around the middle of the 17th century, he tried the experiment of ringing a bell in a jar that was evacuated by means of his air pump, and claimed that he could still hear the sound. As a matter of fact, the first to try the bell in vacuo experiment was apparently the Jesuit, Athanasius Kircher (1602–1680). He described it in his book, Musurgia Universalis, published in 1650, and concluded from his observation that air is not necessary for the transmission of sound [35]. Undoubtedly, the trouble with the investigations of von Guericke and Kircher was the failure to avoid transmission through the walls of the vessel coupled with the rather inadequate vacuum they were able to obtain. In 1660, Robert Boyle [5] (1627–1691) in England repeated the experiment with a much improved air pump and more-careful arrangements, and finally observed the now well-known decrease in the intensity of the sound as the air is pumped out. He definitely concluded that the air is a medium for the transmission of sound, though presumably not the only one. As a matter of fact, this explanation of the experiment, though hallowed by tradition, is a mistaken one. A more careful examination of the situation shows that the ob-

* See Ref. [10], p. 288 ff.
† See Ref. [10], p. 307 f.

served decrease in the intensity of the sound is due not so much to the failure of the low pressure air to transmit sound as to the increasing difficulty of getting the sound out of the bell (or other sound source) into the air and then out again from the air to the glass container. The so-called impedance mismatch between source and surrounding fluid medium becomes greater as the density of the medium decreases. Of course, the experiment does demonstrate a very important connection between the source and the medium in acoustic propagation, and the modern theory of sound implies that without doubt some material medium is necessary for acoustic transmission.

Granted that air is a sound-transmitting medium, the question at once arises: How rapidly does the propagation take place? As early as 1635, Gassendi [46], while in Paris, made measurements of the velocity of sound in air using firearms and assuming that the light of the flash is transmitted instantaneously. His value came out to be 1473 Paris feet per second or 478.4 m/sec in the later terminology of the metric system, which, of course, Gassendi did not use, as it was not introduced until the time of the French Revolution. Somewhat later, by more careful measurement, Mersenne [46] showed that Gassendi's figure was too high; his value was 1380 Paris feet per second or about 450 m/sec. Gassendi did note one important fact; namely, that the velocity is independent of the pitch of the sound, thus discrediting the view of Aristotle, who had taught that high notes are transmitted faster than low notes. On the other hand, Gassendi made the mistake of believing that the wind has no effect on the measured velocity of sound. In 1650, the Italian, G. A. Borelli (1608–1679) and his colleague V. Viviani (1622–1703) of the Accademia del Cimento of Florence tried the same type of experiment [51] and obtained 1077 Paris feet per second or 350 m/sec. All these measurements suffered from lack of reference to the temperature, humidity, and wind velocity. Though the Englishman W. Derham [18] (1657–1735) made extensive measurements of the velocity of sound in the early part of the 18th century (published in 1708) in which he concluded that the velocity is independent of all environmental conditions except wind, his results were definitely shown to be wrong by the Italian, G. L. Bianconi [6,47] (1717–1781), in Bologna, who in 1740 demonstrated definitely that the velocity of sound in air increases with the temperature. The first open-air measurement (no wind) that can be considered at all precise in the modern sense was probably that carried out under the direction of the Academy of Sciences of Paris in 1738, using a cannon as the source of sound. When reduced to 0°C, the result in modern units was 332 m/sec. Careful repetitions during the two succeeding two centuries gave results differing from this value by less than 1%. The best modern value (1942) is indeed 331.45±0.05 m/sec under stand-

11

ard conditions of temperature and pressure [28]. This is a tribute to the care with which those Paris academicians carried out their work. Actually, very few early physical measurements have stood the test of time as well as these of the velocity of sound in air.

The problem of the measurement of the velocity of sound in solid media was tackled by Chladni, whose investigations of the nodal lines in vibrating plates have already been mentioned. He used similar means to study propagation in metal rods and by measuring internodal distances was able to calculate sound velocity in such specimens. In 1808, the French physicist, J. B. Biot [3] (1774–1862), most famous for his work in optics, made actual measurements of the velocity of sound in an iron water pipe in Paris by direct timing of the sound travel. The pipe was nearly 1000 m long. By comparing the times of arrival of a given sound through the metal and through the air, respectively, he established in particular that the velocity of the compressional wave in the solid metal is many times greater than that in the air. This was indeed to be expected from the very much greater elasticity of metal in comparison with air. Biot's experimental values agreed in order of magnitude with those of Chladni.

The first serious attempt to measure the velocity of sound in a liquid like water was apparently that of the Swiss physicist, Daniel Colladon (1802–1893), who, assisted by the mathematician, Charles Sturm (1803–1855), conducted studies in Lake Geneva in 1826. Some might suppose that it was merely curiosity about the actual velocity of sound in lake water that motivated Colladon. But this was not the case. In 1825, the Academy of Sciences in Paris announced as the subject of its prize competition for 1826 the measurement of the compressibility of the principal liquids. Colladon decided to compete, and actually measured successfully the static compressibility of water and some liquids. He must have been somewhat fascinated by its relatively low value and the correspondingly large magnitude of its reciprocal, the bulk modulus. He was, of course, aware of the theoretical relation between the compressibility and the speed of sound. The suggestion was therefore immediate that a check on the accuracy of his compressibility measurements could be sought in the acoustic velocity. The tests were carried out in Lake Geneva in November 1826, and the results appeared in the famous "Mémoire sur la compression des liquides et la vitesse du son dans l'eau", which was awarded the Grand Prize of the Academy [12]. The compressibility of water as computed from the velocity of sound turned out to be very close to the statically measured value. Actually, the whole story of this study of Colladon and Sturm is found set forth in fascinating fashion in Colladon's autobiography [11] (Souvenirs et mémoires: Autobiographie de J.- Daniel Colladon). Here, the man who did the work and indeed became in later life a well-known physicist and engineer tells informally how he did it, including such homely details as the

troubles that he encountered in carrying the powder needed for his light flashes across the frontier between Switzerland and France. The average velocity found in this measurement was 1435 m/sec at 8°C.

Though the propagation of sound through air had already been compared, as we have seem above, with the motion of ripples on the surface of water, the first attempt to theorize seriously in mathematical form about a wave theory of sound was apparently made by Sir Isaac Newton [53] (1642–1727), who in the second book (1687) of his famous Principia mathematica (propositions 47–49) compares the transmission of sound with pulses produced when a vibrating body moves the adjacent portions of the surrounding medium and these in turn move those next adjacent to themselves and so on. Newton then went on to make some rather arbitrary assumptions, the principal one being that, when a pulse is propagated through a fluid, the particles of the fluid always move in simple harmonic motion, or, as he puts it: "are always accelerated or retarded according to the law of the oscillating pendulum." He proves indeed the theorem that, if this is true for one particle, it must be true for all adjacent ones. It is not necessary to go through the complete demonstration, which has always been difficult to follow. The end result is that the velocity of sound is equal to the square root of the ratio of the atmospheric pressure to the density of the air.

As was to be expected, Newton's "derivation" was subjected to much criticism by the natural philosophers of continental Europe, among whom were Euler, John Bernoulli (younger brother of Daniel), and Lagrange. We have already commented on L. Euler's treatise, "Physical Dissertation on Sound" [23] of 1727. In this remarkable work, he set forth his ideas of the physical principles underlying sound propagation as well as sound production with great clarity, and attacked Newton's method as being entirely too specialized. He presented an expression for the velocity of sound in air that is very close to that of Newton, though it must be admitted that his method is not clear. In a later treatise of 1749, Euler developed Newton's theory in much clearer analytic form and obtained Newton's result.

It gradually became clear that the problem of sound propagation would never be completely solved until the wave equation for sound waves in a fluid could be set up and solved. It will be recalled that d'Alembert [16] had first derived this equation for a continuous string in 1747; at that time, he commented on the fact that it should be possible to apply the same equation to sound waves. However, he did not get far with the details; these were worked out by Euler, who seems in this case, as in other problems in both solid and fluid dynamics, to have "known it all."

The claims and counterclaims, the epistolary criticisms of one another by those zealous 18th-century savants, make the task of the historian of science a fascinating but a difficult one. Though Euler [20] seems to have laid the foundation for the theory of the propagation of sound waves in air in three great memoirs to the Berlin Academy in 1759, it was Lagrange who, in memoirs to the Turin Academy at about the same time, revised Newton's reasoning and generalized it to the case of sound waves of arbitrary character as distinct from simple harmonic waves. Actually, of course, he arrived at the same result as Newton for the velocity of sound in air. The reader is entitled to look upon this as evidence of either Newton's genius or else his good luck! This matter of the theoretical calculation of the velocity of sound in a gas forms a celebrated chapter in the history of physics. It is well known that, when the relevant data for air at 60°F ($= 16.25$°C) are substituted into the Newtonian equation for the velocity $c = \sqrt{p/\rho}$ (where p is the gas pressure and ρ the corresponding density), the result is 945 ft/sec or about 288 m/sec. Though this is definitely lower than the early Paris experimental results already mentioned, Newton at first considered that the order of magnitude agreement was satisfactory. However, when more and more accurate measurements confirmed the higher value, Newton evidently became worried, and in the second edition (1713) of his *Principia mathematica* revised his theory to try to produce better agreement with experiment [53]. What he did is not at all clear, but he evidently felt that some correction must be made for the impurity of the actual air. It would be an interesting exercise in the history of science to try to reconstruct the line of thought of a great mind like that of Newton in grappling with this problem.

Nothing more seems to have been done on the problem of the velocity of sound until Pierre Simon Laplace (1749–1827) in 1816 suggested that in the Newtonian and Lagrangian determinations an error had been made in using for the volume elasticity of the air (the reciprocal of the compressibility) the pressure itself, which is equivalent to assuming that the elastic motions of the air particles take place at constant temperature [44]. In view of the rapidity of the motions involved in the passage of the sound wave, however, it seemed to Laplace more reasonable to suppose that the compressions and rarefactions follow the adiabatic law; i.e., heat does not have a chance to flow out of the compressed region before compression gives way to rarefaction. But in this case the adiabatic elasticity is higher than the isothermal value in the ratio γ, where γ is the ratio of the specific heat of the gas at constant pressure to that at constant volume. According to this line of reasoning, the formula of Newton should be changed to $c = (\gamma p/\rho)^{\frac{1}{2}}$; since $\gamma > 1$ always, the newly calculated velocity of sound would necessarily be greater than the old and, therefore, closer to the experimental value. In 1816, when Laplace put forth his theory, though the existence of two specific heats of a gas was recognized, the value of γ was not known very precisely. Using the value 1.5 for air as obtained by the experimentalists LaRoche and Berard, Laplace found that $c = 345.9$ m/sec at 6°C, compared with the best experimental

value then available of 337.18 m/sec for this temperature. This was close enough for Laplace to feel he was on the right track. He returned to the problem later and included a chapter on the velocity of sound in his famous *Mécanique céleste* (1825) [43]. By that time, the well-known heat experimentalists, Clement and Desormes, had performed the classical experiment on the determination of γ (1819) and had found the value 1.35, leading to 332.9 m/sec for the sound velocity at 6°C. Some years later, the more accurate value 1.40 led to complete agreement between Laplace's theory and experiment. This theory is now so well established that it is common practice to determine γ for various gases by precision measurements of the velocity of sound.

The latter half of the 18th century and the first quarter of the 19th witnessed numerous attempts to theorize about waves in continuous media, based largely on the general solution for the wave equation (the equation that says in effect that the second time derivative of the quantity that "waves" is equal to the second space derivative of this same quantity multiplied by the square of the wave velocity) discovered, as noted above, by d'Alembert in 1747. Much attention, for example, was paid to waves on the surface of liquids like water. This work had value in connection with acoustics only to the extent that it led to increased confidence in the applicability of the wave equation to sound propagation in fluids. By 1800, the solution of the equation for aerial sound propagation in tubes subject to the boundary conditions at the ends had been pretty well established, and the predicted harmonic frequencies (normal modes or partials) checked with experiment with reasonable accuracy. There were indeed puzzling discrepancies in detail, not to be fully cleared up until the "end corrections" were understood some half-century later. Experimental techniques for sound measurement in tubes stayed rather crude for a long time; it was not until 1866 that A. Kundt (1839–1894) developed his simple but effective method of dust figures for studying experimentally the propagation of sound in tubes and in particular for the measurement of sound velocity in air and other gases from standing-wave patterns (nodes and loops) [40].

In the meantime, the much more difficult problem of the propagation of a compressional wave in a three-dimensional fluid medium had been attacked by S. D. Poisson [59] in a celebrated memoir of 1820. Three years before, in a similar lengthy memoir, Poisson had given the most elaborate theory up to that time of the transmission of sound in tubes, including the theory of stationary air waves in tubes of finite length, both open and closed [57]. He even considered the possibility of an end correction to take care of the fact that the condensation (the fractional change in density owing to the sound wave) cannot be considered as precisely zero at the open end, with the result that the observed resonance frequencies correspond to a length slightly greater than the actual geometrical length of the tube.

It remained, however, for Hermann von Helmholtz to give in 1860 a more thorough treatment of this whole problem. The special case of an abrupt change in cross section was also studied by Poisson, along with the reflection and transmission of sound at normal incidence on the boundary of two different fluids. Much modern work of practical significance was anticipated in this great memoir of Poisson.

The more difficult problem of the reflection and transmission of a plane sound wave incident obliquely on the boundary of two different fluids was solved by the self-taught Nottingham genius, George Green (1793–1841), in 1838. His memoir placed emphasis on the refraction of sound, and served to stress both the similarities and differences between the reflection and refraction of sound and light [29]. It will be recalled that sound waves in ideal fluids, being strictly compressional, are longitudinal, whereas light waves are transverse. Hence, light waves can be polarized, whereas sound waves in fluids cannot, in the ordinary sense. On the other hand, elastic waves in an extended solid can be both longitudinal and transverse, or more accurately irrotational and solenoidal. This was realized by Poisson in his 1829 study of isotropic elastic media [58]. At the time; this did not seem to have much significance for acoustics, though it had a very important bearing on the elastic solid theory of light, which was actively pursued during the middle decades of the 19th century. This early work, however, has taken on a new and tremendously great significance in the middle of the 20th century, owing to the interest in the propagation of elastic waves in structures like airplane fuselages and space missiles. The connection with modern geophysics (seismological waves) is also obvious.

So far, in this historical résumé of the propagation of sound, it has been tacitly assumed that the disturbance in the material medium being propagated as a sound wave (e.g., the excess density or pressure in a fluid) is very small as compared with the equilibrium value. In this case, the equation for wave propagation is linear. This is the type of equation to which the 18th-century investigators in acoustics gave their full attention. That its solution gives only an approximation to the actual sound transmission for relatively large disturbances was finally realized in the 19th century. However, Euler had already come close to the so-called finite-amplitude wave equation [20] in his famous 1759 memoir, "On the Propagation of Sound." Here, he worked out the equation of motion of a thin slice of air subject to pressure forces on its two sides. He would have obtained the precise result of 19th-century research had he not made an unaccountable algebraic error. His physics was in this case impeccable, but his mathematics went astray. Even Homer nods! At any rate, he realized that the normal linear wave equation, containing only second derivatives of the wave displacement function with respect to space and time, must be corrected by the inclusion of a nonlinear term whenever the gradient of the displacement is an appreciable fraction of unity.

It does not appear that nonlinear acoustic-wave propagation was taken up again seriously until around the middle of the 19th century, when the German mathematician, Georg F. B. Riemann (1826–1866), and the British mathematician and physicist, S. Earnshaw (1805–1888), more or less independently investigated certain special cases [19,61]. In particular, the results showed that in nonlinear propagation the velocity of propagation depends on the amplitude, so that it is only under very special conditions that a nonlinear wave of permanent type can be realized. It may be well to point out that some understanding of this situation had previously been reached by Poisson [56]. All this work led up to the theory of shock waves developed by G. G. Stokes, J. Challis, W. J. M. Rankine, H. Hugoniot, and Lord Rayleigh, among others.* Nonlinear acoustics has assumed great importance in the 20th-century development of the subject.

III. RECEPTION OF SOUND

In the story of acoustics up to very recent times, by far the most important sound receiver of interest has been the human ear, and the reception of sound was for a long time studied largely in connection with the behavior of this organ. In this respect, the human ear has had greater influence on the development of acoustics than human speech. The ear is remarkably versatile and sensitive. It has been established that its normal threshold of hearing is about 10^{-16} W/cm^2 sound intensity or 10^{-9} erg/cm^2 sec. If one takes the area of the normal eardrum as about 0.66 cm^2, this means that an average mechanical energy flow of only 6.6×10^{-10} erg/sec can produce the sensation of sound. A harmonic sound in the audible range of frequency will be identified if its duration is of the order of 0.1 sec. Thus, acoustic energy of the order of 6.6×10^{-11} erg is sufficient to excite identifiable sound in the ear. Energywise, the ear turns out to be fully as sensitive as the eye.

Many elaborate investigations of the anatomy of the ear have been made over the past century or so, and its acoustical behavior has been studied intensively. However, in spite of all this work, no completely acceptable theory of audition has emerged. Precisely how we hear still remains a puzzling problem in modern psychophysics.

After the relation between pitch and frequency had been established, it became an interesting task to determine the frequency limits of audibility. The French physicist, F. Savart (1791–1841), using fans and rotating toothed wheels in investigations around 1830, placed

* For detailed bibliographical references, see Ref. [13], p. 438.

the minimum audible frequency at 8 vibrations/sec and the upper limit at 24 000 vibrations/sec (Ref. [65]) (usually now referred to as cycles per sec or Hertz, after the famous German physicist, Heinrich Hertz, whose studies on electromagnetic waves were epoch making). Other investigators, such as L. F. W. A. Seebeck (1805–1849) (not to be confused with the discoverer of the thermoelectric effect), J. B. Biot (1774–1802), K. R. Koenig (1832–1901), and H. L. F. von Helmholtz (already mentioned earlier), obtained values for the lower limit [39,63,76] ranging from 16 to 32 cps. This serves to emphasize the rôle that individual differences play in hearing, even more noticeable indeed in the case of the upper frequency limit of audibility. The latter may not only vary considerably from individual to individual, but for each person usually decreases with age. Of course, the values in each case depend on the intensity, as has become clearer only in comparatively recent times. The most elaborate studies on audibility made during the 19th century were those of Koenig [38], who devoted a lifetime to the design and production of precision sources of sound of controlled frequency, such as tuning forks, rods, strings, and pipes. This was before the era of electroacoustic sources. Koenig was also responsible for the electrically driven fork.

The closely related problem of minimum sound intensity necessary for audibility (the auditory threshold) was apparently first studied jointly by A. Toepler (1836–1912) and L. Boltzmann (1841–1906) around 1870. By an ingenious use of optical interference, they were able to measure the maximum change in density (or effectively the maximum condensation) in a just-audible sound wave. Their experimental results lead to an audible threshold of about 10^{-11} W/cm², considerably in excess of that obtained by modern methods [69], but at any rate suggestive of the great sensitivity of the human ear. This collaboration of the great Viennese theoretical physicist with Toepler on an acoustic project is interesting in view of Boltzmann's principal fame as one of the creators of the statistical theory of gases.

In 1843, Georg Simon Ohm (1787–1854), the author of the famous law of electric currents, put forward a theory of audition according to which all musical tones arise from simple harmonic vibrations of definite frequency, and the particular quality or timbre of actual musical sounds is due to combinations of simple tones of commensurable frequencies [54]. He held, moreover, that the human ear is able to analyze any complex note into a set of simple harmonic tones, in terms of which it may be expanded mathematically by means of Fourier's theorem. This theorem, dating from 1822 and already mentioned earlier in this historical sketch, states that, subject to certain specific mathematical restrictions, which it is not necessary to discuss here, any arbitrary function of a variable t, say the time, can be expanded in a convergent series of circular functions whose arguments are integral multiples of a fundamental fre-

quency. If the arbitrary function is itself periodic in time, it can be represented in this way for all values of the variable t, whereas if the function is not periodic it can be so represented only over a finite time interval. This theorem has been of great value in the so-called analysis of sounds of all sorts.

Ohm's law stimulated a host of researches in what has come to be called physiological and psychological acoustics, i.e., the acoustics of hearing. The greatest of these in the 19th century were undoubtedly those of von Helmholtz, whose treatise, *Die Lehre von den Tonempfindungen als physiologische Grundlage für die Theorie der Musik* [75], already has been mentioned. It was translated into English by A. J. Ellis under the title *Sensations of Tone*, in 1895. Here, the author gave the first elaborate theory of the mechanism of the ear, the so-called resonance theory, in accordance with which the various elements of the basilar membrane in the cochlea resonate to certain frequencies in the sound falling on the ear. By this theory, he was able to justify theoretically the law of Ohm. von Helmholtz became greatly interested in the mechanical phenomenon of resonance, and in the course of his investigations invented the special type of sound resonator, since known by his name. This is simply a spherical chamber with an orifice. When a harmonic source of sound of appropriate frequency is brought close to the opening, if the sizes of the chamber and the orifice are just right, the sound will be very much amplified by the vigorous oscillatory motion of the air in the orifice. A large chamber resonates to a low frequency or low pitch tone and conversely. Such resonators have had wide use in modern acoustical research and applications. von Helmholtz showed that, when two tones of different frequencies are directed at an asymmetrical vibrator, the resulting vibration will contain frequencies that are the sum and difference of the original ones, and indeed many other linear combinations of the original frequencies will occur. He speculated that the eardrum is such an asymmetrical vibrator and, hence, predicted human ability to detect such summation and difference tones. This prediction has been verified. von Helmholtz's pioneer researches laid the groundwork for all subsequent research in the field of audition. One of the greatest physicists of the 19th century, he touched no field that he did not enrich with his experimental and theoretical genius.

Since the reception of sound by the ear in enclosed spaces like rooms, churches, theaters, and auditoriums in general is a common experience, it is proper that some attention should be paid here to the development of what has come to be called room or architectural acoustics. It was early recognized that some rooms are not satisfactory for good hearing and various devices were used to overcome the difficulties. These were at first simply geometrical contrivances such as sounding boards and other reflectors. A Boston physician, J. B. Upham [73], in 1853 wrote several papers indicating a much clearer grasp of the more important matter involved: namely, the reverberation or multiple reflection of the sound from all the surfaces of the room. He also showed how the reverberation time could be reduced by the installation of fabric curtains and upholstered furnishings. In 1856, Joseph Henry (1797–1878), the distinguished American physicist, who became the first Secretary of the Smithsonian Institutuion in Washington, made a study of auditorium acoustics that reflects a clear understanding of all the factors involved, though his suggestions were all of a qualitative character [31]. In spite of these early moves in what we now realize was the right direction, the subject was completely neglected by architects and during the latter half of the 19th century, attempts were often made to correct gross acoustical defects in rooms by such absurdly inadequate devices as stringing wires, etc. The modern quantitative foundation of architectural acoustics dates from the work of the Harvard University physicist, Wallace C. Sabine (1868–1919), who in 1900 hit upon the law connecting the reverberation time in a room (i.e., the time taken for any initially built-up sound intensity to decay to any arbitrarily chosen fraction of its original value, say the one millionth part) with the volume of the room and the amount of acoustic absorbing material in it [62]. This made applied architectural acoustics possible in the sense that any room can be designed for satisfactory hearing of speech, and to a certain extent for music also, though here subtle psychological factors enter, which provide trouble.

Special devices for the amplification of sound received by the ear have a long history. Horns, for example, are obviously of great antiquity. It is uncertain just when the suggestion arose that they might be used to improve the reception of sound. At all events, the Jesuit, Athanasius Kircher, mentioned previously [35], in 1650 designed a parabolic horn as a hearing aid as well as a speaking trumpet, and evidently realized the importance of the flare in the amplification of both received and emitted sound. Robert Hooke, who, in connection with his duties at the Royal Society, was forever trying out new ideas, experimented with ear trumpets and is even supposed to have suggested the possibility of a device to magnify the sounds of the body [30]. But it seems to have been reserved to the French physician, René Laënnec (1781–1826), actually to invent and employ the stethoscope for clinical purposes (1817–1819)*. The English physicist, Sir Charles Wheatstone (1802–1875), in 1827 developed a similar instrument that he termed a microphone [78], a name now applied to an electromechanical device (i.e., one in which motion produced by sound is made to induce electric currents) for the reception of sound. Electroacoustics had then hardly been thought of. Without it, modern experimental acoustics could not exist.

---

* See *Revue scientifique* (*Paris*), 4th Ser., 9, 42 (1898), for full references to Laënnec's work on the stethoscope.

All through the historical development of physics, one can detect a tendency to reduce the observation of physical phenomena and particularly experimental measurements to something that can be *seen*. Practically all physical measurements involve this principle and employ a pointer or a spot of light moving on a scale. It was inevitable that attempts would be made to study sound phenomena visually; this was, of course, especially necessary for the investigation of sounds whose frequencies lie above the range of audibility of the ear, the so-called ultrasonic radiation. One of the first moves in this direction was the observation by the American physicist, John LeConte (1818–1891), that musical sounds can produce jumping in a gas flame if the pressure is properly adjusted (1858) [45]. The sensitive flame, as it later came to be called, was developed to a high pitch of excellence by the English physicist, John Tyndall (1820–1893), who used it for the detection of high-frequency (inaudible) sounds and for the study of the reflection, refraction, and diffraction of sound waves [71,72]. It still provides a very effective lecture demonstration, but, for practical applications, it has been superseded in the 20th century by various types of electrical microphones coupled to the cathode-ray oscilloscope.

In the endeavor to make visible the form of a sound wave, Koenig [37] about 1860 invented the manometric flame device, which consists of a box through which gas flows to a burner. One side of the box is a flexible membrane. When sound waves strike on the membrane, the alternating changes in pressure produce corresponding fluctuations in the flame that can be made visible by reflecting the light of the flame from a rapidly rotating mirror. Another attempt to visualize sound waves was made by the 19th-century French proofreader, editor, and amateur scientist, Edouard-Leon Scott de Martinville, in 1857 in his "phonautograph" in which a flexible diaphragm at the throat of a receiving horn was attached to a stylus, which in turn touched a smoked rotating-drum surface and traced out a curve corresponding to the incident sound [17]. This was the precursor of the phonograph. An equally ambitious attempt along similar lines was made by Eli Whitney Blake [4] (1836–1895), the first Hazard Professor of Physics in Brown University, who in 1878 made a microphone by attaching a small metallic mirror to a vibrating disk at the back of a telephone mouthpiece. By reflecting a beam of light from the mirror, Blake succeeded in photographing the sounds of human speech. Such studies were much advanced by the American physicist, D. C. Miller (1866–1941), who invented a similar instrument in the "phonodeik," and made very elaborate photographs of sound-wave forms [52]. These various devices of Scott, Blake, and Miller were, of course, the predecessors of the cathode-ray oscilloscope, so useful in modern acoustical research.

16

IV. LORD RAYLEIGH AND MODERN ACOUSTICS

The publication of Lord Rayleigh's *Theory of Sound* [67] in 1877 marks in a sense both the end of what may be called the classical era in acoustics and the beginning of the modern age of sound. Rayleigh was a product of the Cambridge University mathematical school of the mid nineteenth century and graduated there as Senior Wrangler in the Mathematical Tripos of 1865. He was, therefore, well equipped to handle analytically problems encountered in his reading of von Helmholtz's treatise *Sensations of Tone* [75,76]. This led him to conceive the desirability of writing a substantial and authoritative treatise on the whole field of acoustics to bring together in one place material scattered for the most part in articles in learned society journals. At the same time, he took occasion to present in detail some of his own contributions. The result was a work that has long stood as a monument of physical literature, with a tremendous influence on the subsequent development of the science of acoustics, particularly on the analytical side.

The work naturally divides itself in two parts, of which the first relates to mechanical-vibration phenomena of all kinds, including the oscillations of strings, bars, membranes, and plates. Motions of such structures are, of course, closely connected with the production of sound, as has been stressed earlier in this historical review. A valuable feature of the treatise is the insistence on the establishment of general principles as well as applications to special cases of practical significance. Rayleigh was an accomplished applied mathematician and developed helpful techniques for the solution of difficult vibration problems. One of these, the basis for the so-called Rayleigh–Ritz method, has had wide modern applications, not merely in studying the vibrations of solid structures, but also in quantum mechanics. Rayleight never lost sight of the physical meaning inherent in natural phenomena and his analysis always has the merit of being applicable in practice.

The second part of Rayleigh's "Sound" deals principally with acoustical propagation through fluid media. Here he had to handle such difficult matters as, for example, the diffraction of sound waves around obstacles and the general scattering sound suffers when passing through a medium with many suspended particles, e.g., bubbles in water. Acoustic diffraction and scattering form mathematically a much more difficult subject than the corresponding phenomena in light, since the wavelength of ordinary audible sound is of the order of magnitude of the dimensions of the obstacles themselves. Much attention is also paid to the geometrical characteristics of the acoustic radiation from vibrating objects, e.g., pulsating spheres or oscillating disks, so as to produce "beams" of sound. So thoroughly were problems of this kind treated and so completely and clearly did Rayleigh summarize the work of previous investigators on such things as the attenuation of sound in fluids by various dissipative mechanisms, that, when the second enlarged and revised edition of *Theory of Sound* [67] appeared between 1894 and 1896, it was widely felt that the whole subject of acoustics as a branch of physics was now complete and that there was nothing more to learn. There were undoubtedly many who considered that the future of the field lay in the hands of the engineers, as indeed was already proving to be the case with the basic theory of electricity and magnetism developed by Ampere, Faraday, and Maxwell, whose scientific efforts laid the foundations of electrical engineering. There was certainly some justification for this feeling about acoustics at that time, not so much indeed because there were no more interesting acoustical phenomena to investigate, but because the experimental means by which these investigations could be practically carried out were not yet available. For example, the work of Rayleigh and his contemporaries strongly suggested that a host of interesting properties would be associated with sound waves of frequency well above the audible limit. But, when Rayleigh's "Sound" was published, the only practical source of such radiation was a bird whistle!

As one looks back into the physics of the 19th century, it seems almost incredible that electromechanical effects were not used earlier as sources of sound of a wide range of frequency. The main difficulty, of course, lay in the inadequacy of the means of producing electrical oscillations as well as the coupling of these oscillations to solid vibrators. It is true that the piezoelectric effect was discovered by the brothers Pierre and Jacques Curie [15] in 1880, and this suggested that, if there were some way in which to produce alternating positive and negative electric charges on the opposite faces of a properly cut quartz crystal, one could make it vibrate. The successful exploitation of this effect as a source and receiver of sound, however, had to await the invention of the vacuum-tube oscillator and amplifier, which did not come until the work of Fleming and DeForest in the first two decades of the 20th century.

As a matter of fact, the dawn of the 20th century saw many fundamental problems in acoustics unsolved. On the biological side, the nature of hearing was by no means wholly understood, since the relation between the anatomy of the ear and associated nervous system on the one hand and the observed phenomena of audition on the other was by no means clear. Detailed studies of speech had as yet been impractical owing to inadequate means of speech analysis. On the physical side—though there existed a theory for the attenuation of sound in fluid media like the atmosphere in terms of the effect of the transport properties viscosity and heat conduction—it was realized that the results predicted by such effects were not in agreement with experiment, being in general much smaller than the relevant observed values. This was recognized indeed by Lord Rayleigh, who made a shrewd suggestion of a plausible solution of the difficulty in terms of so-called relaxation effects. These were not fully explored until the century was one-quarter over [66].

As has already been suggested, perhaps the biggest obstacle to the further development of acoustics, both as a science and as a branch of technology, around 1900 and for some years thereafter, was the lack of appropriate sources and receivers of sound or transducers, as they came to be called. It was recognized that a great many physical phenomena—e.g., mechanical, thermal, electrical, magnetic—can give rise to sound radiation and conversely can act to transform sound into other physical effects. It was also clear that such transformation or transduction could be of enormous practical importance. One obvious example is the telephone, by which sound can be transferred without attenuation over much greater distances than would be possible through the normal sound-transmitting medium itself. This problem was first solved (according to the decision of the courts in the United States of America after lengthy patent litigation) by Alexander Graham Bell around 1876. The transducing mechanism in this case was, of course, electromagnetic in character. The influence of this invention on the future development of acoustics has been incalculable. In order to improve the telephone, huge amounts of money have been expended on research into every aspect of human communication. As a result of the work at the Bell Telephone Laboratories, for example, we now know much more than might ever have been reasonably expected from mere human curiosity about the way that the human being hears and speaks. We also know how to produce sounds of all frequencies from a few cycles per second up to several thousand million cycles per second and can study efficiently the behavior of solid, liquid, and gaseous media exposed to such sounds. At the same time, it has become possible to study the interaction between high-frequency radiation, now called ultrasonic, and other physical effects such as electric, magnetic, high and low temperature, large ranges of pressures, and the like.

It is scarcely possible to do justice in a brief account to the advances in acoustics in the past half-century, since it has shared in the fantastically increasing pace of the progress of science as a whole. To the layman, the most obvious developments have been in the technological field rather than in the pure science category. Who is not acquainted with radio, talking motion pictures, and television, and with the rôle that acoustics plays in all of them? To these must be added the whole realm of the recording and reproducing of sound, the production of phonograph disks and hi-fi sets, not to mention tape recorders. Public-address systems and other forms of sound-reinforcement devices have brought the voices of human speakers as well as music to large audiences. Less well known but important for the national security are the transducers designed for transmitting sound underwater in the detection of submarines and other objects. In this field, as in the others, the interplay be-

tween pure and applied acoustics has been particularly evident. As the power, sensitivity, and efficiency of underwater transducers have increased, greater knowledge of the acoustical properties of sea water has become necessary in order to make the use of this more sophisticated instrumentation really worthwhile. Hence a new field of acoustical research has been opened up as a branch of physical oceanography. In similar fashion, the improved instrumentation for the study of speech and hearing has stimulated the creation of wholly new branches of physiology and psychology. The construction of high-power ultrasonic transducers has introduced a new tool for medical research both on the diagnostic and therapeutic levels. Even music is being influenced by the comtemporary ways of producing and amplifying sound. The list could be extended to every form of human activity in which sound enters, and there are few for which this is not true.

However, it is important to point out that there exist many as yet unsolved problems in basic acoustics. These involve particularly the interactions of high-frequency sound radiation with matter in its various phases. Successful attack on them awaits the solution of the more purely technological problem of the extension of the practical frequency limit of ultrasonic radiation. The present limit (1966) is about 10^{10} cps. If frequencies of the order of 10^{11}–10^{14} cps could be realized, our understanding of the nature of the solid, liquid, and gaseous states would be much enhanced. Recent research in which the optical laser has been used as a source of ultrasonic radiation holds out great promise. This is not the place for details. They can be found in the pages of *The Journal of the Acoustical Society of America* or in the literature referenced therein [1,34].

In fact, the whole history of acoustics during the past third of a century will be found in the articles in the JOURNAL or in its bibliography.

The writer desires to make grateful acknowledgments to all the sources from which the material of this review has been garnered. He owes particular acknowledgment to D. C. Miller's *Anecdotal History of the Science of Sound"* [51], F. V. Hunt's *Electroacoustics* [33], as well as C. Truesdell's historical references [70] to the works of Euler, already referred to in the text. The author also wishes to express deep appreciation, for valuable assistance in connection with the Bibliography, rendered by the staff of the Physical Sciences Library of Brown University, in particular Antigone C. Coolidge.

It has been asserted by some physicists who are carried away by the glamor of high-energy physics and the properties of the solid state that the future of a so-called "classical" field of physics like acoustics lies wholly in its technological applications and that as physical science it is "played out." This is an error, as the history of science in the last half-century amply shows and the remarks of the preceding paragraph document. There is no ground for assuming that man will ever run out of questions about acoustics any more than

642   volume 39   number 4   1966

18

he will run out of questions about the nucleus and the theoretical particles that inhabit it or can be created from it. What is, of course, true is that as investigation proceeds the boundary lines between the various types of natural phenomena that mankind has artificially erected for purposes of convenience are becoming fuzzier and more unrealistic. The aim of the science of the future is a meaningful synthesis. In this, acoustics will claim its rightful place.

Bibliography

1. Acoustics Research Survey Committee, "A Survey of the Needs and Opportunities for Research in Acoustics," J. Acoust. Soc. Am. 37, 392–395 (1965).
2. D. Bernoulli, *Réflexions et éclaircissements sur les nouvelles vibrations des cordes, exposées dans les mémoires de l'Académie, de 1747 et 1748* (Royal Academy, Berlin, 1755), p. 147 ff.
3. J. B. Biot, Ann. Chim. Phys. 13, 5 ff (1808).
4. E. W. Blake, "A Method of Recording Articulate Vibrations by Means of Photography," Am. J. Sci. Arts, Ser. 3, 16, 54 ff (1878).
5. R. Boyle, *New Experiments Physico–Mechanical, Touching the Air* (Miles Fleshner, London, 1682), 3rd ed., Expt. 27, p. 103.
6. E. Cherbuliez, "Geschichtliche Übersicht der Untersuchungen über die Schallfortpflanzungsgeschwindigkeit in der Luft," Mitt. Naturforsch. Ges. Bern (1870), p. 141 ff; (1871), p. 1 ff.
7. E. F. F. Chladni, *Die Akustik* (Breitkopf & Hartel, Leipzig, 1802).
8. E. F. F. Chladni, *Entdeckungen über die Theorie des Klanges* (Weidmann, Erben & Reich, Leipzig, 1787).
9. R. F. A. Clebsch, *Theorie der Elasticität fester Körper* (B. G. Teubner, Leipzig, 1862).
10. M. R. Cohen and I. E. Drabkin, *A Source Book in Greek Science* (McGraw-Hill Book Co., Inc., New York, 1948), pp. 286–310.
11. J.-D. Colladon, *Souvenirs et mémoires: Autobiographie de J.-Daniel Colladon* (Aubert-Schuchardt, Geneva, 1893).
12. J.-D. Colladon and J. K. F. Sturm, "Mémoire sur la compression des liquides et la vitesse du son dans l'eau," Ann. Chim. Phys. 36, 113 ff, 225 ff (1827).
13. R. Courant and K. O. Friedrichs, *Supersonic Flow and Shock Waves* (Interscience Publishers, Inc., New York, 1948), p. 438.
14. P. Curie, *Œuvres de Pierre Curie* (Gauthier-Villars, Paris, 1908).
15. P. Curie and J. Curie, Compt. Rend. 91, 294 ff, 383 ff (1880).
16. J. Le R. d'Alembert, *Recherches sur le courbe que forme une corde tendue mise en vibration* (Royal Academy, Berlin, 1747), p. 214 ff.
17. E.-L. S. de Martinville, "Inscription automatique des sons de l'air au moyen d'une oreille artificielle," Compt. Rend. 53, 108 ff (1861).
18. W. Derham, "Experimenta et observationes de soni motu aliisque ad id attinentibus," Phil. Trans. Roy. Soc. (London) 26, 1–35 (1708).
19. S. Earnshaw, "On the Mathematical Theory of Sound," Phil. Trans. Roy. Soc. (London) 150, 133 ff (1858).
20. L. Euler, "De la Propagation du son," Mém. Acad. Sci. Berlin 15, 185–209 (1759); or cf. *Leonhardi Euleri Opera Omnia* (III) (B. G. Teubner, Leipzig, 1926), Vol. 1, pp. 428–451.
21. L. Euler, "De Motu corporum flexibilium," Opusculi 3, 88 ff (1751). Presentation date 5 November 1744 [see *Leonhardi Euleri Opera Omnia* (II) (Orell Füssli, Zürich, 1947)], Vol. 10, p. 177.
22. L. Euler, "De Motu vibratorio tympanorum," Acad. Sci. St. Petersburg 10, 243 ff (1746). Presented to the Academy in 1761 and 1762.
23. L. Euler, "Dissertatio physica de sono," Basel, 1727, in *Leonhardi Euleri Opera Omnia* (III) (B. G. Teubner, Leipzig, 1926), Vol. 1, p. 182 ff.
24. L. Euler, "Eclaircissements plus détailles sur la generation et la propagation du son, et sur la formation de l'echo," Mém. Acad. Sci. Berlin 21, 335–363 (1765); or cf. *Leonhardi Euleri Opera Omnia* (III) (B. G. Teubner, Leipzig, 1926), Vol. 1, pp. 540–567.
25. L. Euler, *Remarques sur les mémoires précedents de M. Bernoulli* (Royal Academy, Berlin, 1753), pp. 1–196.
26. J. B. J. Fourier, *La Théorie analytique de la chaleur* (F. Didot, Paris, 1822).
27. Galileo Galilei, *Dialogues concernng Two New Sciences* (1638), H. Crew and A. DeSalvio, translators (Northwestern University Press, Evanston, Ill., 1939), pp. 95–108.
28. D. E. Gray et al., Eds., *American Institute of Physics Handbook* (McGraw-Hill Book, Co., Inc., New York, 1963), 2nd ed., 3–64, 3–65.
29. G. Green, Trans. Cambridge Phil. Soc. 6, 403 ff (1838).
30. R. T. Gunther, *Early Science in Oxford*, Vol. 6: *The Life and Work of Robert Hooke* (Oxford University Press, London, 1930), pp. 330–331.
31. J. Henry, "Acoustics Applied to Public Buildings," in *Annual Report of the Smithsonian Institution* (Smithsonian Institution, Washington, D. C., 1856), p. 221 ff.
32. R. Hooke, "A Description of Helioscopes and Some Other Instruments," reprinted in R. T. Gunther, *Early Science in Oxford* (Oxford University Press, London, 1931), Vol. 8, pp. 119–152.
33. F. V. Hunt, *Electroacoustics* (John Wiley & Sons, Inc., New York, 1954).
34. F. V. Hunt, T. F. Hueter, R. W. Morse, T. Litovitz, R. O. Fehr, W. J. Fry, M. Lawrence, and G. A. Miller, in *Symposium on Unsolved Problems in Acoustics*, J. Acoust. Soc. Am. 30, 375–398 (1958).
35. A. Kircher, *Musurgia universalis* (Francisci Corbelletti, Rome, 1650).
36. G. R. Kirchhoff, "Über das Gleichgewicht und die Bewegung einer elastischen Scheibe," Crelle's J. 40, 51 ff (1850).
37. K. R. Koenig, "Die manometrische Flammen," Poggendorf's Ann. Physik 146, 161 ff (1872).
38. K. R. Koenig, *Quelques expériences d'acoustique* (A. Lahure, Paris, 1882).
39. K. R. Koenig, Ann. Physik 69, 626 ff, 721 ff (1899).
40. A. Kundt, Poggendorf's Ann. Physik 127, 497 ff (1866).
41. J. L. Lagrange, *Mécanique analytique* (Desaint, Paris, 1788).
42. J. L. Lagrange, "Recherches sur la nature et la propagation du son," in *Œuvres de Lagrange* (Gauthier-Villars, Paris, 1867), Vol. 1, p. 39 ff.
43. P. S. Laplace, *Mécanique céleste* (J. B. M. Duprat, Paris, 1825).
44. P. S. Laplace, Ann. Chim. Phys. 3, 238 ff (1816).
45. J. LeConte, "On the Influence of Musical Sounds on the Flame of a Jet of Coal Gas," Phil. Mag. 25, 235 ff (1858).
46. L. M. A. Lenihan, "Mersenne and Gassendi. An Early Chapter in the History of Sound," Acustica 2, 96–99 (1951).
47. J. M. A. Lenihan, "The Velocity of Sound in Air," Acustica 2, 205–212 (1952).

48. **R. B. Lindsay,** "Pierre Gassendi and the Revival of Atomism in the Renaissance," Am. J. Phys. **13,** 235–242 (1945).

49. **M. Mersenne,** *Harmonicorum Libri XII* (G. Baudry, Paris, 1635).

50. **M. Mersenne,** *Harmonie universelle* (S. Cramoisy, Paris, 1636). For English translation, see J. Hawkins, *General History of the Science and Practice of Music* (J. A. Novello, London, 1853), pp. 600–616, 650 ff.

51. **D. C. Miller,** *Anecdotal History of the Science of Sound* (The Macmillan Co., New York, 1935), p. 20.

52. **D. C. Miller,** *The Science of Musical Sounds* (The Macmillan Co., New York, 1916), p. 78 ff.

53. **Sir Isaac Newton,** *Philosophiae naturalis principia mathematica* (Joseph Streater, for the Royal Society, London, 1687; Cambridge: 2nd ed., 1713; London: 3rd ed., 1726). The usually accepted standard English version is the revision of F. Cajori of the translation by A. Motte (1729) (University of California Press, Berkeley, 1946). Truesdell does not consider this "altogether reliable."

54. **G. S. Ohm,** Poggendorf's Ann. Physik 59, 497 ff (1843).

55. **Sir Richard Paget,** *Vowel Resonances* (International Phonetic Association, London, 1922).

56. **S. D. Poisson,** "Mémoire sur la théorie du son," J. Ecole Polytech. 14ème Cah. **7,** 319 ff (1808).

57. **S. D. Poisson,** "Sur le Mouvement des fluides élastiques dans les tuyaux cylindriques, et sur la théorie des instruments à vent," Mém. Acad. Roy. Sci. Inst. France, Année 1817, 2, 305 ff (1819).

58. **S. D. Poisson,** "Mémoire sur l'équilibre et le mouvement des corps élastiques," Mém. Acad. Roy. Sci. Inst. France (2) **8,** 357 ff (1829).

59. **S. D. Poisson,** "Sur l'Intégration de quelques équations linéaires aux differences partielles, et particulièrement de l'équations générale du mouvement des fluides élastiques," Mém. Acad. Roy. Sci. Inst. France, Année 1817, **3,** 121 ff (1820).

60. **O. Reynolds,** "Memoir of James Prescott Joule," Lit. & Phil. Soc., Manchester (1892), p. 101 ff.

61. **G. F. B. Riemann,** "Über die Fortpflanzung ebener Luftwellen von endlichen Schwingungsweite," Göttingen Abh. 8, 43 ff (1858–1859).

62. **W. C. Sabine,** *Collected Papers on Acoustics* (Harvard University Press, Cambridge, Mass., 1927); reprinted by Dover Publications, Inc., New York, 1964, with a new introduction by F. V. Hunt).

63. **L. F. W. A. Seebeck,** Poggendorf's Ann. Physik **53,** 417 ff (1841).

64. **J. Sauveur,** *Système général des intervalles des sons* (L'Académie Royale des Sciences, Paris, 1701), p. 297 ff.

65. **F. Savart,** Poggendorf's Ann. Physik 20, 290 ff (1830).

66. **J. W. Strutt Lord Rayleigh,** "On the Cooling of Air by Radiation and Conduction and on the Propagation of Sound," Phil. Mag. **47,** 308 ff (1899).

67. **J. W. Strutt Lord Rayleigh,** *Theory of Sound* (MacMillan & Co. Ltd., London, 1877; 2nd ed., revised and enlarged 1894, reprinted 1926, 1929). First American ed. with a historical introduction by R. B. Lindsay (Dover Publications, Inc., New York, 1945). The author would like to make grateful acknowledgment to Dover Publications, Inc., for kind permission to use in the present sketch some of the material in the historical introduction just noted.

68. **B. Taylor,** "De Motu nervi tensi," Phil. Trans. Roy. Soc. (London) **28,** 11–21 (1713).

69. **A. Toepler and L. Boltzmann,** Poggendorf's Ann. Physik **141,** 321 ff (1870).

70. **C. Truesdell,** "The Rational Mechanics of Flexible or Elastic Bodies; 1638–1788," Introduction to *Leonhardi Euleri Opera Omnia* (II), Vols. 10, 11 (Orell Füssli, Zürich, 1960). A mine of detailed information on the history of acoustics in the 17th and 18th centuries.

71. **J. Tyndall,** *Sound* (D. Appleton & Co., New York, 1867), Chap. 6

72. **J. Tyndall,** Phil. Mag. **33,** 92 ff, 375 ff (1867).

73. **J. B. Upham,** "A Consideration of Some of the Phenomena and Laws of Sound and Their Application in the Construction of Buildings Designed Especially for Musical Effects," Am. J. Sci. Arts. Ser. I, **65,** 215 ff, 348 ff (1853); 66, 21 ff (1853).

74. **O. von Guericke,** *Nova experimenta de vacuo spatio* (Janssonicus à Waesberge, Amsterdam, 1672).

75. **H. L. F. von Helmholtz,** *Die Lehre von den Tonempfindungen als physiologische Grundlage für die Theorie der Musik* (F. Vieweg & Son, Braunschweig, 1865), 2nd ed.

76. **H. L. F. von Helmholtz,** *Sensation of Tone,* English transl. of 3rd German ed. by A. J. Ellis (Longmans Green Co., London, 1895), pp. 11, 13, 162, 182, 372.

77. **J. Wallis,** "Dr. Wallis's Letter to the Publisher concerning a New Musical Discovery, written from Oxford, March 19, 1676," Phil. Trans. Roy. Soc. (London) **13,** No. 134, 839 ff (1677).

78. **Sir Charles Wheatstone,** *The Scientific Papers of Sir Charles Wheatstone* (The Physical Society, London, 1879).

79. **W. Whewell,** *History of the Inductive Sciences* (D. Appleton & Co., New York, 1874), 3rd ed., p. 24.

20

2

De Anima
De Audibilibus

ARISTOTLE

Aristotle (384–322 B.C.), the celebrated Greek philosopher and founder of the Lyceum in Athens, wrote widely on scientific subjects, though he is best known for his work on logic and other branches of philosophy. He dealt rather sparingly with acoustics, at least in such of his writings that survive. His great book *Physica* (On Physics) does not mention sound, though it deals extensively with the meaning and properties of motion. Aristotle's principal references to acoustics are found in one of the books of his treatise *De Anima* (On the Soul) and in the treatise *De Audibilibus*. Extracts from these are presented here. They show that Aristotle had some notion that sound is connected with the motion of air but it is by no means clear that he grasped the wave concept. He understood the nature of the echo. He was also interested in the biological aspects of sound, particularly speech.

The following selections have been reprinted from "A Source Book of Greek Science" by M. R. Cohen and I. E. Drabkin (Harvard University Press).

Aristotle, *On the Soul* II. 8.   Translation of R. D. Hicks

There are two sorts of sound, one a sound which is operant, the other potential sound.   For some things we say have no sound, as sponge, wool; others, for example, bronze and all things solid and smooth, we say have sound, because they can emit sound, that is, they can produce actual sound between the sonorous body and the organ of hearing.   When actual sound occurs it is always of something on something and in something, for it is a blow which produces it.   Hence it is impossible that a sound should be produced by a single thing, for, as that which strikes is distinct from that which is struck, that which sounds sounds upon something.   And a blow implies spatial motion.   As we stated above, it is not concussion of any two things taken at random which constitutes sound.   Wool, when struck, emits no sound at all, but bronze does, and so do all smooth and hollow things; bronze emits sound because it is smooth, while hollow things by reverberation produce a series of concussions after the first, that which is set in motion being unable to escape.   Further, sound is heard in air, and though more faintly, in water.[1]   It is not the air or the water, however, which chiefly determine the production of sound: on the contrary, there must be solid bodies colliding with one another and with the air: and this happens when the air after being struck resists the impact and is not dispersed.   Hence the air must be struck quickly and forcibly if it is to give forth sound; for the movement of the striker must be too rapid to allow the air time to disperse: just as would be necessary if one aimed a blow at a heap of sand or a sandwhirl, while it was in rapid motion onwards.

Echo is produced when the air is made to rebound backwards like a ball from some other air which has become a single mass owing to its being within a cavity which confines it and prevents its dispersion.   It seems likely that echo is always produced, but is not always distinctly audible: since surely the same thing happens with sound as with light.   For light is always being reflected; else light would not be everywhere, but outside the spot where the sun's rays fall there would be darkness.   But it is not always reflected in the same way as it is from water or bronze or any other smooth surface; I mean, it does not always produce the shadow, by which we define light.

---

[1] As a matter of fact, water is better than air as a medium for sound.   [Edd.]

Void[1] is rightly stated to be the indispensable condition of hearing. For the air is commonly believed to be a void, and it is the air which causes hearing, when being one and continuous it is set in motion. But, owing to its tendency to disperse, it gives out no sound unless that which is struck is smooth. In that case the air when struck is simultaneously reunited because of the unity of the surface; for a smooth body presents a single surface.

That, then, is resonant which is capable of exciting motion in a mass of air continuously one as far as the ear. There is air naturally attached to the ear. And because the ear is in air, when the external air is set in motion, the air within the ear moves.

[Aristotle], *De Audibilibus* 800a1–*b*3, 803*b*18–804*a*8.[2] Translation of T. Loveday and
E. S. Forster (Oxford, 1913)

All sounds, whether articulate or inarticulate, are produced by the meeting of bodies with other bodies or of the air with bodies, not because the air assumes certain shapes, as some people think, but because it is set in motion in a way in which, in other cases, bodies are moved, whether by contraction or expansion or compression, or again when it clashes together by an impact from the breath or from the strings of musical instruments. For, when the nearest portion of it is struck by the breath which comes into contact with it, the air is at once driven forcibly on, thrusting forward in like manner the adjoining air, so that the sound travels unaltered in quality as far as the disturbance of the air manages to reach. For, though the disturbance originates at a particular point, yet its force is dispersed over an extending area, like breezes which blow from rivers or from the land. Sounds which happen for any reason to have been stifled where they arise, are dim and misty; but, if they are clear, they travel far and fill all the space around them.

We all breathe in the same air, but the breath and the sounds which we emit differ owing to structural variations of the organs at our disposal, through which the breath must travel in its passage from within—namely the windpipe, the lungs, and the mouth. Now the impact of the breath upon the air and the shapes assumed by the mouth make most difference

---

[1] Used here not in the sense of a vacuum but of a medium like air. [Edd.]

[2] The connection of sound with motion and impulse was made very early in Greek acoustical theory. The importance of the air, or other medium, in transmitting the impulse was also appreciated (see, e.g., the selection from Aristotle, *De Anima*, quoted above.) In the *De Audibilibus*, which seems to be a fragment of a longer work, the explanation is somewhat different from Aristotle's. There is a definite approach to a wave theory, the portions of air communicating the impulse to proximate portions. Further development of a wave theory of sound is found in Boethius (p. 291, below), who in all probability draws upon a Greek source.

The authorship of *De Audibilibus* is doubtful. Some have ascribed the work to Strato of Lampsacus, other to Heraclides of Pontus. [Edd.]

to the voice. This is clearly the case; for indeed all the differences in the kinds of sounds which are produced proceed from this cause, and we find the same people imitating the neighing of horses, the croaking of frogs, the song of the nightingale, the cries of cranes, and practically every other living creature, by means of the same breath and windpipe, merely by expelling the air from the mouth in different ways. Many birds also imitate by these means the cries of other birds which they hear.

As to the lungs, when they are small and inexpansive and hard, they cannot admit the air nor expel it again in large quantities, nor is the impact of the breath strong and vigorous. For, because they are hard and inexpansive and constricted, they do not admit of dilatation to any great extent, nor again can they force out the breath by contracting after wide distension; just as we ourselves cannot produce any effect with bellows, when they have become hard and cannot easily be dilated and closed.

Voices are thin, when the breath that is emitted is small in quantity. Children's voices, therefore, are thin, and those of women and eunuchs, and in like manner those of persons who are enfeebled by disease or over-exertion or want of nourishment; for owing to their weakness they cannot expel the breath in large quantities. The same thing may be seen in the case of stringed instruments; the sounds produced from thin strings are thin and narrow and "fine as hairs," because the impacts upon the air have only a narrow surface of origin. For the sounds that are produced and strike on the ear are of the same quality as the source of movement which gives rise to the impacts; for example, they are spongy or solid, soft or hard, thin or full. For one portion of the air striking upon another portion of the air preserves the quality of the sound, as is the case also in respect of shrillness and depth; for the quick impulsions of the air caused by the impact, quickly succeeding one another, preserve the quality of the voice, as it was in its first origin. Now the impacts upon the air from strings are many and are distinct from one another, but because, owing to the shortness of the intermittence, the ear cannot appreciate the intervals, the sound appears to us to be united and continuous. The same thing is the case with colours; for separate coloured objects appear to join, when they are moved rapidly before our eyes. The same thing happens, too, when two notes form a concord; for owing to the fact that the two notes overlap and include one another and cease at the same moment, the intermediate constituent sounds escape our notice. For in all concords more frequent impacts upon the air are caused by the shriller note, owing to the quickness of its movement; the result is that the last note strikes upon our hearing simultaneously with an earlier sound produced by the slower impact. Thus, because, as has been said, the ear cannot perceive all the constituent sounds, we seem to hear both notes together and continuously.

3

Acoustics of the Theater

Vitruvius

Marcus Vitruvius Pollio (commonly called Vitruvius), Roman engineer and architect, flourished about 25 B.C., but his precise dates are unknown, as are indeed the details of his life. In connection with his engineering activities he wrote a treatise in ten books, *De Architectura (On Architecture)*, which has survived. He was much interested in the acoustics of the classical theater. We present a few extracts from his treatise in which he can be said to have provided the first ideas on architectural acoustics.

The following selections are reprinted from Vitruvius' "Ten Books on Architecture" (Dover Publications Inc.).

HARMONICS

1. HARMONICS is an obscure and difficult branch of musical science, especially for those who do not know Greek. If we desire to treat of it, we must use Greek words, because some of them have no Latin equivalents. Hence, I will explain it as clearly as

I can from the writings of Aristoxenus, append his scheme, and
define the boundaries of the notes, so that with somewhat careful attention anybody may be able to understand it pretty easily.

2. The voice, in its changes of position when shifting pitch,
becomes sometimes high, sometimes low, and its movements are
of two kinds, in one of which its progress is continuous, in the
other by intervals. The continuous voice does not become stationary at the "boundaries" or at any definite place, and so the
extremities of its progress are not apparent, but the fact that
there are differences of pitch is apparent, as in our ordinary
speech in *sol, lux, flos, vox;* for in these cases we cannot tell at
what pitch the voice begins, nor at what pitch it leaves off, but
the fact that it becomes low from high and high from low is apparent to the ear. In its progress by intervals the opposite is the
case. For here, when the pitch shifts, the voice, by change of
position, stations itself on one pitch, then on another, and, as it
frequently repeats this alternating process, it appears to the
senses to become stationary, as happens in singing when we produce a variation of the mode by changing the pitch of the voice.
And so, since it moves by intervals, the points at which it begins
and where it leaves off are obviously apparent in the boundaries
of the notes, but the intermediate points escape notice and are
obscure, owing to the intervals.

3. There are three classes of modes: first, that which the
Greeks term the enharmonic; second, the chromatic; third, the diatonic. The enharmonic mode is an artistic conception, and therefore execution in it has a specially severe dignity and distinction.
The chromatic, with its delicate subtlety and with the "crowd-
ing" of its notes, gives a sweeter kind of pleasure. In the diatonic, the distance between the intervals is easier to understand,
because it is natural. These three classes differ in their arrangement of the tetrachord. In the enharmonic, the tetrachord consists of two tones and two "dieses." A diesis is a quarter tone;
hence in a semitone there are included two dieses. In the chromatic there are two semitones arranged in succession, and the

27

third interval is a tone and a half. In the diatonic, there are two consecutive tones, and the third interval of a semitone completes the tetrachord. Hence, in the three classes, the tetrachords are equally composed of two tones and a semitone, but when they are regarded separately according to the terms of each class, they differ in the arrangement of their intervals.

4. Now then, these intervals of tones and semitones of the tetrachord are a division introduced by nature in the case of the voice, and she has defined their limits by measures according to the magnitude of the intervals, and determined their characteristics in certain different ways. These natural laws are followed by the skilled workmen who fashion musical instruments, in bringing them to the perfection of their proper concords.

5. In each class there are eighteen notes, termed in Greek φθόγγοι, of which eight in all the three classes are constant and fixed, while the other ten, not being tuned to the same pitch, are variable. The fixed notes are those which, being placed between the moveable, make up the unity of the tetrachord, and remain unaltered in their boundaries according to the different classes. Their names are proslambanomenos, hypate hypaton, hypate meson, mese, nete synhemmenon, paramese, nete diezeugmenon, nete hyperbolaeon. The moveable notes are those which, being arranged in the tetrachord between the immoveable, change from place to place according to the different classes. They are called

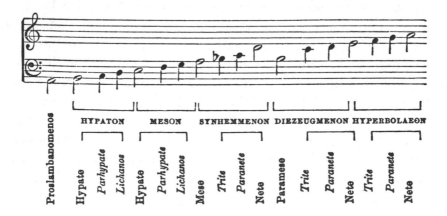

parhypate hypaton, lichanos hypaton, parhypate meson, lichanos meson, trite synhemmenon, paranete synhemmenon, trite diezeugmenon, paranete diezeugmenon, trite hyperbolaeon, paranete hyperbolaeon.

6. These notes, from being moveable, take on different qualities; for they may stand at different intervals and increasing distances. Thus, parhypate, which in the enharmonic is at the interval of half a semitone from hypate, has a semitone interval when transferred to the chromatic. What is called lichanos in the enharmonic is at the interval of a semitone from hypate; but when shifted to the chromatic, it goes two semitones away; and in the diatonic it is at an interval of three semitones from hypate. Hence the ten notes produce three different kinds of modes on account of their changes of position in the classes.

7. There are five tetrachords: first, the lowest, termed in Greek ὕπατον; second, the middle, called μέσον; third, the conjunct, termed συνημμένον; fourth, the disjunct, named διεζευγμένον; the fifth, which is the highest, is termed in Greek ὑπερβόλαιον. The concords, termed in Greek συμφωνίαι, of which human modulation will naturally admit, are six in number: the fourth, the fifth, the octave, the octave and fourth, the octave and fifth, and the double octave.

8. Their names are therefore due to numerical value; for when the voice becomes stationary on some one note, and then, shifting its pitch, changes its position and passes to the limit of the fourth note from that one, we use the term "fourth"; when it passes to the fifth, the term is "fifth." [1]

9. For there can be no consonancies either in the case of the notes of stringed instruments or of the singing voice, between two intervals or between three or six or seven; but, as written above, it is only the harmonies of the fourth, the fifth, and so on up to the double octave, that have boundaries naturally corresponding to those of the voice: and these concords are produced by the union of the notes.

---

[1] The remainder of this section is omitted from the translation as being an obvious interpolation.

CHAPTER V

SOUNDING VESSELS IN THE THEATRE

1. In accordance with the foregoing investigations on mathematical principles, let bronze vessels be made, proportionate to the size of the theatre, and let them be so fashioned that, when touched, they may produce with one another the notes of the fourth, the fifth, and so on up to the double octave. Then, having constructed niches in between the seats of the theatre, let the vessels be arranged in them, in accordance with musical laws, in such a way that they nowhere touch the wall, but have a clear space all round them and room over their tops. They should be set upside down, and be supported on the side facing the stage by wedges not less than half a foot high. Opposite each niche, apertures should be left in the surface of the seat next below, two feet long and half a foot deep.

2. The arrangement of these vessels, with reference to the situations in which they should be placed, may be described as follows. If the theatre be of no great size, mark out a horizontal range halfway up, and in it construct thirteen arched niches with twelve equal spaces between them, so that of the above mentioned "echea" those which give the note nete hyperbolaeon may be placed first on each side, in the niches which are at the extreme ends; next to the ends and a fourth below in pitch, the note nete diezeugmenon; third, paramese, a fourth below; fourth, nete synhemmenon; fifth, mese, a fourth below; sixth, hypate meson, a fourth below; and in the middle and another fourth below, one vessel giving the note hypate hypaton.

3. On this principle of arrangement, the voice, uttered from the stage as from a centre, and spreading and striking against the cavities of the different vessels, as it comes in contact with them, will be increased in clearness of sound, and will wake an harmonious note in unison with itself.

But if the theatre be rather large, let its height be divided into four parts, so that three horizontal ranges of niches may be marked out and constructed: one for the enharmonic, another for the chromatic, and the third for the diatonic system. Beginning with the bottom range, let the arrangement be as described above in the case of a smaller theatre, but on the enharmonic system.

4. In the middle range, place first at the extreme ends the vessels which give the note of the chromatic hyperbolaeon; next

to them, those which give the chromatic diezeugmenon, a fourth below; third, the chromatic synhemmenon; fourth, the chromatic meson, a fourth below; fifth, the chromatic hypaton, a fourth below; sixth, the paramese, for this is both the concord of the fifth to the chromatic hyperbolaeon, and the concord[1] of the chromatic synhemmenon.

5. No vessel is to be placed in the middle, for the reason that there is no other note in the chromatic system that forms a natural concord of sound.

In the highest division and range of niches, place at the extreme ends vessels fashioned so as to give the note of the diatonic hyperbolaeon; next, the diatonic diezeugmenon, a fourth below; third, the diatonic synhemmenon; fourth, the diatonic meson, a fourth below; fifth, the diatonic hypaton, a fourth below; sixth, the

---

[1] Codd. *diatessaron*, which is impossible, paramese being the concord of the fourth to the chromatic meson, and identical with the chromatic synhemmenon.

proslambanomenos, a fourth below; in the middle, the note mese, for this is both the octave to proslambanomenos, and the concord of the fifth to the diatonic hypaton.

6. Whoever wishes to carry out these principles with ease, has only to consult the scheme at the end of this book, drawn up in accordance with the laws of music. It was left by Aristoxenus, who with great ability and labour classified and arranged in it the different modes. In accordance with it, and by giving heed to these theories, one can easily bring a theatre to perfection, from the point of view of the nature of the voice, so as to give pleasure to the audience.

7. Somebody will perhaps say that many theatres are built every year in Rome, and that in them no attention at all is paid to these principles; but he will be in error, from the fact that all our public theatres made of wood contain a great deal of boarding, which must be resonant. This may be observed from the behaviour of those who sing to the lyre, who, when they wish to sing in a higher key, turn towards the folding doors on the stage, and thus by their aid are reinforced with a sound in harmony with the voice. But when theatres are built of solid materials like masonry, stone, or marble, which cannot be resonant, then the principles of the "echea" must be applied.

8. If, however, it is asked in what theatre these vessels have been employed, we cannot point to any in Rome itself, but only to those in the districts of Italy and in a good many Greek states. We have also the evidence of Lucius Mummius, who, after destroying the theatre in Corinth, brought its bronze vessels to Rome, and made a dedicatory offering at the temple of Luna with the money obtained from the sale of them. Besides, many skilful architects, in constructing theatres in small towns, have, for lack of means, taken large jars made of clay, but similarly resonant, and have produced very advantageous results by arranging them on the principles described.

CHAPTER VIII

ACOUSTICS OF THE SITE OF A THEATRE

1. ALL this having been settled with the greatest pains and skill, we must see to it, with still greater care, that a site has been selected where the voice has a gentle fall, and is not driven back with a recoil so as to convey an indistinct meaning to the ear. There are some places which from their very nature interfere with the course of the voice, as for instance the dissonant, which are termed in Greek κατηχοῦντες; the circumsonant, which with them are named περιηχοῦντες; again the resonant, which are termed ἀντηχοῦντες; and the consonant, which they call συνηχοῦντες. The dissonant are those places in which the first sound uttered that is carried up high, strikes against solid bodies above, and, being driven back, checks as it sinks to the bottom the rise of the succeeding sound.

2. The circumsonant are those in which the voice spreads all round, and then is forced into the middle, where it dissolves, the case-endings are not heard, and it dies away there in sounds of indistinct meaning. The resonant are those in which it comes into contact with some solid substance and recoils, thus producing an echo, and making the terminations of cases sound double. The consonant are those in which it is supported from below, increases as it goes up, and reaches the ears in words which are distinct and clear in tone. Hence, if there has been careful attention in the selection of the site, the effect of the voice will, through this precaution, be perfectly suited to the purposes of a theatre.

The drawings of the plans may be distinguished from each other by this difference, that theatres designed from squares are meant to be used by Greeks, while Roman theatres are designed from equilateral triangles. Whoever is willing to follow these directions will be able to construct perfectly correct theatres.

4

Concerning the Principles of Music

BOETHIUS

Translation by R. Bruce Lindsay
from "De Institutione Musicae"
(Latin edition of G. Friedrich, Leipzig, 1867),
Sections 1, 3, 8, 9, 10, 11

Anicius Manlius Saverinus Boethius (480–524) was a Roman philosopher whose chief scientific fame came from his translations of and commentaries on the work of Aristotle. In this way he served as a vehicle for bringing Greek science to the attention of the Roman west. His best known original work is his *Consolations of Philosophy*. The extracts translated and presented here on sound in music and early experiments in acoustics were undoubtedly taken from Greek sources. They are of interest in their foreshadowing of the wave theory of sound.

1. Preface: Music and the
Adornment of Our Manners

The perception of sensations is so clearly in the power of living things that without this no organism could be understood. But knowledge of these things has not through inquiry been clearly connected with the perception of the soul. For it has not been established what sense we bring to the perception of things that are capable of being perceived. The nature of those very senses by which we perform our actions as well as the particular properties of perceptible things cannot be obvious or explicable except through suitable investigation. The faculty of sight is present in all mortals. This is stimulated by images reaching the eye or by rays emitted to the things seen. Among the learned there is some uncertainty about this. The common folk are assailed by the same doubt. On the other hand when one looks at a triangle or a square he certainly recognizes what is looked at by the eyes. But to understand the real nature of a square or rectangle it is necessary to ask a mathematician. The same can be said of other perceptions and this is particularly true of the judgment of the ears. It is the faculty of the ear to catch sounds, not only indicating their actual existence and recognizing the differences among various sounds, but also to provide pleasure when the sounds are of sweet quality. The hearer can, however, be distressed if scattered or incoherent sounds strike the ear. There are four mathematical disciplines of which music is one; the other three are concerned mainly with the investigation of rational truth, but music concerns not only speculation but also human behavior. For there is nothing more characteristic of human beings than their ability to be relaxed by pleasant melodies and indeed to suffer the opposite by unpleasant sounds. This holds not only for older people or those devoted to a certain profession, but for all including infants, and young people as well as the aged. There is indeed no period in life in which one is not affected by the delights of song. Hence we can understand what was said, not in vain, by Plato, that the soul of the world is knit together by the harmony of music.

3. On Sound and the Elements of Music

Consonance, which governs every modulation of music, cannot take place before sound is present. Sound indeed is not produced without some prior impulse or percussion, and such impulses and percussions cannot exist without being preceded by motion. For if all things are motionless, no one of them can strike another and hence set the other in motion. With all things motionless no sound can be produced. Hence sound is defined as due to a percussion of the air, which persists undissipated until it is heard. Some of these motions are indeed faster and others slower; some of them are less frequent and others more frequent. For if any one considers a continuous motion, he necessarily apprehends only its speed or its slowness; but if one moves his hand, he will move it either at greater or less frequency. And if the motion takes place slowly and at low frequency, it is necessary that, because of this same slowness and infrequency, the resulting sound should be of low pitch. If on the other hand the motion is quick and frequent, the sound emitted will be of high pitch. Therefore a string if stretched tightly gives a high pitch sound, whereas if it is stretched loosely it gives a low pitched sound. For when the string is more taut it produces a faster impulse and goes back and forth more frequently. It also strikes the air more frequently. But the string which is looser exerts weak and slow impulses and through this very weakness produces low-frequency sounds and does not vibrate so long. For it should not be thought that when a string is struck only one sound is produced, or that there is only one percussion in such cases. The air is moved as often as the string strikes it in its vibrations. Inasmuch as the velocities of the sounds are joined together one after another, no interruption is perceived by the ears and one sound appears to strike the sense organ, of either low or high pitch, even though each consists of several sounds, the low pitched ones from the less frequent and the high pitched ones from the more frequent. It is similar to the case of the cone, called a spinning top. If one paints a single strip of this with red paint or some other color, and then spins it as fast as possible, the whole cone seems to be colored red. This does not mean the top is actually all red, but that the velocity of the red strip seizes upon the uncolored parts and does not allow them to be seen. But more of this later on!

It therefore follows that inasmuch as high pitch sounds are produced by more frequent and faster motions, and low pitch sounds by slower and less frequent ones, it is clear that by a certain increase in motions a high pitch sound can be produced out of a low pitch one, and by cutting out motions, a low pitch can be produced from a high pitch. For high pitch corresponds to more motions than low. On the other hand, in cases in which a plurality made the difference, it is essential that a very large number be involved. Indeed the relation between paucity and plurality involves numerical comparison. Of those things that are compared numerically some are equal and others are unequal. Therefore sounds also are in some cases equal and in others separated from each other. But for sounds which are separated by no inequality, there is no consonance. [Editor's note: This can mean only that they are in unison.] For consonance is the harmony of dissimilar sounds reduced somehow to one.

8. The Nature of Musical Sound, Interval, and Consonance

A musical sound is a coming together of ordinary sounds in time, that is, a joining of them in one tune. We do not indeed wish to define sound in general, but rather that which is called in Greek *phthongos* (a tone). An interval is the distance between a high pitch and a low pitch tone. Consonance is the combination of high and low pitch tones which strike the ear uniformly and agreeably. Dissonance is due to the harsh and unpleasant impact on the ear of a mixture of tones. For while neither wishes to be mixed with the other and each strives to reach the ear by itself and so to speak oppose the other, each transmits sound unpleasantly to the ear.

9. Errors of the Senses

On this subject we must state that not every sense perception is precisely reliable, though of all these, the sense of hearing plays the principal role. For if there were no hearing there could be no talk of sound at all. But audition holds the first place as a way of communication. The final power of gaining knowledge consists in reasoning, which can never be carried out without error, even when adhering to definite rules. For concerning the errors of the senses it should have been said long since that the power of perceiving is not the same for all men nor is it the same at all times for the same individual. Anyone who wishes to investigate anything with accuracy should not rely on his own judgment alone. For this reason the followers of Pythagoras felt constrained to follow the middle road. For they did not allow the ear the final decision with respect to any perception, yet at the same time there were many things which were investigated by them only through the agency of hearing. They measure indeed by the ear those very harmonics which are separated from each other. They introduce rules through which the judgment of the ear may be tested so that each sense organ, as it were, has to be obedient to these rules, leaving reason dominant. It is admitted that changes of all kinds that occur in life are made evident through the agency of the senses. Nevertheless the latter can provide no completely reliable judgment and there can be no necessary truth in them if reason is absent. For the sense organ itself can be handicapped by both very great and very small effects. Very small things cannot be adequately sensed because of the very smallness of the corresponding sensations and are often confused with greater ones. Thus in the case of sound, faint sounds can be heard only with great difficulty, whereas very strong sounds produce a deafening affect.

10. How Pythagoras Investigated Consonance

It was then principally for the reasons set forth in the previous section that Pythagoras abandoned the judgment of the ears and transferred attention to measuring scales, having no faith in the human ear, which can suffer change in part through its own nature and in part through external accidents. It can also vary with age. He had no confidence either in musical instruments from which are often produced great variation and instability. If, for example, you wish to consider strings, more humid air will weaken the vibrations while dry air strengthens. them. The large size of the string will produce a tone of lower pitch, while a thinner string will produce a tone of higher pitch. Or in some other way the original state of uniformity may change. Since the same situation prevailed with all other instruments, Pythagoras thought all these unworthy of consideration and had little faith in othem. So for a long time he sought assiduously for other means by which judgments concerning consonance could be firmly established. In the meantime, while he was passing a smith's shop, by the pleasure of the gods he heard the hammers when struck produce in some way out of the diverse sounds a musical harmony. Astonished at this, which had long been a subject of inquiry to him, he went into the shop and after long consideration decided that the diversity of sounds was due to the force of the blows. In order that he might solve this problem decisively he ordered the men to exchange hammers. But it was found that the properties of the sounds did not depend on the strength of the men, but the same properties were found to exist with the interchanged hammers. When he had observed this he examined the weight of the hammers. Of five hammers, two were found with weights in ratio of 2 to 1 and these produced sounds an octave apart. He found that the one which was double the weight of the other had a weight four-thirds that of still another and produced a sound higher by a fourth. One hammer, which had a weight three-halves that of another, produced the consonance a fifth above. The two hammers to which the previously mentioned hammers had been shown to have the ratio 4 to 3 and 3 to 2, respectively, were found to have to each other the ratio of 9 to 8. Even before Pythagoras the musical consonance of octave, fourth, and fifth were recognized, but Pythagoras was the first to find by the way just described the proportions associated with these musical harmonies. In order to make clarer what has been just said, let us, for example, assume that the four hammers (the fifth being disregarded) have weights represented by the numbers 12, 9, 8, 6, respectively. Then the hammers with weights 12 and 6 were found to be an octave apart. The hammers with weights 12 and 9 (ratio 4 to 3) are a fourth apart, and the same is true of the hammers with weights 8 and 6, respectively. The hammers with weights 9 and 6, respectively, are a fifth apart.

[Editor's note: The story of the hammers is apocryphal. It was undoubtedly taken by Boethius from much earlier sources. The statements have no basis in fact, since the weight of a vibrator by itself has nothing to do with its frequency. It is the geometrical dimensions which, for the same line density and tension (or elastic coefficient) are decisive. The story of how Pythagoras reached this conclusion is explained by Boethius in Section 11.]

11. The Ways in Which the Various Proportions of Harmonic Sounds Were Studied by Pythagoras

On his return home from the smith's shop Pythagoras attempted in various ways to find out whether the whole theory of consonant sounds resides in these proportions. He now turned to strings attaching equal weights to them, and judged their consonances by ear. On the other hand he also varied the lengths of reeds by doubling and halving them and by choosing other proportions, and thus by differing observations developed a complete faith in his results. . . . Led on by these earlier results he examined the length and thickness of strings. And thus he invented the monochord, concerning which we shall have something to say later. The monochord acquired this designation [Latin: *regula*] not merely because of the wooden scale by which we measure the dimensions of strings and the corresponding sounds, but because any particular investigation of this kind made with a monochord [regula] is so firmly established that no investigation can any longer be misled by doubtful evidence.

5

Dialogues on Music and Acoustics

GALILEO GALILEI

Galileo Galilei (1564–1642), the Italian natural philosopher who was one of the founders of modern physics, is usually known in the literature by his first name. His greatest contribution to science was probably in the field of mechanics, specifically in the motion of falling bodies and projectiles. His discoveries in astronomy are more popularly known. Evidently Galileo was fascinated by the phenomena of sound. In his *Dialogues Concerning Two New Sciences* (1638) at the end of the discussion of the First Day, he introduces his ideas on acoustics with particular reference to the importance of the concept of frequency. Many of the points brought out are probably not wholly original with Galileo and owe a good deal to his predecessors and contemporaries, e.g., Mersenne. His clear handling of the early problems of acoustics justify us in considering his work as a landmark in the development of the subject.

MATHEMATICAL DISCOURSES

CONCERNING

Two New Sciences

RELATING TO

Mechanicks and Local Motion,

IN

FOUR DIALOGUES.

I. Of the Refiſtance of Solids againſt Fraction.	III. Of Local Motion, *viz.* Equable, and naturally Accelerate.
II. Of the Cauſe of their Coherence.	IV. Of Violent Motion, or of PROJECTS.

By GALILEO GALILEI,

Chief Philoſopher *and* Mathematician *to the Grand Duke of* TUSCANY.

With an APPENDIX concerning the Center of Gravity of SOLID BODIES.

Done into *Engliſh* from the *Italian,*

By THO. WESTON, *late Maſter, and now pubiiſh'd by* JOHN WESTON, *preſent Maſter, of the Academy at* Greenwich.

LONDON:

Printed for J. HOOKE, at the *Flower-de-Luce,* over-againſt St. *Dunſtan's* Church in *Fleet-ſtreet.* M. DCC. XXX.

I come now to the other Queſtions relating to *Pen-*
dulums, Matters which to ſome would ſeem very frivolous,
and more eſpecially to thoſe Philoſophers who are continually employ'd in the more profound Queſtions of natural
Philoſophy : Yet I am far from contemning them, en-
couraged by *Ariſtotle's* Example, in whom I can't help
admiring this, *viz.* that he hath left nothing unhandled,
as one may ſay, that's worth one's Conſideration.   And
as to your Queſtions, I think I can give you a Conceit of
my own, relating to ſome Problems concerning *Muſick,*
a noble Subject, of which ſo many great Men, and even
Ariſtotle himſelf, have written ; and touching which he
consiſders

confiders many curious Problems : And if I, from eafy and fenfible Experiments, can deduce Reafons of the wonderful Accidents of Sounds, I may hope my Difcourfe may be grateful to you.

SAGR. Not only grateful, but by me in particular moft defireable ; for I take great Delight in all Mufical Inftruments : And altho' I have thought much concerning Confonances, yet I never could comprehend whence it arifes, that one fhould more pleafe me, and give me greater Delight than another ; that fome others do not only not delight me, but, on the contrary, highly offend me. And again, that common Problem concerning two Strings fet to an Unifon, one of which actually foundeth upon ftriking the other, I, as yet, cannot folve ; nor do I clearly underftand the Forms, and fome other Particulars relating to Confonances.

SALV. I will fee whether from thefe our *Pendulums* I can deduce any thing to anfwer all thefe Difficulties.

And firft as to your Queftion, *viz.* Whether the fame *Pendulum* doth really and exactly perform all its Vibrations, greateft, mean, and leaft, in Times precifely equal, I refer myfelf to that which I have long fince learnt of our *Academic*, who plainly demonftrates, that a Moveable defcending along the Chord fubtending any Arch, paffes them all in equal Times, *i. e.* as well the Subtenfe of 180°, (*i. e.* the whole Diameter) as the Subtenfe of 100°, or 60, 10, 2, ½ a Degree, or of four Minutes, fuppofing them all to terminate in the loweft Point, which touches the horizontal Plane : Again, concerning Defcendents by

T 2

the

the Arches of the fame Chords, elevated above the Horizon, Experience likewife fhews that they all are paffed over in equal Times, but in fhorter than the Times in which the Chords are run thro' ; an Effect fo wonderful, that at firft View one would think the contrary fhould happen ; for the Terms of the Beginning and Ends of the Motion being common, and a right Line being the fhorteft of thofe which are comprehended between the fame Terms, it feems reafonable that the Motion made thro' is fhould be perform'd in the fhorter Time ; which yet it quite otherwife, and the fhorteft Time, and confequently the fwifteft Motion, is that which is made thro' the Arch of which that right Line is the Chord.

In the next Place : As to the Times wherein the Vibrations of Moveables, fufpended by Strings of different Lengths are performed, they are in fubduple Proportion of the Lengths of the Strings ; or, if you will, the Lengths are in a duplicate Ratio of the Times, *i. e.* as the Squares of the Times ; fo that fuppofing, for Example, the Time of a Vibration of one *Pendulum*, to be double the Time of a Vibration of another *Pendulum*, it neceffarily follows, that the Length of the String of that *Pendulum* be quadruple of the Length of the String of this : And then in the Time of one Vibration of that, another will make three Vibrations, when the String of that is nine times as long as that of the other : From whence it follows, that the Lengths of the Strings have the fame Proportion to one another, as have the Squares of the Numbers of the Vibrations made in the fame Time.

SAGR. So

SAGR. So that, if I underſtand you aright, I may eaſily know the Length of a String hanging from any never ſo great a Height, tho' I can't ſee its upper End, by which it is faſten'd on high, but only its lower End : For if I faſten Weight enough to the ſaid String here below, and ſet it a vibrating to and fro, letting ſome Friend number ſome of its Vibrations.   Then if I at the ſame time count the Vibrations of another Moveable hung by a String exactly a Yard long ; then from the Number of Vibrations which thoſe *Pendulums* make in the ſame time, I ſhall find the Length of that String : Thus, for Example's Sake, ſuppoſe that in the time that my Friend hath counted 20 Vibrations of the longer String, I have counted 240 of my String that is a Yard long : The Numbers 20 and 240 being ſquar'd, which are 400 and 57600, I can pronounce the longer String to contain 57600 of thoſe Meaſures whereof my String contains 400 ; and becauſe this is but one Yard in Length, I will divide 57600 by 400, and the Quotient will be 144 : And ſo many Yards I affirm that String is in Length.

SALV. And you'll not be miſtaken one Inch, eſpecially if you take a ſufficient Number of Vibrations.

SAGR. You give me frequent Occaſion to admire the Riches, and withal the extraordinary Bounty of Nature, whilſt from Things ſo common and mean, you deduce, one way or other, many curious and new Notions, and ſuch oftentimes as are remote from all Imagination.   I have very often carefully conſider'd the Vibrations in particular

ticular of the Lamps, which in some Churches hang by very long Ropes, and which have been by Chance stirr'd or mov'd by any one: But the most that I inferr'd from that Observation was this, that their Opinion was very improbable, who will have that such Motions are maintain'd and continu'd by the *Medium*, *i. e.* by the Air: For I should think the Air endued with great Judgment, and with the greatest Ease, to cause an hanging Weight to vibrate for Hours together with so much Regularity: But I should never have learnt from hence, that if the same Moveable, suspended by a String a hundred Yards long, now elevated above the lowest Point ninety Degrees, now only one, or half a one, that, I say, it should spend as much time in moving thro' the least Arch, as it does in moving thro' the greatest; I don't think this would ever have entered into my Head, and, to be plain, I still think it next to impossible. But now I stand prepar'd to hear how from these the most simple and plain Things can be assign'd me such Reasons concerning those Musical Problems, as in some measure to give me Satisfaction about them.

SALV. Above all things you must know that every *Pendulum* hath the Time of its Vibrations so limited and prefixed, that it is impossible to make it vibrate in any other Period than that which is natural to it: For let any one take the String the Weight is fasten'd to in his Hand, and let him try to increase or lessen the Number of its Vibrations, and he shall find his Labour to be in vain: But we may, on the contrary, to a *Pendulum*, tho' it be heavy, and at Rest, give a Motion by only blowing upon it,

†

it, and, repeating the Blaſt, a Motion conſiderable ; but this ſecond Blaſt muſt be made at at Inſtant ſuitable to the Motion of the Vibrations ; ſo that if at the firſt Blaſt we move it half an Inch from the Perpendicular, and then again, if when it has returned to us, and is beginning its ſecond Vibration, we add a ſecond Blaſt, we ſhall confer a new Motion to it ; and thus by ſucceſſively repeating the Blaſts at proper Times, but not when the *Pendulum* is moving towards us, (for ſo we ſhould check and not help the Motion) we may confer upon it ſuch an *Impetus*, that a much greater Force than our Breath will be required to ſtop it again.

SAGR. I have obſerv'd, when I was a Boy, that one Man alone by ſuch Impulſes, given at right Times, has been able to raiſe a very great Bell ; and when four or ſix have taken hold of the Rope to ceaſe the Bell from Motion, they have been all raiſed from the Ground ; they all being not able to withſtand the *Impetus*, which one Man alone, with regular Pulls, had conferred upon the Bell.

SALV. An Example that declares my Meaning as fully as my foregoing one is ſuitable to render the Reaſon of that admirable Problem of the Strings of the Viol or Lute, which moves and makes not that only to ſound, which is tun'd to an Uniſon with it, but that alſo which is ſet to an Eighth, and to a Fifth : The String being touch'd or ſtruck, its Vibrations begin and continue ſo long as its Sound laſts; theſe Vibrations give a Vibration and Tremor to the adjacent Air, whoſe Tremors and Circulations, extending

extending themſelves a great Way, ſtrike upon all the Strings of the ſame Inſtrument, as alſo upon the Strings of any others near it : That String that is ſet to an Uniſon with that touched, being diſpoſed to make its Vibrations in the ſame Time, at the firſt Impulſe begins to move a little ; and upon the Addition of a ſecond, a third, and twentieth, and many more, all made in fit and periodic Times, it receives at laſt the ſame Tremulation with that firſt touched, and, as plainly may be ſeen, it dilates its Vibrations exactly according to the Dilatation of thoſe of its **Mover.**

This Undulation that extends itſelf thro' the Air, moves and makes to vibrate, not only the Strings, but likewiſe any other Body diſpoſed to Tremulation, and to vibrate in the ſame time with the trembling String ; wherefore if upon the Sides of the Inſtrument, ſeveral ſmall Particles of Hair, or Briſtles, or any other flexible Matter, be laid, ſounding the Viol, we ſhall ſee now this, now that Corpuſcle tremble, as this or that String is ſtruck, whoſe Vibrations are made in the ſame Time ; but the others will not move at the Striking or Sound of this String ; nor will that Corpuſcle move at the Sounding of any other String.

If the Baſe String of a Viol be ſmartly ſtruck with a Bow, and a thin and ſmooth Drinking-Glaſs be ſet by it ; if the Tone of the String be an Uniſon with the Tone of the Glaſs, the Glaſs will tremble and ſenſibly reſound : Again, we plainly ſee the Circulation of the *Medium* about the reſounding Body, to diffuſe to a large Space, by making a Glaſs to ſound, that has ſome Water in it, by rubbing the Rim or Edge of it with the Tip of one's
Finger :
†

Finger : For we fhall thus fee the Water in the Glafs to undulate in a moft regular Order ; which Effect will yet more clearly be feen, if we put the Foot of the Glafs in the Bottom of a Veffel of reafonable Bignefs, and fill it with Water nearly to the Glafs's Rim, and then make it found by rubbing it round, as before, with the Tip of one's Finger, for then we fhall fee the Circulations in the Water to be moft regular, and with great Velocities to fpread to a great Diftance round about the Glafs : Nay, I have often happen'd to fee, in making a pretty big Glafs, almoft full of Water, to found as before, the Waves form'd with an exact Equality ; but the Tone of the Glafs happening fometimes to rife an Eighth higher, I have feen at that very Inftant every one of the faid Waves to divide themfelves into two : which Accident moft plainly proves the Form of the Octave to be the Double.

SAGR. The fame Thing has happen'd to me more than once, to my great Delight and Advantage too : For I ftood a long time in Doubt concerning the Forms of Confonance, not thinking the Reafons commonly brought by the learned Authors, who have hitherto wrote of Mufick, fufficiently demonftrative.   They tell us that the *Diapafon*, that is the *Octave*, is contain'd by the *Double* ; and that the *Diapente*, which we call *the Fifth*, is contain'd by the *Sefquialter* : For if a String, ftretch'd upon the marine Trumpet, be founded open, and afterwards placing a Bridge under the Midft of it, its half only be founded, you'll hear an Eighth : And if the Bridge be placed under one third of the String, and you then ftrike the two thirds open, it foundeth a Fifth to that of the whole String ftruck

U   when

49

when open : whereupon they infer that the Eighth is con-tain'd between two and one, and the Fifth between three and two. But I don't think we can conclude from hence, that the *Double* and *Sefquialteral* can naturally affign the Forms of the *Diapafon* and *Diapente* : And my Reafon for't is this : There are three Ways by which we may fharpen the Tone of a String, *viz.* by fhortning it, by ftretching it, or by making it thinner : If now, retaining the fame Tenfion and Thicknefs, we would hear an Eighth, we muft make it fhorter by half, *i. e.* we muft firft found the whole String, and then its half. But if keeping the fame Length and Thicknefs, we would have it rife to an Eighth from its prefent Tone, by ftretching it, or fcrewing it higher, 'tis not fufficient to ftretch it with a Double, but with four times the Force : Thus, if at firft it was ftretch'd by a Weight, fuppofe of one Pound, we muft hang a four Pound Weight to it, in order to raife its Tone an Eighth. And laftly, if keeping the fame Length and Tenfion, we would have a String to found an Eighth, this String muft be but one fourth of the Thicknefs of the other it muft found an Eighth to. And this that I fay of the Eighth, *i. e.* that its Form taken from the Tenfion or Thicknefs of the String in a duplicate *Ratio* to that which it receiveth from the Length, I would have underftood of all other Mufical Intervals ; becaufe if that which the Length gives us in a *Sefquialter* Proportion, *i. e.* when firft by ftriking it open, and then its two thirds, we would do the fame from the Tenfion, or Tenuity, or thinning the String, we muft double the *Sefquialter* Proportion, by taking the double *Sefquiquartan* : And if the grave String be ftretch'd by a four Pound Weight to the fharp one, we muft hang not

fix

fix but nine Pounds : And as to the Thickneſs ; we muſt make the grave String thicker than the acute, in the Proportion of nine to four, to gain the Fifth.

These Things being really ſo in Fact, I ſaw no Reaſon why theſe ſage Philoſophers ſhould rather conſtitute the Form of the *Eighth Double* than *Quadruple* ; and the Form of the *Fifth* rather *Seſquitertian*, than the *double Seſquiquartan:* But becauſe 'tis impoſſible to number the Vibrations of the String, which, while it ſounds, are made very ſwiftly ; I ſhould always have been in Doubt whether or no it were true, that the String ſounding the higher *Eighth* did perform double the Number of Vibrations, in the ſame Time to thoſe of the graver String, unleſs the Undulations, which may be continued as long as you pleaſe, by keeping the Glaſs ſounding and vibrating, had ſenſibly ſhewn me, that in the very Inſtant wherein ſometimes we hear a Tone to jump to an *Eighth*, there are ſeen to ariſe other leſſer Undulations, which, with infinite Accuracy divide each of the former into two.

SALV. An excellent Obſervation this ! whereby we can diſtinguiſh one by one the Undulations produced by the Tremor of the ſounding Body ; which are thoſe which, diffuſing themſelves thro' the Air, cauſe a Titillation on the Drum of our Ear, whence in our Soul is produc'd a Sound. But ſince we can't ſee and obſerve thoſe Undulations in the Water, any longer than that Rubbing with the Tip of the Finger laſteth, and they being not at a Stay, but in conſtant Motion all that Time, wou'd'nt it be a great Thing to contrive ſuch as we might exquiſitely

U 2   obſerve,

obferve, as long as we pleafe, for Months if we will, nay, for Years, whence we fhould be able exactly to meafure, and eaſ̄ to count them?

SAGR. I affure you I fhould highly efteem fuch an Invention.

SALV. The Difcovery was merely accidental ; the Obfervation and applying it to Ufe was my own : I own I like the Thing, and have often confider'd it with great Satisfaction, tho' the Experiment at firft View may feem to be of little or no Account or Worth.

Scraping a Copper Plate with an Iron Chizel to take out fome Spots in it, upon moving the Chizel quick to and fro, amongft the many Attritions or Rubbings, I, more than once, heard it fend forth a whiftling Noife or Sound ; and then looking upon the Plate, I efpied a long Row of fmall Streaks, parallel to one another, and exactly equidiftant.   Returning to my Scraping again, I found, by feveral Trials, that the Chizel left thofe Streaks upon the Plate, at thofe Scrapings, and thofe only that yielded the aforefaid Noife : But when the Attrition was made without fuch Whiftling, there appear'd not the leaft Sign of any fuch Streaks : Repeating the Experiment, for my farther Satisfaction, feveral times afterwards, fcraping one while with greater, another with lefs Velocity, the Whiftling I found to have one while a fharper Tone, and another a graver ; and I obferv'd the Marks made at the fharper Sounds, to be more, and clofer together ; and fewer, and farther afunder, at the grave Sounds : And furthermore, as the Swiftnefs towards the End of the felf-fame Scrape grew

grew greater than that at the Beginning, the Sound was heard continually to grow sharper, and the Streaks were observ'd to stand thicker, but always made with the greatest Regularity, and exactly equidistant from one another: And furthermore, when in scraping such Sound was heard, I felt the Chizel to tremble in my Hand, and a certain Shivering to run along my Arm; and, in short, I observ'd the very same Thing effected by this Tool, which may be observ'd in ourselves when we whisper, and afterwards speak aloud; for when we send forth our Breath without forming a Sound, we do not perceive any Motion in the Throat or Mouth, in Comparison of that Tremor which we feel in the Wind-Pipe and Jaws, upon sending forth the Voice, especially if the Tones emitted be grave and strong. I have, moreover, at such a time observ'd, amongst the Strings of the Viol, two to sound Unisons, to two of the Sounds made by scraping after the Manner aforesaid, and exact Fifths to one another: then measuring the Intervals of the Streaks of both the Attritions or Scrapes, I found forty-five Spaces of the one to be equal to thirty of the other, which, indeed, is the Form attributed to the *Diapente.*

But now, before I proceed any farther, I must tell you, that of the three Ways of making a Tone more acute, that which you refer to the Slenderness or Fineness of the String, may with more Truth be attributed to its Weight: The Alteration taken from the Thickness, answers indeed when the Strings are of the same Matter, and so a String of Gut to sound an Eighth, must be four times as thick as the other Gut-string, below which it is to sound an Eighth; and one of Wire also four times thicker than another

another of Wire : but if betwixt two Strings, one of Wire, and the other of Gut, I would get an Eighth, I muſt not make it four times thicker, but four times heavier ; ſo that, as to the Thickneſs, the Wire one ſhall not be four times thicker, but four times heavier than the other ; for then the Wire one will be finer than that of Gut which ſounds an Eighth above it.   Whence it comes to paſs, that ſtringing one Inſtrument with Gold Strings, and another with Braſs ones, provided the Strings of the one Inſtrument be of the ſame Length, Thickneſs, and Tenſion with thoſe of the other, the Tone of the Gold Strings ſhall be about a Fifth lower than that of the Braſs ones, Gold being nearly double the Weight of Braſs.   And here I would have you obſerve, that the Gravity of the Moveable more reſiſts the Velocity of the Motion, than the Thickneſs does, contrary to what one at firſt would think ; for, indeed, in Appearance, 'tis more reaſonable that the Velocity ſhould be retarded, by the Reſiſtance of the *Medium*, againſt dividing or opening, more in a thick and light Moveable, than in a heavy and ſmall one : and yet in this Caſe it happens quite contrary.   But to purſue our firſt Purpoſe ; I ſay, that 'tis neither the Length, nor the Tenſion, nor the Thickneſs, that conſtitutes the neareſt immediate Reaſon of the Forms in Muſical Intervals, but the Proportion of the Number of Vibrations and Percuſſions in the Undulations of the Air, ſtriking the Drum of our Ear, which itſelf alſo doth tremulate under the ſame Meaſures of Time.

Having eſtabliſh'd this Point, we may perhaps be able to aſſign a juſt Reaſon whence it comes to paſs, that of Sounds differing in Tone, ſome Pairs are heard with great
Delight,

†

Delight, others with lefs ; and that others are very offenfive to the Ear.   This we may do, I fay, by feeking the Reafon of the more or lefs perfect Confonances, and alfo of Diffonances.   The Offence thefe give, proceeds, I believe, from the difcordant and jarring Pulfations of two different Tones, which, without any Proportion, ftrike the Drum or the Ear : And the Diffonances will be extreme harfh, in cafe the Times of the Vibrations are incommenfurable.   And of fuch this is one ; When one of two Strings, Unifons, is founded with fuch a Part of another, as is the Side of a Square of its Diagonal; which Diffonance is like to the *Tritone* or *Semi-diapente*.

Thofe Pairs of Sounds fhall be Confonances, and will be heard with Pleafure, which ftrike the *Timpanum* in fome Order ; which Order requires, in the firft Place, that the Percuffions made in the fame Time be commenfurable in Number, that the Cartilage of the *Timpanum* or Drum may not be fubject to a perpetual Torment of bending itfelf two different Ways, in Submiffion to the ever difagreeing Percuffion.

The firft and moft grateful Confonance will be the *Eighth*, fince for every Stroke wherewith the grave String ftrikes the Ear, the acute one gives two ; fo that *both*, in every fecond Vibration of the fharp String, ftrike at once, or together ; wherefore half the whole Number of Strokes agree in ftriking together; but the Strokes of the Strings that are *Unifons*, are always ftruck together, and therefore are but as one only String, and make no Confonance.

The *Fifth* alfo is delightful, where, for every two Strokes of the grave String, the fharp one gives three ; whence it follows, that if we. count the Vibrations of the acute String,

String, we fhall find the third Part of the whole Number to agree, and ftrike together ; that is, between each Pair of Confonances, there are two fingle Sounds : and in the *Diateffcron* there are three fingle ones interpos'd.   In the fecond, that is, in the *Sefquioctave* Tone in nine Pulfations there's only one that ftrikes in Concord with the other of the graver String : and all the reft are Difcords, and fall upon the *Timpanum* irregularly and troublefomely, and thence by the Ear are efteem'd as Diffonances.

SIMP. I wifh you would explain this Difcourfe a little more.

SALV. Let the Line A B be the Space and Dilatation of one Vibration of the grave String, and the Line C D of the acute one, which are Eighths to each other : Bifect A B in E ; then 'tis manifeft that if the Strings begin to

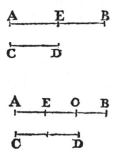

move in the Terms A and C, that when the acute Vibration fhall be come to the Term D, the other will be extended only to its half, or to its middle Point E, which not being the Term or Bound of the Motion, it doth not yet ftrike, but yet a Stroke is made in D.   Then whilft

the
t

the Vibration returns from D to C, the other passes from E to B, upon which the two Percussions of B and C now strike both together upon the *Timpanum ;* and because they continue to repeat in like manner the subsequent Vibrations, we may conclude that the Union of the Percussions of the Line CD, with the Percussions of the Line AB happen every other Time : But the Pulsations of the Terms A B are always accompanied, and are always the same with one of CD ; which is manifest from hence, that supposing A and C to strike together, whilst A passes to B, C goes on to D, and goes back to C ; so that the Strokes B and C are made together.

But now let AB and CD be two Vibrations which produce the *Diapente,* whose Times are in a *Sesquialteral* Proportion, and divide the Line AB of the grave String into three equal Parts, by the Points E and O ; and suppose the Vibrations to begin at the same Time from the Terms A and C ; then its manifest, that when the Stroke is made in D, the Vibration of AB is got only to O ; wherefore the *Timpanum* receives the Pulsation D only. Then whilst it returns from D to C, the other Vibration is pass'd from O to B, and return'd to O, first making a Pulsation in B, which is also made alone, and in Countertime (a Thing to be consider'd) for since we suppos'd the first Pulsations to be made at the same Moment in the Terms A and C, the second which was made alone from the Term D was made so long after as is the Time of Transition thro' CD, that is AO : but the following one which is in B, is distant from the other, as much as is the Transition OB only, which is the half : but it goes on from O to A, in the Time the other passes from C to D,

X   and

and thus are two Pulfations made together in A and D: There afterwards follow other Periods like thefe, that is, with the Interpofition of two fingle Pulfations of the acute String by itfelf; and one fingle Pulfation of the grave String, alone or by itfelf, between thofe two fingle Strokes ·of the acute one.,

Wherefore fuppofing the Time divided into Moments, *i. e.* into its leaft equal Particles, and fuppofing that in the two firft Moments, the concordant Pulfations made in A and C, to pafs to O and D, and that in D a Stroke is made; and that in the third and fourth Moment, one pafTes from D to C, and ftrikes in C; and the other goes on from O to B, and returns to O, ftriking in B: and in the fifth and fixth Moment the one pafTes from C to D, and the

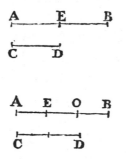

other from O to A; then ftriking together we fhall receive the Pulfations on the *Timpanum* diftributed in this Order, that fuppofing the Pulfations of the two Strings to be made in the fame Moment, two Moments after it fhall receive one folitary Percuffion, in the third Moment another, alfo folitary, in the fourth, another; and two Moments after, that is, in the fixth, two together; and here ends the Period, and *Anomaly*, if I may fo fay; which Period is often, after the fame Manner, repeated.

SAGR. I
†

SAGR. I can hold no longer! but muft loudly exprefs the Pleafure I have to hear fuch appofite Reafons affigned of Effects, concerning which I have been fo long in the Dark. Now I underftand how it comes to pafs, that an *Unifon* differs not from a fingle Tone: Now I fee why the *Octave* is the principal Confonance, and fo like to the *Unifon*, that it is taken and conjoin'd with others as an *Unifon*. It refembles an *Unifon*, becaufe, as the Pulfations of Strings fet to an *Unifon* all ftrike together, the Pulfations of the grave String, which founds an Eighth, are all accompany'd with Beats of the acute String, and of thefe one folitary one is interpos'd, and in equal Diftances, and in a certain Order, without any Variety ; whence fuch Confonance, without the leaft Allay, becomes moft melodious.

But the *Fifth*, with thofe its Counter-times, and which the Interpofitions between the Pairs of conjoin'd Pulfa-tions of two folitary ones of the acute String, and one folitary one of the grave one, and thefe three in fuch Interval of Time, as is its half, produces fuch a Titillation upon the Cartilage of the *Timpanum*, that, allaying the Sweet-nefs by a Mixture of Tartnefs, it feems at one and the fame Time to kifs and bite.

SALV. Since I fee you are fo pleas'd with thefe Novel-ties, I'll fhew you a Method whereby not the Ear only, but the Eye alfo, may be recreated, in beholding the fame Sports the Ear feels or hears.

Hang Balls of Lead, or other heavy Matter, by three Strings of different Lengths, in fuch Manner, that in the Time the longeft makes two Vibrations, the fhorteft makes

X 2   four,

four, and the middle one three; which will happen when the longeft contains fixteen Feet, or other Meafures, fuch as the middle one contains nine, and the fhorteft four. And removing them all at once from Perpendicularity, and then letting them go, you'll fee a various Intermixture of thofe *Pendulums* with various Accidents, but fo that at every fourth Vibration of the longeft, all the three will concur together in one and the fame Term; which they afterwards leave and then again repeat the fame Period; And this Mixture of Vibrations is the fame with that which being made in Strings of Inftruments, prefents to the Ear an *Eighth*, with an intermediate *Fifth*. And if we order the Length of other Strings in the like Difpofition, fo that their Vibrations anfwer to thofe of other mufical confonant Intervals, we fhall ftill fee new Connections, and thofe always fuch as all the Strings (be they three or four) will, in determinate Times, and after determinate Numbers of Vibrations, concur in the fame Moment, at the Term of their Vibrations, and thence they again begin another like Period ; But if the Vibrations of two or more Strings, are either incommenfurable, fo that by vibrating to and fro, they never harmonioufly perform a determin'd Number of Vibrations; or if they are not incommenfurable, and yet return not till after a long Time, and after a great Number of Vibrations, then the Sight is confounded by the irregular and confus'd Order of irregular Intermixtures, as the Ear with Regret receives the intemperate Impulfes of the Air's Tremulations, which, without Order or Rule, fucceffively ftrike its *Timpanum* or *Drum*.

But whither, Gentlemen, have we been tranfported for fo many Hours, by various Problems, and unlook'd for Difcourfes ?

†

Difcourfes? We have made it Night, and yet have handled few or none of the Points firft propos'd ; but, on the contrary, have fo loft our Way, that I hardly remember where we begun, and what that fmall Introduction was, which, by Way of *Hypothefis*, we laid down as a Principle to build upon in the Demonftrations which were to follow.

SAGR. True, it will be convenient, therefore, that we put an End to this Day's Conference, and refrefh our Minds by the Night's Repofe, and To-morrow (if you pleafe fo far to favour us) we will refume the Difcourfes defir'd, and at firft chiefly intended.

SALV. I'll not fail to be here at the ufual Hour, to enjoy your Company, and do you what Service I can.

The END of the FIRST DAY'S CONFERENCE.

GALILÆUS

6

On the Velocity of Sound in Air

MARIN MERSENNE

Marin Mersenne (1588–1648), the French natural philosopher, mathematician, and theologian, was a member of the religious order of Minims. He was a friend of Descartes and supported the Cartesian philosophy. He is perhaps best known for his work in mathematics, but he was also interested in physical problems and devoted attention to the problem of the vibrating string as well as to the velocity of sound in air. The following extracts are from his Latin treatise. The second one is of particular interest since it suggests that Mersenne had grasped the fundamental idea of gun ranging by sound.

F· MARINI
MERSENNI
MINIMI
COGITATA
PHYSICO
MATHEMATICA.

In quibus tam naturæ quàm artis effectũs
admirandi certiſſimis demonſtra-
tionibus explicantur.

PARISIIS,
Sumptibus ANTONII BERTIER, viâ Iacobeâ

M. DC. XLIV.
CVM PRILEGIO REGIS,

6

On the Velocity of Sound in Air

MARIN MERSENNE

Translated from the Latin by
R. Bruce Lindsay

Harmoniae, Liber Primus: Proposition 5

The sound of organs and stringed instruments travels at the rate of 1380 feet per second.

From the fact just stated those who are particularly impressed with harmonic sounds can learn how much sooner such sounds are enjoyed with satisfaction when they are near instead of being far off from the source and also how much earlier the motion of a bell or bell hammer is seen than the sound is heard. We have previously said much about the velocity of sound in our Latin work on harmonic sounds as well as in French works and more recently in the tract on ballistics (Proposition 35).

You will find it easy to make experiments on this matter in very long colonnades such as are found in the royal domain we call the Tuileries. For when you hear a stringed instrument, an organ or a bell at a distance of 230 six-foot intervals you perceive with your eyes one second earlier the motion of the producer of the sound (e.g., the hammer, etc.), if indeed you can distinguish clearly enough at that distance. Similarly the sound is heard after 2, 3, 4 or more seconds if the instruments are placed twice as far, three times as far, or four times as far away, etc. With 11 seconds the sound in going through a distance of one league practically vanishes. If, say, a horn were sounded at some point on the earth and it could be heard after passing once around the earth, it would be perceived 10 hours later. For in that time the sound will complete a semicircumference of a great circle both to the right and the left and indeed in every other direction (provided it is intense enough).

However, this magnitude or intensity of sound should be at least 360 times greater than that of the sound from the greater military ordnance, which can be heard clearly only up to ten leagues. Other details may be seen in the first three books of our greater work "Harmonicorum Libri XII" (1636).

Ballistica: Proposition 35

The velocity of sound is greater than the velocity of cannon balls and equals 230 six-foot intervals per second.

Whoever wishes to measure the velocity of sound under various condition, by night, by day, in valleys, in woods or mountains, either with or against the wind, in fair or rainy weather, in all these circumstances experiment always leads to the same velocity of sound. After you have observed the sound at a distance of 230 six-foot intervals from the source and retreat another space of 230 six-foot intervals so as to be at 460 six-foot intervals from the source the same or equal sound will take 2 seconds to cover the distance. If we multiply the original distance by 5 so that we hear the noise of a cannon at a distance of 1150 six-foot intervals from the source, the fire at the mouth of the cannon will always be seen at night 5 seconds before the sound is heard. When we do the experiment with a Gallic league of 2500 six-foot intervals (in terms of which the earth's circumference is 7200 leagues) it is easy to conclude the time it will take the sound to travel a whole league or any number of leagues. For the velocity of sound is not diminished by its attenuation in intensity. With the hearing of perceptible sound the last part strives to equal the velocity of the first.

The noise of a cannon covers one league in the space of 11 seconds, since a league is equivalent to 11 times 230 six-foot intervals (the 230 six-foot intervals being the distance sound travels in one second) minus 30 six-foot intervals, which hardly needs to be considered here since it corresponds only to the distance traveled by sound in about one-seventh of a second.

From these facts we are permitted to bring together several results. In the first place, a soldier watching the firing of a gun at 100 six-foot intervals whose fire he has already seen is able to dodge the shot. This I demonstrate thus: it is established from observation that a cannon ball takes one second to go 100 six-foot intervals. Therefore half of that second at most will be used up in the transfer of the sound. Therefore the soldier will have a whole second from the time of the flash (assuming he saw it at the instant it occurred) in which he can take three or four steps, before the ball gets there. For that reason there remains half a second from the time when he heard the sound before the arrival of the ball; though I would not advise anyone to try this unless protected by armor of every kind, so as to be prepared for every eventuality. But with a wall placed in between it would be possible for someone to find out which came first, the sound or the fall of the shot on the ground.

In the second place, from the sound and the observation of the cannon flash it is easy to find out how far apart the cannons are; so that from this even very ingenious men cannot fail to promote their art. In the third place, from the sound of thunder and the preceding flash of lightning it can be learned how far away the discharge was, provided that the place of the flash does not change. If you observe how many seconds (using pulse beats which take place precisely every second, if a pendulum or some other time measuring instrument is not available) elapse between flash and sound, there are just as many space intervals each equal to a 230 six-foot interval between you and the discharge; if the time is 5 seconds, the distance is about half a league, if 10 seconds, a league; it makes no difference whether the spatial separation is vertical or lateral or oblique.

In the fourth place, if through circular motions of the air, sound is generated like the circles which we see formed in water by the finger or by the dropping of a little stone, it is permitted through measurement of the velocity of bodies and motions to estimate their density and weight. Water is said to be 1380 times denser than air. It follows that the radius of the circle which is formed in water disturbed by sound in the transmission time of one second scarcely exceeds one foot, whereas in the corresponding case in air the radius is 1380 feet (that is 230 six-foot intervals) because the density of water relative to air is approximately in this proportion, which we conclude from Proposition 29 of the *Pneumatica*. One doubt indeed is cast on this, namely the observation in Proposition 25 of this book (on Ballistics) which seems to indicate that the sounds of the greater guns travel more slowly. When one mathematician at the fort of Theodo observed the sounds of these some 13 or 14 seconds after seeing the flash he was hardly half a league distant from the gun and the sound traveled through a whole league and more in this time. Because of this the sound of these large guns ought to be studied until we can draw the same conclusions about sounds of this kind as we observe for sounds of speech, horns, etc. In the preface of this book you will find many other things which you will read not without wonder and should meditate on.

7

Air as the Medium
for the Transmission of Sound

ROBERT BOYLE

   Robert Boyle (1627–1691), the celebrated British natural philosopher, is best remembered for his experiments on the elastic properties of air and his well-known law connecting the pressure and volume of a gas at constant temperature. His possession of an adequate air pump enabled him to do experiments with partial vacuums which were not possible for his immediate predecessors. Among these experiments is the famous one on the decrease in the intensity of the sound produced by a bell ringing in a bell jar from which the air is continuously pumped out. This was long considered as a decisive experiment demonstrating that the air or some equivalent material medium is necessary for the propagation of sound. We now know that the conventional explanation of the experiment is inadequate. The decrease in intensity of the sound is due to the increasing difference in acoustic impedance between the vibrating object and the surrounding medium.

   The following extract is taken from an early edition of one of Boyle's books.

Reprinted from "New Experiments, Physico-Mechanical, Touching the Spring of the Air," by Rbt. Boyle, 2nd Edition, T. Robinson (Oxford) 105–110 (1662)

7

EXPERIMENT XXVII.

THat the Air is the medium whereby founds are convey'd to the Ear, hath been for many Ages, and is yet the common Doctrine of the Schools. But this Received Opinion hath been of late oppos'd by fome Philofophers upon the account of an Experiment made by the Induftrious *Kircher*, and other Learned Men: who have (as they affure us) obferv'd, That if a Bell, with a Steel Clapper, be fo faften'd to the infide of a Tube, that upon the making the Experiment *De Vacuo* with that Tube, the Bell remain'd fufpended in the deferted fpace at the upper end of the Tube: And if alfo a vigorous Load-ftone be apply'd on the outfide of the Tube to the Bell, it will attract the Clapper, which upon the Removal of the Load-ftone falling back, will ftrike againft the oppofite fide of the Bell, and thereby produce a very audible found; Whence divers have concluded, That 'tis not the Air, but fome more fubtle Body that is the medium of founds. But becaufe we conceiv'd that, to invalidate fuch a confequence from this ingenious Experiment (though the moft luciferous, that could well be made without fome fuch Engine as ours) fome things might be fpecioufly enough alleadg'd; we thought fit to make a tryal or two, in order to the Difcovery of what the Air doth in conveying of founds, referving divers other Experiments tryable in our Engine concerning founds, till we can obtain more

[P] leafure

68

leafure to profecute them. Conceiving it then the beſt way to make our tryal with ſuch a noiſe as might not be loud enough to make it difficult to diſcern ſlighter variations in it ; but rather might be, both laſting, (that we might take notice by what degrees it decreaſ'd ;) and ſo ſmall, that it could not grow much weaker without becoming imperceptible. We took a Watch, whoſe Caſe we open'd, that the contain'd Air might have free egreſs into that of the Receiver. And this Watch was ſuſpended in the cavity of the Veſſel onely by a Pack-thred, as the unlikelieſt thing to convey a ſound to the top of the Receiver : And then cloſing up the Veſſel with melted Plaiſter, we liſten'd near the ſides of it, and plainly enough heard the noiſe made by the ballance. Thoſe alſo of us, that watch'd for that Circumſtance, obſerv'd, that the noiſe ſeem'd to come directly in a ſtraight Line from the Watch unto the Ear. And it was obſervable to this purpoſe, that we found a manifeſt diſparity of noiſe, by holding our Ears near the ſides of the Receiver, and near the Cover of it : which difference ſeem'd to proceed from that of the Texture of the Glaſs, from the ſtructure of the Cover (and the Cement) through which the ſound was propagated from the Watch to the Ear. But let us profecute our Experiment. The Pump after this being imploy'd, it ſeem'd that from time to time the ſound grew fainter and fainter ; ſo that when the Receiver was empty'd as much as it uſ'd to be for the foregoing Experiments, neither we, nor ſome ſtrangers that chanc'd to be then in the room, could, by applying our Ears to the very ſides, hear any noiſe from within ; though we could eaſily perceive that by the moving of the hand which mark'd the ſecond minutes, and by that of the ballance, that the Watch neither ſtood ſtil, nor remarkably varied from its wonted motion. And to ſatisfie our ſelves farther that it was indeed the abſence of the Air about the Watch that hinder'd us from hearing it, we let in the external Air at the Stop-cock, and then, though we turn'd the Key and ſtopt the Valve, yet we could plainly hear the noiſe made by the ballance, though we held our Ears ſometimes at two Foot diſtance from the outſide of the

the Receiver. And this Experiment being reiterated in another place, succeeded after the like manner. Which seems to prove, that whether or no the Air be the onely, it is at least, the principal medium of Sounds. And by the way it is very well worth noting, that in a Vessel so well clos'd as our Receiver, so weak a pulse as that of the ballance of a Watch should propagate a motion to the Ear in a Physically straight Line, notwithstanding the interposition of so close a Body as Glass, especially Glass of such thickness as that of our Receiver: since by this it seems that the Air imprison'd in the Glass, must, by the motion of the ballance, be made to beat against the concave part of the Receiver, strongly enough to make its convex part beate upon the contiguous Air, and so propagate the motion to the Listners Ears. I know this cannot but seem strange to those, who, with an eminent Modern Philosopher, will not allow that a Sound, made in the cavity of a Room, or other place so clos'd, that there is no intercourse betwixt the external and internal Air, can be heard by those without, unless the sounding Body do immediately strike against some part of the inclosing Body. But not having now time to handle Controversies, we shall onely annex, That after the foregoing Experiment, we took a Bell of about two Inches in Diameter at the bottom, which was supported in the mid'st of the cavity of the Receiver by a bent stick, which by reason of its Spring press'd with its two ends against the opposite parts of the inside of the Vessel: in which, when it was clos'd up, we observed that the Bell seemed to sound more dead then it did when just before it sounded in the open Air. And yet, when afterwards we had (as formerly) emptyed the Receiver, we could not discerne any considerable change (for some said they observ'd a small one) in the loudness of the sound. Whereby it seemed that though the Air be the principal medium of sound, yet either a more subtile matter may be also a medium of it, or else an ambient Body that contains but very few particles of Air, in comparison of those it is easily capable of, is sufficient for that purpose. And this, among other things, invited us to

[P 2]

con-

conſider, whether in the above mentioned Experiment made with the Bell and the Load-ſtone, there might not in the deſerted part of the Tube remain Air enough to produce a ſound: ſince the Tubes for the Experiment *De Vacuo* (not to mention the uſual thinneſs of the Glaſs) being ſeldom made greater then is requiſite, a little Air might beare a not inconſiderable proportion to the deſerted ſpace: And that alſo, in the Experiment *De Vacuo*, as it is wont to be made, there is generally ſome little Air that gets in from without ; or at leaſt ſtore of bubbles that ariſe from the Body of the Quick-ſilver, or other Liquor it ſelf, Obſervations heedfully made have frequently informed us: And it may alſo appear, by what hath been formerly delivered concerning the *Torricellian* Experiment.

On the occaſion of this Experiment concerning ſounds, we may adde in this place, That when we tryed the Experiment formerly mentioned, of firing Gun powder with a Piſtol in our evacuated Receiver, the noiſe made by the ſtriking of the Flint againſt the Steel, was exceeding languid in compariſon of what it would have been in the open Air. And on divers other occaſions it appeared that the ſounds created within our exhauſted Glaſs, if they were not loſt before they reach'd the Ear, ſeem'd at leaſt to arrive there very much weaken'd. We intended to try whether or no the Wire-ſtring of an Inſtrument ſhut up into our Receiver, would, when the ambient Air was ſuck'd out, at all tremble, if in another Inſtrument held cloſe to it, but without the Receiver, a ſtring tun'd (as Muſicians ſpeak, how properly I now examine not) to an Uniſon with it, were briskly toucht, and ſet a Vibrating. This, I ſay, we purpoſ'd to try to ſee how the motion made in the Air without, would be propagated through the cavity of our evacuated Receiver. But when the Inſtrument wherewith the tryal was to be made came to be imploy'd, it prov'd too big to go into the Pneumaticall Veſſel: and we have not now the conveniency to have a fitter made.

We thought likewiſe to convey into the Receiver a long and ſlen-

slender pair of Bellows, made after the fashion of those usually employ'd to blow Organs, and furnished with a small Musical instead of an ordinary Pipe. For we hop'd, that by meanes of a string fastned to the upper part of the Bellows, and to the moveable stopple that makes a part of the Cover of our Receiver, we should, by frequently turning round that stopple, and the annexed string, after the manner already often recited, be able to lift up and distend the Bellows; and by the help of a competent weight fastned to the same upper part of the Bellows, we should likewise be able, at pleasure, to compress them: and by consequence, try whether that subtler matter then Air (which, according to those that deny a *Vacuum*, must be suppos'd to fill the exhausted Receiver) would be able to produce a sound in the Musical Pipe; or in a Pipe like that of ordinary Bellows, to beget a Wind capable to turne or set on moving some very light matter, either shap'd like the Sails of a Wind-Mill, or of some other convenient form, and exposed to its Orifice. This Experiment, I say, we thought to make, but have not yet actually made it for want of an Artificer to make us such a pair of Bellows as it requires.

We had thoughts also of trying whether or no, as Sounds made by the Bodies in our Receiver become much more languid then ordinary, by reason of the want of Air: so they would grow stronger, in case there were an unusual quantity of Air crouded and shut up in the same Vessel. Which may be done (though not without some difficulty) by the help of the Pump, provided the Cover and Stopple be so firmly fasten'd (by binding and Cement, or otherwise) to the Glass and to each other, that there be no danger of the condensed Airs blowing of either of them away, or its breaking through the junctures. These thoughts, My Lord, as I was saying, we entertain'd; but for want of leasure, as, of as good Receivers as ours, to substitute in its place, in case we should break it before we learn'd the skill of condensing the Air in it, we durst not put them in practice: Yet, on this occasion, give me leave to advertise Your Lordship once for all,

That

That though for the reasons newly intimated, we have, One-ly in the seventeenth Experiment, taken notice, that by the help of our Engine the Air may be condens'd as well as rarified; yet there are divers other of our Experiments, whose *Phænomena* it were worth while to try to vary, by means of the compres-sion of the Air.

8

Theoretical Derivation of the Velocity of Sound in Air

ISAAC NEWTON

Sir Isaac Newton (1642–1727), one of the giants of natural philosophy, is most famous for his work in gravitation, cosmology, light, and mathematics. In his famous treatise *Principia Mathematica* (Mathematical Principles of Natural Philosophy), however, in Book II in connection with his investigations of wave motion in fluids, he made what is usually considered to be the first attempt at a theoretical derivation of the velocity of sound in a fluid like air. This section is reproduced here as one of the classics of acoustics. It must be confessed that Newton's derivation was not considered completely clear by his contemporaries or successors in the 18th century. His work was criticized by such men as Euler and Lagrange, as will be seen in other papers in this volume. Newton's value falls definitely short of the experimental value. Later and more careful theoretical derivations only served to confirm Newton's value. The discrepancy was only cleared up by Laplace, whose famous 1816 paper is reproduced in this volume. It is interesting to note that modern physical research has shown that Newton's expression is the correct one for gases for which the ratio of frequency to pressure is very large.

Reprinted from "Mathematical Principles of Natural Philosophy," translated by Andrew Motte, first American edition (D. Adee, New York) 356–357 (1848)

SECTION VIII.

Of motion propagated through fluids.

PROPOSITION XLI. THEOREM XXXII.

A pressure is not propagated through a fluid in rectilinear directions unless where the particles of the fluid lie in a right line.

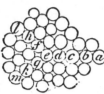

If the particles *a, b, c, d, e,* lie in a right line, the pressure may be indeed directly propagated from *a* to *e;* but then the particle *e* will urge the obliquely posited particles *f* and *g* obliquely, and those particles *f* and *g* will not sustain this pressure, unless they be supported by the particles *h* and *k* lying beyond them; but the particles that support them are also pressed by them; and those particles cannot sustain that pressure, without being supported by, and pressing upon, those particles that lie still farther, as *l* and *m*, and so on *in infinitum.* Therefore the pressure, as soon as it is propagated to particles that lie out of right lines, begins to deflect towards one hand and the other, and will be propagated obliquely *in infinitum;* and after it has begun to be propagated obliquely, if it reaches more distant particles lying out of the right line, it will deflect again on each hand; and this it will do as often as it lights on particles that do not lie exactly in a right line. Q.E.D.

Cor. If any part of a pressure, propagated through a fluid from a given point, be intercepted by any obstacle, the remaining part, which is not intercepted, will deflect into the spaces behind the obstacle. This may be demonstrated also after the following manner. Let a pressure be propagated from the point A towards any part, and, if it be possible, in rectilinear

directions; and the obstacle
NBCK being perforated in BC,
let all the pressure be intercepted
but the coniform part APQ passing through the circular hole BC.
Let the cone APQ be divided
into frustums by the transverse
planes, *de, fg, hi*. Then while
the cone ABC, propagating the
pressure, urges the conic frustum
degf beyond it on the superficies

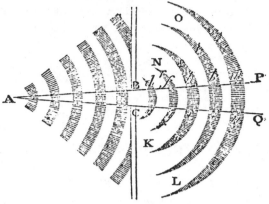

de, and this frustum urges the next frustum *fgih* on the superficies *fg*, and
that frustum urges a third frustum, and so *in infinitum;* it is manifest
(by the third Law) that the first frustum *defg* is, by the re-action of the
second frustum *fghi*, as much urged and pressed on the superficies *fg*, as
it urges and presses that second frustum. Therefore the frustum *degf* is
compressed on both sides, that is, between the cone A*de* and the frustum
fhig; and therefore (by Case 6, Prop. XIX) cannot preserve its figure,
unless it be compressed with the same force on all sides. Therefore with
the same force with which it is pressed on the superficies *de, fg*, it will
endeavour to break forth at the sides *df, eg;* and there (being not in the
least tenacious or hard, but perfectly fluid) it will run out, expanding itself, unless there be an ambient fluid opposing that endeavour. Therefore,
by the effort it makes to run out, it will press the ambient fluid, at its sides
df, eg, with the same force that it does the frustum *fghi;* and therefore,
the pressure will be propagated as much from the sides *df, eg*, into the
spaces NO, KL this way and that way, as it is propagated from the superficies *fg* towards PQ.   Q.E.D.

PROPOSITION XLII.   THEOREM XXXIII.

*All motion propagated through a fluid diverges from a rectilinear pro-
gress into the unmoved spaces.*

CASE 1. Let a motion be
propagated from the point A
through the hole BC, and, if it
be possible, let it proceed in the
conic space BCQP according to
right lines diverging from the
point A. And let us first suppose this motion to be that of
waves in the surface of standing
water; and let *de, fg, hi, kl*, &c.,
be the tops of the several waves,
divided from each other by as
many intermediate valleys or hollows.   Then, because the water in the

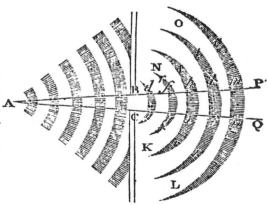

ridges of the waves is higher than in the unmoved parts of the fluid KL, NO, it will run down from off the tops of those ridges, e, g, i, l, &c., d, f, h, k, &c., this way and that way towards KL and NO; and because the water is more depressed in the hollows of the waves than in the unmoved parts of the fluid KL, NO, it will run down into those hollows out of those unmoved parts. By the first deflux the ridges of the waves will dilate themselves this way and that way, and be propagated towards KL and NO. And because the motion of the waves from A towards PQ is carried on by a continual deflux from the ridges of the waves into the hollows next to them, and therefore cannot be swifter than in proportion to the celerity of the descent; and the descent of the water on each side towards KL and NO must be performed with the same velocity; it follows that the dilatation of the waves on each side towards KL and NO will be propagated with the same velocity as the waves themselves go forward with directly from A to PQ. And therefore the whole space this way and that way towards KL and NO will be filled by the dilated waves *rfgr, shis, tklt, vmnv*, &c. Q.E.D. That these things are so, any one may find by making the experiment in still water.

CASE 2. Let us suppose that *de, fg, hi, kl, mn*, represent pulses successively propagated from the point A through an elastic medium. Conceive the pulses to be propagated by successive condensations and rarefactions of the medium, so that the densest part of every pulse may occupy a spherical superficies described about the centre A, and that equal intervals intervene between the successive pulses. Let the lines *de, fg, hi, kl*, &c., represent the densest parts of the pulses, propagated through the hole BC; and because the medium is denser there than in the spaces on either side towards KL and NO, it will dilate itself as well towards those spaces KL, NO, on each hand, as towards the rare intervals between the pulses; and thence the medium, becoming always more rare next the intervals, and more dense next the pulses, will partake of their motion. And because the progressive motion of the pulses arises from the perpetual relaxation of the denser parts towards the antecedent rare intervals; and since the pulses will relax themselves on each hand towards the quiescent parts of the medium KL, NO, with very near the same celerity; therefore the pulses will dilate themselves on all sides into the unmoved parts KL, NO, with almost the same celerity with which they are propagated directly from the centre A; and therefore will fill up the whole space KLON. Q.E.D. And we find the same by experience also in sounds which are heard through a mountain interposed; and, if they come into a chamber through the window, dilate themselves into all the parts of the room, and are heard in every corner; and not as reflected from the opposite walls, but directly propagated from the window, as far as our sense can judge.

CASE 3. Let us suppose, lastly, that a motion of any kind is propagated from A through the hole BC. Then since the cause of this propagation is that the parts of the medium that are near the centre A disturb and agitate those which lie farther from it; and since the parts which are urged are fluid, and therefore recede every way towards those spaces where they are less pressed, they will by consequence recede towards all the parts of the quiescent medium; as well to the parts on each hand, as KL and NO, as to those right before, as PQ; and by this means all the motion, as soon as it has passed through the hole BC, will begin to dilate itself, and from thence, as from its principle and centre, will be propagated directly every way. Q.E.D.

PROPOSITION XLIII. THEOREM XXXIV.

Every tremulous body in an elastic medium propagates the motion of the pulses on every side right forward; but in a non-elastic medium excites a circular motion.

CASE. 1. The parts of the tremulous body, alternately going and returning, do in going urge and drive before them those parts of the medium that lie nearest, and by that impulse compress and condense them; and in returning suffer those compressed parts to recede again, and expand themselves. Therefore the parts of the medium that lie nearest to the tremulous body move to and fro by turns, in like manner as the parts of the tremulous body itself do; and for the same cause that the parts of this body agitate these parts of the medium, these parts, being agitated by like tremors, will in their turn agitate others next to themselves; and these others, agitated in like manner, will agitate those that lie beyond them, and so on *in infinitum.* And in the same manner as the first parts of the medium were condensed in going, and relaxed in returning, so will the other parts be condensed every time they go, and expand themselves every time they return. And therefore they will not be all going and all returning at the same instant (for in that case they would always preserve determined distances from each other, and there could be no alternate condensation and rarefaction); but since, in the places where they are condensed, they approach to, and, in the places where they are rarefied, recede from each other, therefore some of them will be going while others are returning; and so on *in infinitum.* The parts so going, and in their going condensed, are pulses, by reason of the progressive motion with which they strike obstacles in their way; and therefore the successive pulses produced by a tremulous body will be propagated in rectilinear directions; and that at nearly equal distances from each other, because of the equal intervals of time in which the body, by its several tremors produces the several pulses. And though the parts of the tremulous body go and return in some certain and determinate direction, yet the pulses propagated from thence through the medium will dilate themselves towards the sides, by the foregoing Proposition; and will be propagated on all sides from that tremulous body, as from a common centre, in superficies nearly spherical and concentrical. An example of this we have in waves excited by shaking a .finger in water, which proceed not only forward and backward agreeably to the motion of the finger, but spread themselves in the manner of concentrical circles all round the finger, and are propagated on every side. For the gravity of the water supplies the place of elastic force.

Case 2. If the medium be not elastic, then, because its parts cannot be condensed by the pressure arising from the vibrating parts of the tremulous body, the motion will be propagated in an instant towards the parts where the medium yields most easily, that is, to the parts which the tremulous body would otherwise leave vacuous behind it. The case is the same with that of a body projected in any medium whatever. A medium yielding to projectiles does not recede *in infinitum*, but with a circular motion comes round to the spaces which the body leaves behind it. Therefore as often as a tremulous body tends to any part, the medium yielding to it comes round in a circle to the parts which the body leaves; and as often as the body returns to the first place, the medium will be driven from the place it came round to, and return to its original place. And though the tremulous body be not firm and hard, but every way flexible, yet if it continue of a given magnitude, since it cannot impel the medium by its tremors any where without yielding to it somewhere else, the medium receding from the parts of the body where it is pressed will always come round in a circle to the parts that yield to it. **Q.E.D.**

Cor. It is a mistake, therefore, to think, as some have done, that the agitation of the parts of flame conduces to the propagation of a pressure in rectilinear directions through an ambient medium. A pressure of that kind must be derived not from the agitation only of the parts of flame, but from the dilatation of the whole.

PROPOSITION XLIV. THEOREM XXXV.

If water ascend and descend alternately in the erected legs KL, MN, of a canal or pipe; and a pendulum be constructed whose length between the point of suspension and the centre of oscillation is equal to half the length of the water in the canal; I say, that the water will ascend and descend in the same times in which the pendulum oscillates.

I measure the length of the water along the axes of the canal and its legs, and make it equal to the sum of those axes; and take no notice of the resistance of the water arising from its attrition by the sides of the canal. Let, therefore, AB, CD, represent the mean height of the water in both legs; and when the water in the leg KL ascends to the height EF, the water will descend in the leg MN to the height GH. Let P be a pendulous

79

body, VP the thread, V the point of suspension, RPQS the cycloid which

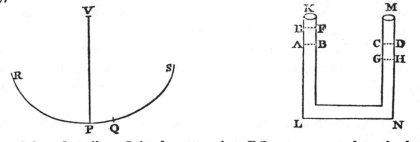

the pendulum describes, P its lowest point, PQ an arc equal to the height AE. The force with which the motion of the water is accelerated and retarded alternately is the excess of the weight of the water in one leg above the weight in the other; and, therefore, when the water in the leg KL ascends to EF, and in the other leg descends to GH, that force is double the weight of the water EABF, and therefore is to the weight of the whole water as AE or PQ to VP or PR. The force also with which the body P is accelerated or retarded in any place, as Q, of a cycloid, is (by Cor. Prop. LI) to its whole weight as its distance PQ from the lowest place P to the length PR of the cycloid. Therefore the motive forces of the water and pendulum, describing the equal spaces AE, PQ, are as the weights to be moved; and therefore if the water and pendulum are quiescent at first, those forces will move them in equal times, and will cause them to go and return together with a reciprocal motion.   Q.E.D.

COR. 1. Therefore the reciprocations of the water in ascending and descending are all performed in equal times, whether the motion be more or less intense or remiss.

COR. 2. If the length of the whole water in the canal be of 6½ feet of *French* measure, the water will descend in one second of time, and will ascend in another second, and so on by turns *in infinitum;* for a pendulum of $3\frac{1}{18}$ such feet in length will oscillate in one second of time.

COR. 3. But if the length of the water be increased or diminished, the time of the reciprocation will be increased or diminished in the subduplicate ratio of the length.

PROPOSITION XLV. THEOREM XXXVI.

The velocity of waves is in the subduplicate ratio of the breadths.

This follows from the construction of the following Proposition.

PROPOSITION XLVI. PROBLEM X.

To find the velocity of waves.

Let a pendulum be constructed, whose length between the point of suspension and the centre of oscillation is equal to the breadth of the waves;

and in the time that the pendulum will perform one single oscillation the waves will advance forward nearly a space equal to their breadth.

That which I call the breadth of the waves is the transverse measure lying between the deepest part of the hollows, or the tops of the ridges. Let ABCDEF represent the surface of stagnant water ascending and descending in successive waves; and let A, C, E, &c., be the tops of the waves; and let B, D, F, &c., be the intermediate hollows. Because the motion of the waves is carried on by the successive ascent and descent of the water, so that the parts thereof, as A, C, E, &c., which are highest at one time become lowest immediately after; and because the motive force, by which the highest parts descend and the lowest ascend, is the weight of the elevated water, that alternate ascent and descent will be analogous to the reciprocal motion of the water in the canal, and observe the same laws as to the times of its ascent and descent; and therefore (by Prop. XLIV) if the distances between the highest places of the waves A, C, E, and the lowest B, D, F, be equal to twice the length of any pendulum, the highest parts A, C, E, will become the lowest in the time of one oscillation, and in the time of another oscillation will ascend again. Therefore between the passage of each wave, the time of two oscillations will intervene; that is, the wave will describe its breadth in the time that pendulum will oscillate twice; but a pendulum of four times that length, and which therefore is equal to the breadth of the waves, will just oscillate once in that time. **Q.E.I.**

Cor. 1. Therefore waves, whose breadth is equal to $3\frac{1}{18}$ *French* feet, will advance through a space equal to their breadth in one second of time; and therefore in one minute will go over a space of $183\frac{1}{3}$ feet; and in an hour a space of 11000 feet, nearly.

Cor. 2. And the velocity of greater or less waves will be augmented or diminished in the subduplicate ratio of their breadth.

These things are true upon the supposition that the parts of water ascend or descend in a right line; but, in truth, that ascent and descent is rather performed in a circle; and therefore I propose the time defined by this Proposition as only near the truth.

PROPOSITION XLVII. THEOREM XXXVII.

If pulses are propagated through a fluid, the several particles of the fluid, going and returning with the shortest reciprocal motion, are always accelerated or retarded according to the law of the oscillating pendulum.

Let AB, BC, CD, &c., represent equal distances of successive pulses; ABC the line of direction of the motion of the successive pulses propagated from A to B; E, F, G three physical points of the quiescent medium situate in the right line AC at equal distances from each other; Ee, Ff, Gg, equal spaces of extreme shortness, through which those points go and return with a reciprocal motion in each vibration; ε, ϕ, γ, any intermediate places of the same points; EF, FG physical lincolæ, or linear parts of the medium lying between those points, and successively transferred into the places $\varepsilon\phi$, $\phi\gamma$, and ef, fg. Let there be drawn the right line PS equal to the right line Ee. Bisect the same in O, and from the centre O, with the interval OP, describe the circle SIPi. Let the whole time of one vibration; with its proportional parts, be expounded by the whole circumference of this circle and its parts, in such sort, that, when any time PH or PHSh is completed, if there be let fall to PS the perpendicular HL or hl, and there be taken Eε equal to PL or Pl, the physical point E may be found in ε. A point, as E, moving acccording to this law with a reciprocal motion, in its going from E through ε to e, and returning again through ε to E, will perform its several vibrations with the same degrees of acceleration and retardation with those of an oscillating pendulum. We are now to prove that the several physical points of the medium will be agitated with such a kind of motion. Let us suppose, then, that a medium hath such a motion excited in it from any cause whatsoever, and consider what will follow from thence.

In the circumference PHSh let there be taken the equal arcs, HI, IK, or hi, ik, having the same ratio to the whole circumference as the equal right lines EF, FG have to BC, the whole interval of the pulses. Let fall the perpendiculars IM, KN, or im, kn; then because the points E, F, G are successively agitated with like motions, and perform their entire vibrations composed of their going and return, while the pulse is transferred from B to C; if PH or PHSh be the time elapsed since the beginning of the motion of the point E, then will PI or PHSi be the time elapsed since the beginning of the motion of the point F, and PK or PHSk the time elapsed since the beginning of the motion of the point G; and therefore Eε, Fϕ, Gγ, will be respectively equal to PL, PM, PN, while the points are going, and to Pl, Pm, Pn, when the points are returning. Therefore $\varepsilon\gamma$ or EG + Gγ — Eε will, when the points are going, be equal to EG — LN, and in their return equal to EG + ln. But $\varepsilon\gamma$ is the breadth or expansion of the part EG of the medium in the place $\varepsilon\gamma$; and therefore the expansion of that part in its going is to its mean expansion as EG — LN to EG; and in its return, as EG + ln or EG + LN to EG. Therefore since LN is to KH as IM to the radius OP, and KH to EG as the circumference PHShP to BC; that is, if we put V for the radius of a circle whose circumference is equal to BC the interval of the pulses, as OP to V; and, *ex æquo*, LN to EG as IM to V; the expansion of the part EG, or of the physical point F in the place $\varepsilon\gamma$, to the mean expansion of the same part in its first place EG, will be as V — IM to V in going, and as V + im to V in its return. Hence the elastic force of the point F in the place $\varepsilon\gamma$ to its mean elastic force in the place EG is as

$$\frac{1}{V - IM} \text{ to } \frac{1}{V} \text{ in its going, and as } \frac{1}{V + im} \text{ to } \frac{1}{V} \text{ in its return.}$$ And by the same reasoning the elastic forces of the physical points E and G in going

are as $\dfrac{1}{V - HL}$ and $\dfrac{1}{V - KN}$ to $\dfrac{1}{V}$; and the difference of the forces to the

mean elastic force of the medium as $\dfrac{HL - KN}{VV - V \times HL - V \times KN + HL \times KN}$

to $\dfrac{1}{V}$; that is, as $\dfrac{HL - KN}{VV}$ to $\dfrac{1}{V}$, or as HL — KN to V; if we suppose (by reason of the very short extent of the vibrations) HL and KN to be indefinitely less than the quantity V. Therefore since the quantity V is given, the difference of the forces is as HL — KN; that is (because HL — KN is proportional to HK, and OM to OI or OP; and because HK and OP are given) as OM; that is, if Ff be bisected in Ω, as $\Omega\phi$. And for the same reason the difference of the elastic forces of the physical points ε and γ, in the return of the physical lineola $\varepsilon\gamma$, is as $\Omega\phi$. But that difference (that is, the excess of the elastic force of the point ε above the elastic force of the point γ) is the very force by which the intervening physical lineola $\varepsilon\gamma$ of the medium is accelerated in going, and retarded in returning; and therefore the accelerative force of the physical lineola $\varepsilon\gamma$ is as its distance from Ω, the middle place of the vibration. Therefore (by Prop. XXXVIII, Book I) the time is rightly expounded by the arc PI; and the linear part of the medium $\varepsilon\gamma$ is moved according to the law abovementioned, that is, according to the law of a pendulum oscillating; and the case is the same of all the linear parts of which the whole medium is compounded. Q.E.D.

Cor. Hence it appears that the number of the pulses propagated is the same with the number of the vibrations of the tremulous body, and is not multiplied in their progress. For the physical lineola $\varepsilon\gamma$ as soon as it returns to its first place is at rest; neither will it move again, unless it receives a new motion either from the impulse of the tremulous body, or of the pulses propagated from that body. As soon, therefore, as the pulses cease to be propagated from the tremulous body, it will return to a state of rest, and move no more.

PROPOSITION XLVIII. THEOREM XXXVIII.

The velocities of pulses propagated in an elastic fluid are in a ratio compounded of the subduplicate ratio of the elastic force directly, and the subduplicate ratio of the density inversely; supposing the elastic force of the fluid to be proportional to its condensation.

CASE 1. If the mediums be homogeneous, and the distances of the pulses in those mediums be equal amongst themselves, but the motion in one medium is more intense than in the other, the contractions and dilatations of the correspondent parts will be as those motions; not that this proportion is perfectly accurate. However, if the contractions and dilatations are not exceedingly intense, the error will not be sensible; and therefore this proportion may be considered as physically exact. Now the motive elastic forces are as the contractions and dilatations; and the velocities generated in the same time in equal parts are as the forces. Therefore equal and corresponding parts of corresponding pulses will go and return together, through spaces proportional to their contractions and dilatations, with velocities that are as those spaces; and therefore the pulses, which in the time of one going and returning advance forward a space equal to their breadth, and are always succeeding into the places of the pulses that immediately go before them, will, by reason of the equality of the distances, go forward in both mediums with equal velocity.

CASE 2. If the distances of the pulses or their lengths are greater in one medium than in another, let us suppose that the correspondent parts describe spaces, in going and returning, each time proportional to the breadths of the pulses; then will their contractions and dilatations be equal; and therefore if the mediums are homogeneous, the motive elastic forces, which agitate them with a reciprocal motion, will be equal also. Now the matter to be moved by these forces is as the breadth of the pulses; and the space through which they move every time they go and return is in the same ratio. And, moreover, the time of one going and returning is in a ratio compounded of the subduplicate ratio of the matter, and the subduplicate ratio of the space; and therefore is as the space. But the pulses advance a space equal to their breadths in the times of going once and returning once; that is, they go over spaces proportional to the times, and therefore are equally swift.

CASE 3. And therefore in mediums of equal density and elastic force, all the pulses are equally swift. Now if the density or the elastic force of the medium were augmented, then, because the motive force is increased in the ratio of the elastic force, and the matter to be moved is increased in the ratio of the density, the time which is necessary for producing the same motion as before will be increased in the subduplicate ratio of the density, and will be diminished in the subduplicate ratio of the elastic force. And therefore the velocity of the pulses will be in a ratio compounded of the subduplicate ratio of the density of the medium inversely, and the subduplicate ratio of the elastic force directly. Q.E.D.

This Proposition will be made more clear from the construction of the following Problem.

PROPOSITION XLIX. PROBLEM XI.

The density and elastic force of a medium being given, to find the velocity of the pulses.

Suppose the medium to be pressed by an incumbent weight after the manner of our air; and let A be the height of a homogeneous medium, whose weight is equal to the incumbent weight, and whose density is the same with the density of the compressed medium in which the pulses are propagated. Suppose a pendulum to be constructed whose length between the point of suspension and the centre of oscillation is A: and in the time in which that pendulum will perform one entire oscillation composed of its going and returning, the pulse will be propagated right onwards through a space equal to the circumference of a circle described with the radius A.

For, letting those things stand which were constructed in Prop. XLVII, if any physical line, as EF, describing the space PS in each vibration, be acted on in the extremities P and S of every going and return that it makes by an elastic force that is equal to its weight, it will perform its several vibrations in the time in which the same might oscillate in a cycloid whose whole perimeter is equal to the length PS; and that because equal forces will impel equal corpuscles through equal spaces in the same or equal times. Therefore since the times of the oscillations are in the subduplicate ratio of the lengths of the pendulums, and the length of the pendulum is equal to half the arc of the whole cycloid, the time of one vibration would be to the time of the oscillation of a pendulum whose length is A in the subduplicate ratio of the length $\frac{1}{2}$PS or PO to the length A. But the elastic force with which the physical lineola EG is urged, when it is found in its extreme places P, S, was (in the demonstration of Prop. XLVII) to its whole elastic force as HL — KN to V, that is (since the point K now falls upon P), as HK to V: and all that force, or which is the same thing, the incumbent weight by which the lineola EG is compressed, is to the weight of the lineola as the altitude A of the incumbent weight to EG the length of the lineola; and therefore, *ex æquo*, the force

85

with which the lineola EG is urged in the places P and S
is to the weight of that lineola as HK × A to V × EG; or
as PO × A to VV; because HK was to EG as PO to V.
Therefore since the times in which equal bodies are impelled
through equal spaces are reciprocally in the subduplicate
ratio of the forces, the time of one vibration, produced by
the action of that elastic force, will be to the time of a vibration, produced by the impulse of the weight in a subduplicate ratio of VV to PO × A, and therefore to the time
of the oscillation of a pendulum whose length is A in the
subduplicate ratio of VV to PO × A, and the subduplicate ratio of PO to A conjunctly; that is, in the entire ratio of V to A.   But in the time of one
vibration composed of the going and returning of the pendulum, the pulse will
be propagated right onward through a
space equal to its breadth BC.   Therefore the time in which a pulse runs over

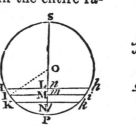

the space BC is to the time of one oscillation composed of
the going and returning of the pendulum as V to A, that is,
as BC to the circumference of a circle whose radius is A.
But the time in which the pulse will run over the space BC
is to the time in which it will run over a length equal to
that circumference in the same ratio; and therefore in the
time of such an oscillation the pulse will run over a length
equal to that circumference.   Q.E.D.

Cor. 1. The velocity of the pulses is equal to that which
heavy bodies acquire by falling with an equally accelerated motion, and in their fall describing half the altitude A.   For the pulse will, in the time of this fall, supposing it to move with the velocity acquired by that fall, run over a
space that will be equal to the whole altitude A; and therefore in the
time of one oscillation composed of one going and return, will go over a
space equal to the circumference of a circle described with the radius A;
for the time of the fall is to the time of oscillation as the radius of a circle
to its circumference.

Cor. 2. Therefore since that altitude A is as the elastic force of the
fluid directly, and the density of the same inversely, the velocity of the
pulses will be in a ratio compounded of the subduplicate ratio of the density inversely, and the subduplicate ratio of the elastic force directly.

9

General System of Sound Intervals
and Its Application to Sounds
of All Systems and All Musical Instruments

JOSEPH SAUVEUR

Joseph Sauveur (1653–1716), the French mathematician and physicist, is credited with the first introduction of the word *acoustics* to represent the science of sound. Though he wrote treatises on fortification, his main interest was in acoustics and in particular acoustics as a means of understanding music. He may well be considered the creator of musical acoustics. The following extract is taken from his famous "Système Général des Intervales du Son," published in 1701. In this he develops the concept of harmonics of a stretched string and introduces the notion of nodes and loops in the vibrating string, showing that he fully understood the nature of standing waves from a phenomenological point of view. The dynamical approach to the problem had to wait for Taylor, d'Alembert, Euler, and Lagrange.

9

General System of Sound Intervals
and Its Application to Sounds
of All Systems and All Musical Instruments

JOSEPH SAUVEUR

Translated from the French by R. Bruce Lindsay
from "Système Général des Intervales du Son,"
Memoirs de l'Academie Royale des Sciences, 297–300, 347–354 (1701)

Preface

The occasion on which I have found myself asked to explain the theory of music to enlightened princes and to persons of profound intellectual qualities leads me to remark that those who are attached to music from a speculative point of view have had in view only a limited number of properties of sound and in particular only the practice of singing in use at their time. With respect to the systems of music they are content to take what others have changed little by little, as taste in music changes. So far as I know, no one has taken a more serious look at the matter or regards it as the object of a science superior to music, in order to abstract from it a part of particular importance to him and which has a natural and simple connection with other points included in the same science.

I have come then to the opinion that there *is* a science superior to music, and I call it *acoustics*; it has for its object sound in general, whereas music has for its object sounds agreeable to the ear.

In order to treat this science in the same way as other sciences, and in particular optics, with which it has a close connection, it becomes necessary to explain the nature of sound, the organ of hearing and in some detail all the properties of sound, and to reach agreement with respect to the consonance and dissonance of sounds which are of importance in music, to the sympathy of sounds, and finally to explain instruments, not only those of musical importance but those for acoustics in general.

As sound is produced by the vibrations of the parts of a sounding body and as the principal property of these vibrations consists in the relation of the number of vibrations of a sound with that of the vibrations of another sound which form the different degrees of sound or sound intervals with respect to high and low pitch, in 1696 I took the step of seeking a common measure of all sound intervals, competent to measure them in their least perceptible differences and of giving names and characters to all sounds, such that one can include among them those that are necessary for ordinary music and which cover in a simple manner all the properties respecting that art without indeed designing to exclude any of the notes to which musicians have for a long time been accustomed.

Afterwards I wrote an essay on acoustics in a treatise on speculative music I indited at the Royal College in 1697. One might expect that I would have had it printed, but I was prevented by the following reasons: 1) the names and characters which I gave to sounds being new, I did not doubt that I would fare ill especially among the musicians, and people who would hold an opposite opinion. I hoped by the objections they raised against me to find occasion for some correction; but since they regarded sound only from the standpoint of their own needs, such small changes as I made I was obliged to make for myself, 2) in working on my treatise on speculative music, I recognized the necessity of having a fixed sound with respect to which one could compare all other sounds of high or low pitch. In 1700 I gave a way in which I imagined one could find this. Since in the Proceedings of the Academy there was shown only the necessity and advantage to be drawn from this, I give here the way of finding it, 3) in thinking about the phenomena of sound, one may remark that especially at night, in the case of long strings, besides the principal sound, one hears other faint sounds of a twelfth or seventeenth of this sound. Similarly trumpets, besides these principal sounds, have others for which the number of vibrations is a multiple of that of the fundamental. I have found nothing to satisfy me in the explanation of the behavior of the marine trumpet. But in looking for the cause of this phenomenon myself, I conclude that in addition to the undulation which the string makes throughout its whole length to form the fundamental note, the string also divides itself into two, three, four, etc. equal undulations to form the octave, the twelfth, the fifteenth, etc. of this note. From this I have concluded the necessity for the existence of knots and bulges [Editor's note: nodes and loops in modern terminology] in these undulations. The manner of perceiving these I have explained in connection with harmonic sounds. 4) This phenomenon has led me to investigate some others connected with the sympathy of sounds, concerning wind instruments and concerning acoustical instruments in general and in particular how one can perfect them to the same degree as optical instruments. And I expect that these things should ultimately lead to a perfect body of acoustics.

Since the branch of acoustics which has for its object the intervals of sounds serves as a basis for all the other branches and since it has had time to be put in order, and since further I have given to my system all the attention one could expect, and I have made a general application to all sorts of systems and musical instruments, and since finally people are beginning to use the intervals of my system, I have concluded it is time to give it to the public. I am doing this with the greatest possible brevity and clarity. I am taking the liberty of introducing some new words, which are necessary for the understanding of my system. I do not give demonstrations here of those ideas which I present, because in addition to the fact that several have already been made in the case of one part, the demonstration of the remaining would demand a complete treatise on acoustics.

I. Concerning Ratios of Sounds and Intervals

Acoustics teaches us that if two sounding bodies make the same number of vibrations in the same time, they are in unison; that if one makes more vibrations than the other in the same time, that which makes the smaller number produces the lower pitch (grave note), whereas that which makes the larger number produces the higher pitch (acute note). Hence the relation between sounds of low and high pitch is exemplified in the ratio of the numbers of vibrations which they both make in the same time. That is why the ratio of two sounds which are said to differ by a fifth is the ratio 2:3. This means that while the sound of lower pitch makes two vibrations, the sound of higher pitch (a fifth above the other) makes three vibrations.

We are going to compare sounds by the ratio of the number of vibrations of one sound to the corresponding number of the other and shall in what follows call this simply the sound ratio. There is another way of comparing sounds and that is by their intervals.

In order to grasp what the interval between two sounds is, let us at first imagine two sounds which are equal or in unison. We further imagine that one of these goes up in pitch more and more toward infinity, that is, departs more and more from the other. It is this deviation or distance of separation between the two sounds which we call in general, the interval between them. The same thing happens when the sound in question decreases steadily in pitch.

These intervals divide themselves initially into the *diapason* or octaves. This happens when the higher pitch sound makes two vibrations compared with one for the lower pitch sound. Thus a sound which goes up in pitch passes through intervals of a first, second, third, fourth octave, etc., when its number of vibrations has the ratio, 2, 4, 8, 16, etc. with the number of vibrations for the lower pitch sound. Similarly when the sound goes down in pitch it passes through successive octaves when the number of vibrations is 1/2, 1/4, 1/8, 1/16 etc. that of the number in the first place.

IX. Concerning Harmonic Sounds

I call the harmonic of a fundamental sound that sound which makes several vibrations while the fundamental makes only one. Thus one sound which is a twelfth of the fundamental is a harmonic because it makes three vibrations while the fundamental makes only one.

[Editor's note: Following the above introductory statement the author provides a table of harmonic sounds covering five octaves, with the accepted nomenclature of the associated musical scale and Sauveur's proposed new names. The section then proceeds as follows.]

After having defined and determined harmonics, it remains to make them perceptible to the ear and even to the eye and to explain their properties. Divide the string of a monochord into equal parts, for example, into five (one can divide a rule of the same length accordingly and lay it along the string). Pluck the string at will. It will give off a sound which I call the fundamental. Place a light obstacle C at

90

one of the division points D, [see Figure 1] like the tip of a feather, if the string is a fine one. The motion of the string takes place on both sides of the resting obstacle. This produces the fifth harmonic, that is to say a seventeenth.

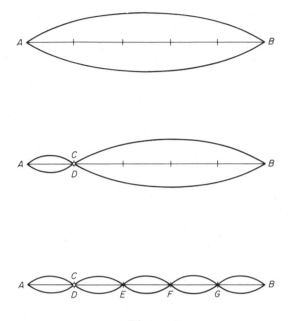

Figure 1.

In order to understand the reason for this result, notice that when one plucks the cord AB as a whole undulations are produced in the whole of its length but when one inserts an obstacle at the first division D of the string, which I have supposed to be divided into five equal parts, the total undulation AB is divided at first into the two AD and DB. Since AD is one-fifth of AB or one-quarter of DB, it performs undulations five times faster than the total AB or four times faster than the other portion DB. Hence the part AD involves the neighboring part DE and forces the latter to follow its motion. Consequently the motion of DE ought to be equal to that of AD, for a larger part would go more slowly. A smaller part would go more rapidly. In turn the part DE forces the neighboring part EF to follow the same motion, and so on to the last. Hence all the parts perform undulations which cross each other at the dividing points D, E, F, etc. Consequently the string produces the fifth harmonic or a seventeenth.

I have called the points A, D, E, F, G, B the *knots* of undulation [Editor's note: *nodes* in modern terminology; in Sauveur's French they are called *noeuds*]; the middles of these undulations will be called the *bulges* of the undulation [Editor's note: *loops* in modern terminology, Sauveur's French reads *ventres*].

If the obstacle is placed at the second dividing point E, the same harmonic will be produced. For, in the first place [see Figure 2] the obstacle C will force the string at first to make the two undulations *AE* and *EB*. In the second place, the undulation *AE* being more rapid than the other, will force the part *EG* which is equal to it to follow its motion. In the third place, the remaining part *GB* which is half of *EG* and hence moves twice as fast, will force its equal *GF* to follow its motion and this in turn will force *FE* to do the same, and so on to the extremity. Hence the whole string will be divided by these undulations into parts equal to the greatest common measure of the parts *AC* and *CB* formed by the light obstacles *C*.

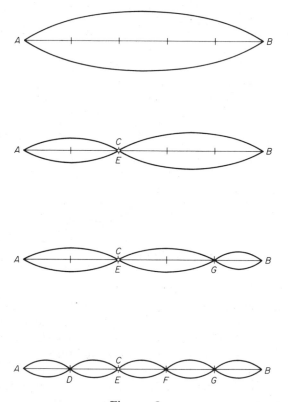

Figure 2.

One will be convinced of the reality of these undulations, first by the ear; for those who have a keen ear will distinguish a harmonic proportional to the parts which form the undulations; one can also assure oneself of this by putting the monochord in unison with this harmonic. In the second place, one can be convinced by the eyes, for if one divides the string into five equal parts and inserts a movable bridge *C* at *D* or *E* and then puts small pieces of black paper [Editor's note: hanging over the string] at the division point *E* and *F* and pieces of white paper at the middle points of these parts, if one strikes the part *AC* [Editor's note: he really should have said *AD*]; one will see the pieces of white paper which are at the loops of the undulations leap off the string, while the pieces of black paper, which are at the nodes, will remain at rest.

From what has just been established, we can draw the following conclusions:

I. If one has produced a harmonic by placing a small obstacle at D, one will continue to produce this harmonic even if the obstacle is removed or if we put an obstacle at any other node or at all nodes.

II. After having formed a harmonic, say the 5th, if one places a light obstacle on the loop of an undulation which divides it, for example, into three parts, it will lead to the production of the third harmonic of the originally formed harmonic, that is to say the 15th harmonic of the fundamental.

III. Without placing a light obstacle on the string, one can still produce in it a harmonic sound; first, if by the side of this string we pluck another which is in unison with one of the harmonics of the untouched string; second, if the first string is not in unison with one of the harmonics of the other, they will share harmonic sounds, which will be the greatest common measures of the fundamentals of the two strings. Thus if the one is a quarter of the other, for which the ratio is 3:4, the smaller will produce the 3rd harmonic and the latter the 4th harmonic.

IV. The harmonic produced by sympathetic vibration with a neighboring string or by the presence of a light obstacle is the more perceptible the greater are its undulations. Thus the third harmonic is more perceptible than the 4th. Those harmonics produced by sympathetic vibration are very soon imperceptible. Those which are produced by light obstacles are unlimited in number, but since the smallest can be distinguished only with the greatest difficulty we shall suppose in what follows that a three-foot long string can be audible up to the 32nd harmonic or the 5th octave, although one can hear above that to the 128th harmonic.

V. According to these suppositions every node which corresponds to a harmonic is distant from the next nearest node corresponding to another harmonic at least one 32nd part of a whole undulation. For example, the third of the string which produces the third harmonic is distant from the closest node of the other harmonics by at least one 32nd part of this third, or, what amounts to the same thing, to a third of the 32nd part of the whole string. For if we suppose that the 32nd harmonic is the last of all, if we divide the 32nd part of the string into three parts, the whole string will be divided into 96 parts. The nodes for both the 3rd harmonic and the 32nd harmonic can only be on some of these 96 division marks; hence they will be distant one from another at least by the amount of one of these 96 parts. Or, if they have a common node, the following node will be distant by at least the same amount. The node of the next harmonic is distant from that of the third harmonic by at least 1/96 of the whole string or by 1/32 of 1/3 of the string, or by 1/3 of 1/32 of the whole string, which amounts to the same thing.

VI. From this it follows, first, that the nodes of the first harmonic, that is to say the extremities of the string, are farther away from each other than the nodes of the next harmonic since the latter are separated by at least 1/32 of the whole string. In the second place the node of the second harmonic is distant from the next by a half of 1/32; etc. Hence the nodes of the lower harmonics are more widely spaced than those for the upper harmonics. This is why the lower harmonics have large amplitudes of motion around their nodes and the upper harmonics correspondingly smaller ones.

VII. It will happen that if the node of a small harmonic sound lies in the neighborhood of two nodes of a greater harmonic, the smaller will be eclipsed by the two larger, so that one will hear the smaller harmonic only if they are of the order of the lower harmonic, i.e., the first, second, third, and fourth, etc.

VIII. If one plucks the whole string as it stands at rest and then slides a light obstacle along the string one will hear a warble of harmonic sounds whose order appears confused; but which nevertheless can be determined by the principles we have established.

IX. We can draw other consequences from the same principles. For example, we can conclude that the small harmonics will displace as much air as the large ones.

X. Experience shows that long strings, when they are good or harmonious produce the first harmonics, principally those which are not an octave apart. Bells and other resonant bodies show the same effect.

XI. Long wind instruments also divide their lengths into kinds of equal undulations. If an undulation of the air which occupies the whole length between the mouthpiece through which the air enters and the first opening through which the air can escape, is forced into very rapid movement, it will divide into two equal undulations and then into three or four, etc., according to the length of the instrument. Hence in blowing slowly into a wind instrument, you hear the fundamental. If one forces the blowing, or in the case of the trumpet, the serpent or the hautboy, one compresses the lips more, the sound can change into the second, third, fourth, etc. harmonics. But in order to discover all the properties of wind instruments, it is necessary to enter upon a detailed examination.

10

Concerning the Motion of a Stretched String

BROOK TAYLOR

Brook Taylor (1685–1731), English mathematician, was a cultured 18th century man with many interests in law, philosophy, art, and religious studies. Educated at St. John's College, Cambridge, he held no university position but, having independent means, devoted his life to scholarly pursuits. He became for a time Secretary of the Royal Society of London. Taylor is now best known for his work in mathematics, where he extended the calculus and applied it successfully to numerous problems in mechanics. Taylor's theorem on the expansion of a function in series is well known. His treatment of the vibratory motion of a stretched string led for the first time to the correct formula for the frequency in terms of the length, tension, and mass. Taylor's method is set forth in the following translation of his paper "De motu nervi tensi" (1713). The reader will be interested to note the way in which Taylor followed Newton's geometrical notation (with the use of fluxions, etc.) in his use of the calculus. An editorial note at the end of the translation shows how his result may be expressed in modern form.

10

Concerning the Motion of a Stretched String

BROOK TAYLOR

Translated by R. Bruce Lindsay
from "De Motu Nervi Tensi,"
Philosophical Transactions of the Royal Society, London, **28**, 26–32 (1713)

Lemma 1

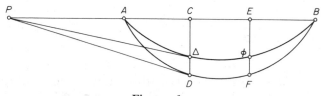

Figure 1.

In Figure 1 let ADFB and A$\Delta\phi$B be two curves, having the following relation that if we draw at will the ordinates $C\Delta D$ and $E\phi F$, we have

$$\frac{C\Delta}{CD} = \frac{E\phi}{EF}$$

Then with the ordinates becoming infinitely smaller, in order that the curves may coincide with the axis AB, I say that in the limit the ratio of the curvature at Δ to that at D must be as $C\Delta$ to CD.

[Ed. note: We omit Taylor's proof and pass on to Lemma 2.]

Lemma 2

Figure 2.

In any case of vibration let the stretched string between two points A and B take on the form of the arbitrary curve $Ap\pi B$ [see Figure 2]. Then I say that whatever be the increment of velocity of any arbitrary point P or its acceleration arising from the force of the strings tension, it will be proportional to the curvature of the string at the same point.

Demonstration

Imagine the string to be made up of rigid, infinitely small but equal particles like pP, $P\pi$, etc. [see Figure 2] and at the point P erect the perpendicular PR equal to the radius of curvature at P. This line PR meets at t the tangents Pt and $t\pi$. It is intersected at s by the lines ps and $s\pi$ parallel to $t\pi$ and pt, respectively, and by the chord $p\pi$ in c. Then, by the principles of mechanics, the absolute force by which both particles pP and $P\pi$ are drawn toward R will be to the force of the string's tension as st is to pt, and the half of this force, which acts on the one particle pP is to the tension of the string as ct is to tp, and this is because of the similar triangles ctp and tpR) as tp or Pp is to Rt or PR. For this reason, on account of the given tension force, the absolute accelerating force is proportional to Pp/PR. But the acceleration generated is directly proportional to the absolute force and inversely proportional to the matter to be moved; the matter to be moved is the particle Pp itself. Therefore the acceleration is as $1/PR$, which is proportional to the curvature at P. For the curvature is proportional to the reciprocal of the radius of the osculating circle. Q.E.D.

Problem 1

To determine the motion of a stretched string

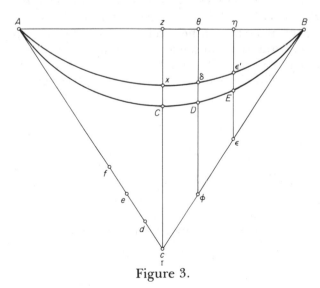

Figure 3.

97

In this problem and following I assume that the string is moved through a very small distance from the axis *AB*, in order that the increase in tension from the inrrease in length may be neglected and that the same may be true of the obliquity of the radii of curvature.

Consulting the figure, assume that the string is originally stretched along *AB*. With the use of a plectrum the midpoint *z* is pulled down to point *c* [Editor's note: so that the string assumes the triangular form shown in the figure]: When the plectrum is removed, the point *c* will first begin to move (by Lemma 2). But then by the bending of the string at the nearby points *d* and *ϕ*, these points will also begin to move and thereafter *ε* and *e*, and so on. Because of the large bending at *c* that point starts to move most quickly, and then by the increased curvature at points *D* and *E*, etc. they will continue to be accelerated to higher velocity. By the same action, due to the undiminished curvature at *C*, that point on the other hand is accelerated to more or less velocity. It then happens that with the forces properly adjusted among themselves, all the motions conspire to bring all points together back to the axis at the same time and simultaneously recede from it, back and forth indefinitely.

But for this to happen, the string must always take on the form of a curve like *ACDEB* whose curvature at any point *E* is proportional to the distance of *E* from the axis, with the velocities of the points *C,D,E*, etc. standing among themselves in the ratio of the distances from the axis, namely *Cz, Dθ, Eη*, etc. For truly in this case, the spaces *Cx, Dδ, Eε* etc., traversed in the same small time interval will be proportional to their velocities, that is, to the spaces *Cz, Dθ, Eη*, etc. to be traversed. Whence the residual spaces *xz, δθ, εη*, etc. will be to each other in the same ratio. Therefore, by Lemma 2 the accelerations will be to each other in the same ratio. This being agreed upon, with the ratio of velocities always kept the same as well as the ratio of the spaces to be traversed, all points reach the axis at the same time and simultaneously return to it; whence the curve *ACDEB* is terminated correctly. Q.E.D.

Besides this if the two curves *ACDEB* and *AxδεB* are compared with each other, the curvatures in *D* and *δ* are as the distances from the axis *Dθ* and *δθ* (Lemma 1). Moreover by Lemma 2 the acceleration of any given point on the string is proportional to its distance from the axis *AB*. Whence (Proposition 51 in Section X of Newton's *Principia*) all vibrations, the large as well as the smallest are performed with the same period and the motion of any point whatever is similar to that of a pendulum.

Corollary

The curvatures are universely proportional to the radii of the osculating circles.

Problem 2

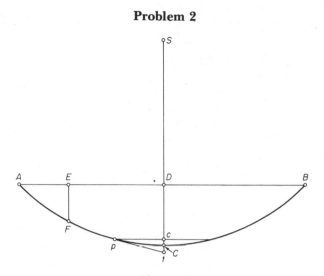

Figure 4.

Being given the length and weight [pondus] of a string, together with the stretching weight, to find the time of one vibration.

Let the string be stretched between A and B [see Figure 4] by the stretching weight P and let the weight of the string itself be N and its length be L. Then let the string be displaced into the position $AFpCB$ and at the midpoint C erect the normal CS, equal to the radius of curvature at C. This normal intersects the axis AB at D. Through an assumed point p near C, draw the normal pc to SC and the tangent pt.

Therefore as in Lemma 2, it holds that the absolute force with which the particle pC is accelerated is to the stretching force P as ct is to pt, that is, as pC is to CS. But the stretching weight P is to the weight of the particle pC itself in the proportion compounded of the ratios P to N and N to the weight of the particle pC; that is, as $P \cdot L$ is to $N \cdot pC$. Therefore, by the composition of these ratios the accelerating force is to the weight as $P \cdot L$ is to $N \cdot CS$. Let us then take a pendulum of length CD. Then (Section X, Proposition 52 of Newton's *Principia*) the period of the string is to the period of this pendulum as $\sqrt{N} \cdot CS$ is to $\sqrt{P} \cdot L$. But for a given gravitational force the lengths of pendulums are in the squared ratio of their periods, whence the length of the pendulum whose vibrations are isochronous with those of the string will be

$$\frac{N \cdot CS \cdot CD}{P \cdot L}$$

or, writing $CS = aa/CD$ (corollary of Problem 1) we can rewrite this

99

$$\frac{N \cdot aa}{P \cdot L}$$

For the introduction of the quantity a, let the abscissa of the curve AE be denoted by z, and the ordinate EF by x. Let the distance along the curve itself be $AF = v$ and let $CD = b$. Then the radius of curvature at F will be aa/x. But in terms of $\dot{v}$ [fluxion or time rate of change of v], the radius of curvature is $\dot{v}\dot{x}/\dot{z}$. Hence $aa/x = \dot{v}\dot{x}/\dot{z}$. This can be written $aa\ddot{z} = \dot{v}x\ddot{x}$. Integrating with respect to time assuming that $\dot{v}$ is constant yields

$$aa\dot{z} = \frac{\dot{v}xx}{2} - \frac{\dot{v}bb}{2} + \dot{v}aa$$

where the quantities $\dot{v}bb/2$ and $\dot{v}aa$ have been added as constants of integration in order that $\dot{z}$ may equal $\dot{v}$ at the middle point C. We then put

$$\dot{v}^2 = \dot{x}^2 + \dot{z}^2$$

and eliminate $\dot{v}$ between this and the previous equation. The result is for $\dot{z}$

$$\dot{z} = \frac{a^2\dot{x} - b^2x/2 + x^2\dot{x}/2}{\sqrt{a^2b^2 - a^2x^2 - x^4/4 - b^4/4 + b^2x^2/2}}$$

Letting b and x vanish with respect to a as the curve comes into coincidence with the axis, we get

$$\dot{z} = \frac{a\dot{x}}{\sqrt{bb - xx}}$$

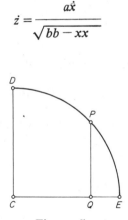

Figure 5.

But with C as center [see Figure 5] I describe the circular arc DPE of radius $CD = b$ and I take $CQ = x$ and erect the normal QP and calling the arc $DP = y$, we get

$$\dot{y} = \frac{b\dot{x}}{\sqrt{bb - xx}} = \frac{b}{a}\dot{z}$$

100

whence

$$y = \frac{b}{a} z \quad \text{and} \quad z = \frac{a}{b} y$$

Making $x = b = CD$ (in which case indeed $y =$ quadrant arc DPE and $z = AD = L/2$), we get

$$\frac{L}{2} = a \cdot \frac{DE}{CD} \quad \text{or} \quad a = \frac{L \cdot CD}{2DE}$$

Now let CD be to $2DE$ (as the diameter of the circle is to its circumference) as d is to c. Then we get

$$aa = LL \cdot \frac{dd}{cc}$$

If we substitute this value of aa above, we get

$$\frac{N}{P} \cdot L \cdot \frac{dd}{cc}$$

as the length of the pendulum isochronous with the string itself. If then D is the length corresponding to period 1 second, we have

$$\frac{d}{c} \sqrt{\frac{N}{P} \cdot \frac{L}{D}}$$

for the period of the vibrating string. The corresponding frequency or number of vibrations of the string in one period of the pendulum (taken as 1 second) is then

$$\frac{c}{d} \sqrt{\frac{P \cdot D}{N \cdot L}}$$

[Editor's note: To see how the formula which Taylor derived for the frequency of the vibrating string agrees with the corresponding formula in modern terminology, we note that from simple geometry

$$\frac{c}{d} = \pi$$

$P =$ tension force in the string, is usually represented by T in modern nomenclature. $N =$ weight of the string $= mg$, where m is the mass and g the acceleration of gravity. $D =$ length of the simple pendulum with period $= 1$ sec and $L =$ length of the stretched string. We then have from elementary physics

$$1 = 2\pi \sqrt{D/g} \quad \text{or} \quad D = g/4\pi^2$$

Substituting into Taylor's formula for the frequency f gives

$$f = \pi \sqrt{\frac{T \cdot g}{mg \cdot 4\pi^2 L}} = \frac{1}{2} \sqrt{\frac{T}{mL}}$$

$$= \frac{1}{2L} \sqrt{\frac{TL}{m}} = \frac{1}{2L} \sqrt{\frac{T}{\sigma}}$$

where σ is the line density of the string or the mass per unit length ($=m/L$). This is the recognized modern formula for the *fundamental* frequency of the string. Note that Taylor's method yields the fundamental only and makes no pretence of calculating the higher harmonics. His whole method is rather specialized but sufficiently ingenious. As C. Truesdell has pointed out ("The Rational Mechanics of Flexible or Elastic Bodies, 1638–1788," Introduction to Leonhardi Euleri Opera Omnia (II), Vols. 10, 11, Orell Füssli, Zurich, 1960) perhaps the most striking thing about Taylor's article historically is the fact that at the end of Lemma 2 he applies Newton's second law for the first time to an element of a continuous body. Moreover he realized the significance of the component of accelerating force acting on such an element in the direction of the motion. If he had written his equation of motion of the element in the modern differential calculus form, it would have taken the form of the conventional wave equation for the string, as later derived and solved by D'Alembert (see the translation of his article in this volume).]

11

Dissertation on Sound

LEONHARD EULER

Leonhard Euler (1707–1783), one of the greatest mathematicians who ever lived, was born in Basel. Though he received his education in Switzerland, he spent the bulk of his professional career in Germany and Russia. His output of research in every known branch of pure and applied mathematics was prodigious. His collected works are still in the process of publication, having reached in 1972 the total of 63 volumes. As a young man Euler became interested in acoustics and presented to the University of Basel in 1727 his "Dissertatio Physica de Sono," after the work of Newton and Mersenne probably the first theoretical treatise on sound in Western Europe. In this he criticized Newton and presented his own version of the nature of sound and its propagation. A translation of this Latin treatise is presented here to give an idea of the grasp of fundamental physics by this 20-year-old student. Editorial notes make clear the relation between Euler's terminology and that of modern acoustics. Unfortunately he does not always explain in detail how he arrived at his results: the article reflects in some measure the brashness of brilliant youth.

Euler's interests in acoustics and related wave propagation problems lasted well into his later years. In particular, in his 40's he became obsessed with the importance of establishing a valid mathematical theory of aerial sound. He paid close attention to the work of his able young contemporary Lagrange, who at a very early age also devoted much attention to acoustics. On October 23, 1759 Euler wrote a letter to Lagrange which, according to Clifford Truesdell (Editor's Introduction to Series II, Vol. 13, of Euler's collected Works, Orell Füssli, Zurich, 1956), marked "the turning point in the theory of aerial propagation." This letter is given later in the volume in English translation.

In the same year (1759) Euler wrote a long memoir on the theory of sound, published later in the Memoirs of the Berlin Academy. In this article, while paying tribute to the somewhat earlier successful work of Lagrange on the same problem, he showed how he could achieve essentially the same results with much simpler analysis than Lagrange had employed. Moreover he carried his investigation further than Lagrange, particularly with respect to standing waves in tubes and the explanation of some paradoxes connected with the propagation of sound waves in finite structures. This memoir is presented further in the volume in its entirety in English translation by the editor of this volume.

Q. F. F. Q. S.

DISSERTATIO PHYSICA

DE SONO,

QUAM

ANNUENTE NUMINE DIVINO

JUSSU MAGNIFICI ET SAPIENTISSIMI PHI-
LOSOPHORUM ORDINIS

PRO

VACANTE PROFESSIONE PHYSICA

Ad d. 18. Febr. A. MDCCXXVII.

In Auditorio Juridico hora 9.

Publico Eruditorum Examini subjicit

LEONHARDUS EULERUS

A. L. M.

Respondente

Præstantissimo Adolescente

ERNESTO LUDOVICO BURCARDO

Phil. Cand.

BASILEÆ,

Typis E. & J. R. THURNISIORUM, Fratrum.

11

Translated from the Latin by R. Bruce Lindsay
from "Dissertatio Physica de Sono" (Basel, 1727)
and also in *Opera Omnia III*, i, 182–198

Chapter One
On the Nature and Propagation of Sound

1) The explanation of sound by the old philosophers was obscure and confused, so far as it is possible to judge from their writings that have come down to us. Some, with Epicurus, have considered sound as something which flows out of vibrating bodies. On the other hand, others and especially the Latin interpreters of Aristotle have placed the nature of sound in the fraction of the air which is emitted in the violent collision of bodies. Among more recent authorities Honore Fabri and Rene Descartes thought that sound consists in the trembling of air, though concerning this trembling their thoughts were equally confused. Newton, a man of most penetrating insight, tried to attack and explain this matter more accurately; he met with little happier success particularly in explaining the true nature of the propagation of sound. I have endeavored to attack and elucidate for the public this difficult matter of sound: I have divided the work into two chapters. In this first chapter we shall discuss what in fact sound consists of and in what fashion it is propagated from one place to another. In the latter chapter three ways of producing sound will be considered.

2) However, before I approach the treatment of sound itself, something ought to be said about air, as subject to sound. I conceive of air as composed of infinitely small globules compressed by the weight of the atmosphere above them and enjoying enough spring or elasticity so that when the compressing force has been removed they are able to resume their natural state. While therefore the weight of the air above compresses that below and forbids the globules of air to expand, the elastic force of the globules is balanced by the weight of the atmosphere. It is therefore

permissible to define by experiment the maximum weight of the atmosphere as equal to that of a column of mercury 2460 scruples high or equivalently 2460 thousands of a Rhenish foot. [Editor's note: A Rhenish foot = 314 mm; hence this comes out to be about 722 mm of Hg.] I shall always adhere to this measure in what follows. If however one chooses to take the atmosphere at its minimum weight, it is found to be equivalent to a column of mercury 2260 scruples high or about 710 mm of Hg. When, on the other hand, the weight of the atmosphere is determined by means of an air pump, it is observed that the specific gravity of native silver is to that of air at its maximum temperature as 12000 is to unity. For the coldest air the ratio is about 10000 to unity. [Editor's note: These values are a bit low, but of correct order of magnitude in comparison with modern measurements.]

3) If we conceive that one of the series of air globules is compressed more than the rest, acting in accordance with its own laws it will expand pushing against the adjacent globules and producing compression in them which push against still others in turn, so that far distant globules feel a certain small amount of compression. And in this way sound is transferred to other places. However the motion with which the original globule of air expanded cannot suddenly cease after this globule has reached the same state as the others; it is carried too far. It is then compressed again by the other globules. This again goes too far. Hence every globule situated not too far away from the first one dilates and then contracts in a trembling motion. However this trembling of the globules of air ought to cease the more readily in view of the infinitely small size of the globules and thence follow the very short time of a single oscillation. It follows therefore that in a finite time there are innumerable oscillations or undulations from a globule of this kind. These, because of the motion of each globule are not able to lead to continual diminution. Since, however, a finite time is required for sound to excite our senses, it is clear that sound cannot consist of this oscillatory motion of the air.

4) So finally sound arises when a globule exposed to an outside force at finite intervals suffers successive compressions. It is necessary, to be sure, for the excitation of sound that a given globule should be alternately contracted and expanded. In truth the times of these oscillations must not be infinitely small but finite in order that the number of these vibrations or oscillations in a given time can be determined. The number of pulses striking the ear in a given finite time must be such that it is possible to express it numerically.

5) Having thoroughly understood the trembling motion which constitutes sound, we find it easy to explain the different properties of sound. Here I call attention only to the principal ones. Sounds are commonly divided into loud and soft. A sound is loud or strong when the compressions of the globules of air are more powerful. A sound is weak or soft when those compressions are weaker. When sound produced by a vibrating globule is propagated by the communication of its compression with the globules arranged in the sphere around it, the number of the latter globules increases as the square of the distance from the given globule; hence the strength or loudness of the sound decreases as the inverse square of the distance from the source, unless the sound receives reinforcement from another source.

6) The distinction among sounds of greatest importance is that between low pitch and high pitch. Low pitch exists when the vibrations of globules of air follow

each other slowly or when in a given time fewer undulations take place. A sound of high pitch, on the other hand, is one whose vibrations have only brief delays between them, so that in the same time more oscillations are completed. Hence sounds, in respect to high and low pitch, stand to each other in the ratio of the number of oscillations performed in a given time.

7) Sound is also either simple or compound. A simple sound is one whose vibrations are all equally spaced one from another and all equally strong. A complex sound consists of many simple sounds all sounding together. This produces either consonance or dissonance. Consonance is perceived in complex sounds if the components perserve a rather simple relation to each other, e.g., one twice the other as in the diapason or octave, or one three-halves the other, as in the fifth in music, etc. Dissonances occur, on the other hand, when the ratio of the components is more complicated, e.g., in the double seventh as in the triton [fish horn].

8) Let us now consider the propagation of sound somewhat more carefully. We hope to find a consistent result if from the theory set forth above we compute the distance which sound is able to traverse in a given time, that is, in a minute, an hour, or a second. It has been observed that sounds of all kinds, whether loud or soft, whether low pitch or high pitch, are transported through a given space in the same time, that is, always with the same velocity. In order to confirm this, it is necessary to ask how much time it takes a compressed globule of air to project its compression at a given distance. This motion can be obtained without difficulty from the laws of communication and the contemplation of the nature of air. I omit the method of deriving this in order to avoid mathematical analysis. However, I quote the result.

9) In order to comprehend the matter in most general terms, let the specific gravity of mercury be to the specific gravity of air as n is to unity, let the height of the mercury in the barometer $= k$, and let the length of the pendulum be f, by whose oscillations it is permitted to measure the time taken by sound to traverse the interval a. Having made these assumptions I find that the time of one oscillation of the pendulum of length f is to the time of sound propagation through the interval a as 1 is to

$$\frac{a}{4\sqrt{nkf}}$$

[Editor's note: By one oscillation Euler meant what in modern terms is a half of a complete oscillation. Hence T = period of a half oscillation $= \pi\sqrt{f/g}$. On this basis, we obtain the velocity v of sound equivalent to the above formula as follows:

$$\frac{\pi\sqrt{f/g}}{a/v} = \frac{4\sqrt{nkf}}{a}$$

or

$$v = \frac{4\sqrt{nkf}}{\pi\sqrt{f/g}} = \frac{4}{\pi}\sqrt{nkg}$$

Now

$$n = \rho_{Hg}/\rho_{air}$$

Hence

$$v = \frac{4}{\pi}\sqrt{\frac{\rho_{Hg}kg}{\rho_{air}}} = \frac{4}{\pi}\sqrt{\frac{p_{air}}{\rho_{air}}}$$

This differs from Newton's formula

$$v = \sqrt{\frac{p_{air}}{\rho_{air}}}$$

only by the factor $4/\pi$. Whereas Newton's formula gave values too small compared with the experimentally measured ones, Euler's formula gives values somewhat too large.]

10) If n and k are given in scruples and we take $\pi\sqrt{f/g}$ = unity, f = 3166 scruples, then we get

$$\frac{a}{4\sqrt{nkf}} = \frac{a}{4\sqrt{3166nk}}$$

Hence the velocity of sound comes out to be

$$4\sqrt{3166nk}$$

in scruples per second. [See above Editor's note.]

11) The following conclusions flow from the above. If nk remains the same, the sound velocity will remain the same also. If the density of the air is proportional to the elasticity, sound proceeds with the same velocity. Indeed in air compressed to its maximum extent sound will not appear to the senses to move faster than in air which is most highly rarefied. And hence sound must move with the same velocity on the highest mountains as in the lowest valleys unless other causes directly affect it.

12) With the increase in the factor nk the sound velocity must increase. If therefore the density of the air remains the same or is diminished, while on the other hand its elasticity (pressure) is increased, the sound velocity will become greater. If, however, on the contrary the density of the air increases, with its elasticity (pressure) remaining the same or decreasing, the sound velocity is diminished. Hence it follows that since both the weight (density) as well as the elasticity of the air surrounding the earth are subject to various changes, the velocity of sound will also vary thereby. Therefore the velocity of sound will be greatest at maximum heat and maximum dryness of

the atmosphere or more accurately when the liquids in the barometer and thermometer are highest. On the other hand in the harshest cold and in a severe storm the velocity of sound should be at a minimum, which will happen when the barometer and thermometer register their lowest.

13) Therefore the maximum velocity of sound will be found for $n = 12000$ and $k = 2460$ scruples, so that the value in scruples/sec becomes

$$4\sqrt{(3166)(12000)(2460)} = 1.224 \times 10^6 \text{ scruples/sec}$$

[Editor's note: This corresponds to a velocity of about 385 m/sec, which in turn implies a temperature of 100°C. It must be remembered, however, that Euler's formula is not the correct one and gives values about 8% higher than the value $\sqrt{\gamma p/\rho}$ where $\gamma = c_p/c_v$, the ratio of the specific heats at constant pressure and constant volume, respectively.]

Thus sound at its maximum velocity should according to my theorem move through an interval of 1222 Rhenish feet per second. [Editor's note: It seems to be more nearly 1224 feet.] The minimum velocity of sound is obtained by taking $n = 10000$ and $k = 2260$, so that in scruples per second the velocity becomes

$$4\sqrt{(3166)(10000)(2260)} \doteq 1.070 \times 10^6 \text{ scruples/sec}$$

or nearly 1070 Rhenish feet per second. [Editor's note: This is about 336 m/sec.] Hence the distances sound will cover in one second are contained within the limits 1224 and 1070 Rhenish feet.

14) If these values are compared with experiment they are found to agree very well, and this confirms my method. For Flamsteed and Derham through very accurately performed experiments measured the velocity of sound to be 1108 feet per second, which lies midway between the limiting values given above. If now we consider what Newton did in this matter, we find that, reduced to our notation, he got for the velocity of sound the value $(p/d)\sqrt{3166nk}$, where p/d = ratio of the periphery of the circle to the diameter, i.e., π. Hence his expression is smaller than ours since he multiplies $\sqrt{3166nk}$ by π and we have 4 in place of π.

15) Hence it is not surprising that the most sagacious Newton found too small a value for the distance sound will travel in one second. He determined it as not greater than 947 feet per second. This is a large discrepancy from the value found experimentally. Nevertheless he took the result as a confirmation of his method, attributing the discrepancy to an impurity in the air, a pure subterfuge. For even if the air is changed with vapor, its elastic force is always equal to the atmospheric pressure and the weight of the atmosphere is not thereby changed. The velocity of sound is not able to undergo any change under these circumstances. Nor does the size of the air molecules have anything to do with this matter. [Editor's note: Euler was wrong here, as modern studies have shown. His criticism of Newton, while to a certain extent just, was too glib and harsh in the light of his own ignorance of the whole story of sound propagation in air.]

Chapter Two
On the Production of Sound

16) For the production of sound it is required that, as I have explained in the preceding chapter, the air is rendered vibratory, so that indeed the globules of air should have contractions and expansions, separated from each other by a finite time. I have been able to conclude that this trembling or vibratory motion of the air takes place in threefold fashion, corresponding to the threefold nature of sounds. For which reason in this chapter words must be invented for the three different ways of producing sound. For the first variety of sounds, I mention the sounds of strings, of drums, of bells, and of instruments constructed with reeds, etc. These are all sounds which owe their origin to a vibrating solid body. In the second class of sounds are included the sound of thunder, of cannons, as well as of tree branches and, due to the violent motion of bodies of this kind, all of which are sounds arising from the sudden restitution of compressed air as well as the strong percussion of air. I enumerate as sounds of the third class those of pipes, the nature of which I shall discuss very carefully, since thus far no one has provided anything very reliable about this matter.

17) Up to the present time, so far as I know, everyone has referred all sounds to the first class of sounds, and it has not been judged possible to produce a sound save by the vibration of a solid body. However, the falsity of this view will soon be made clear when I have explained the two remaining ways of producing sound. Now, however, the first mode in which sound may be excited must be considered more carefully. At the present moment indeed I shall examine strings, how they produce sounds and what kinds of sound are produced, since other sounds from solids can readily be reduced to this type. In order to obtain exact results I shall consider strings stretched by weight while others are stretched by being wrapped around a column which will enable one to measure the force extending the string.

18) Before all, it is to be observed that the same strings produce equal sounds with respect to low and high pitch, with whatever force they are vibrated, though it is admitted that there can be a large discrepancy with respect to loudness and softness. For sounds are loud in proportion to the velocity with which the string cuts through the air, and sounds are equally strong if the air is pushed aside with the same force. Therefore, since both low pitch and high pitch musical sounds ought to be equally strong in order that agreeable harmony should prevail, in the construction of musical instruments it is necessary to pay particular attention to this, namely that sounds are produced equal with respect to strength and loudness, and in order to obtain this the following rules must be diligently observed. These rules have indeed been derived by recent craftsmen in rather crude form from much practice. The truth of them will indeed be made plain in what follows:

I. The lengths of strings are inversely proportional to the sounds, i.e., to the number of vibrations produced in a given time.

II. The thicknesses or transverse sections of the strings are also inversely proportional to the sound (i.e., the frequency) if indeed strings of the same material are in question. On the other hand the effect will be less if as the thickness increases, the density decreases.

These rules can also be applied to instruments made of pipes where here we must take the length or altitude of the pipe in place of the length of the string and the internal size of the pipe in place of the thickness of the string.

19) When a string is oscillated it affects the globules of air, which are compressed since they are unable to move instantaneously. During the duration of the oscillatory motion the globules of air continually suffer new compressions, whence sound is produced. Hence the air acts on the ear or the ear drum as often as the string goes through its oscillatory motion. And so it is possible to find the number of compressions conveyed to the ear by a given sound in given time by investigating the number of oscillations of the string producing this sound in the same time. My solution which agrees exactly with the solutions of John Bernoulli and Brook Taylor is as follows:

20) Let the mass stretching the string = p, let the mass of the string = q, and the length of the string = a. From these three things the number of vibrations in a given time can be found. I find that the number of oscillations per second is

$$\frac{22}{7}\sqrt{\frac{3166p}{aq}}$$

where a is given in scruples. [Editor's note: Euler represents $\pi =$ by 22/7. He takes 3166 scruples $\doteq$ 100 cm nearly as the length of a seconds pendulum with full period 2 seconds or half period 1 second.]

From this formula the frequencies of sounds produced by different strings are to each other as $\sqrt{p/aq}$, that is, the frequency varies directly as the square root of the tension of the string and inversely as the square root of product of the length of the string and its mass. I do not deduce more particular consequences that follow from this, but I inquire into the nature of known sounds and the number of vibrations corresponding to them determined from an experiment which I conducted.

21) I hook a copper wire of the kind whose thickness is designated by the number 18, of length 980 scruples, with mass 49/175,000 pounds, stretched with the weight of a mass of 11/4 pounds. This should yield a sound corresponding to that which in music is called ds. If in the given general formula we substitute 980 in the place of a, 11/4 in place of p, and 49/175,000 for q, it turns out that the frequency of ds is 559 cycles/sec and since $\overline{ds}$ is to c as 6 to 5, c has 466 cycles/sec and thence low C (116 cycles/sec). [Editor's note: To check Euler's formula for the fundamental frequency of the finite stretched string

$$\frac{22}{7}\sqrt{\frac{3166p}{aq}}$$

we recall that for the fundamental the wavelength λ is

$$\lambda = 2l$$

Hence f = frequency = V/λ = velocity/wavelength
But

$$V = \sqrt{\frac{\text{tension}}{\text{line density}}}$$

and line density = mass of string/length = q/a

in Euler's notation. Tension = pg in Euler's notation, where g is acceleration of gravity. Hence

$$V = \sqrt{\frac{pag}{q}}$$

$$f = \frac{\sqrt{(pa/q) \cdot g}}{2a} = \frac{1}{2}\sqrt{(p/aq) \cdot g}$$

Now for the seconds pendulum the period is

$$p = 2 \text{ sec} = 2\pi\sqrt{l/g} \text{ and therefore } g = l\pi^2.$$

But for $P = 2$, $l = 100$ cm or about 3166 scruples. Hence

$$f = \frac{1}{2}\sqrt{\frac{p(3166)\pi^2}{aq}} = \frac{\pi}{2}\sqrt{\frac{3166p}{aq}}$$

This is ½ Euler's value. My only conclusion is that he counted as vibrations what we count as half vibrations. This would still be consistent with his use of the pendulum formula, since the period of a seconds pendulum is 1 second if we count half cycles.

Euler does not say he is calculating the fundamental, but it seems clear that he is. Euler does not provide the above solution, but it seems clear that he must have used something like this method.]

22) To this type of sound production must be referred also the sounds made by vibrating reeds inserted in tubes blown by wind; although the latter also pertain in part to the third mode of sound production. A device of this kind is to be seen in various pneumatic organs, which imitate the sounds of trumpets, bugles, and the human voice, all of which instruments have to be inflated by air (wind) in order to emit sound. The wind, in seeking a passage for itself opens the reed like a valve; however, in opening it stretches it too much so that the valve closes again, attempting to restore its original state; it is then opened again, so that it imposes a vibrating motion on the air passing through it. It is necessary indeed that with the wind flowing evenly the valve shall stay quiet and the sound shall stop. However, taking heed of this, the wind itself, while propelled by bellows strikes the orifices of the device unequally. The wind with the help of the pipe and the inserted valve is rendered vibratory.

23) The human voice is obviously generated in the same way. In the organ of speech the epiglottis takes the place of the reeds. [Editor's note: Euler is probably referring here to the vocal cords, which are attached to the glottis and epiglottis.] This is made vibratory by the air rising through the rough windpipe. This vibratory motion of the air is increased at the head of the windpipe and then is modified in various ways in the oral cavity, by which the low and high pitch voice is modulated and various vowels are formed. These sounds are supplied with consonants by the agency of the lips, the tongue, and the throat. But also by means of the nose, when

air made tremulous by the epiglottis also emerges from the nose, various sounds can be uttered both of low and high pitch which differ from oral sounds in that they are not able to be distinguished clearly from vowels nor to form consonants.

24) The sounds emitted by vibrating reeds, unless they are strengthened by pipes, are so feeble that they are perceived with difficulty, just as is the case with vibrating plates where nearly nothing is perceived by the ears. These sounds are amplified in a wonderful way in pipes as the human voice is by the mouth. And yet a great change in pitch and sharpness is produced by the pipe. This is not the place, indeed, to discuss in detail the intensities and inflections of these sounds. It would be the work of a special chapter to explain these matters in detail where also the marvelous amplification of sound in speaking tubes would be explained as well as the theory of the echo and many other things; but leisure time is not yet available to discuss these matters with care. As for what is contained in the writings of others, so far as I have examined it, it is in part confused and for the most part false.

25) I have put into the second class of sounds those sounds which arise either when a large quantity of compressed air is suddenly released or when air is subject to strong impact. In the latter case the air is also compressed when the struck body (of air) tries to resist leaving its position. After the compression the air when left to itself again expands. Hence the cause of the sounds belonging to the second class is the restoring of the air to its original state after compression. That which permits this restoration to generate sound is that the compressed air in dilating expands too much and hence contracts again, which produces undulation in the air, so that even the smallest globules of air, which go to make up the mass of the air share this trembling motion and consequently produce sound. It should be noted that if the greater part of the air is compressed, the sound produced is lower in pitch and when the smaller part is compressed the sound is of higher pitch. Sounds of this kind should not be able to last long, but ought to stop forthwith, since the air, diffusing into far distant places steadily gives up its tremulous motion.

26) Therefore all causes which are able both to compress and decompress air in such a way that continual relaxation is possible are adaptable to the production of sound. For this reason all very rapid motions of bodies in air ought to be able to lead to sound. For the air because of its own inertia is not able to be compressed freely by the moving bodies but on expanding again induces a trembling motion in the little air globules, thereby producing sound. The sounds of vibrating rods flow more strongly and indeed fastest of all moving bodies. The sounds of breathing and of the wind also owe their origin to this source, for the air in front is compressed by that behind just as by a solid body.

27) Of sounds which are produced by the sudden relaxation of compressed air the strongest are those of missile hurlers [cannons] and thunder. Various experiments carried out with gun powder have showed that the cause of these very loud sounds is the expansion of compressed air, since the air is found to be condensed to the greatest extent at the place where an exit is provided for it by the burning powder, so that it can burst forth with greatest force. Since, however, from the gun powder so many cloud-forming vapors are combined, it is no wonder that with all this material producing a fire, the resulting sounds are stupendous.

28) The third class of sounds is formed by those of flutes. The explanation of the nature of these sounds has bothered investigators in a wonderful way in every age. Many have thought that by the blowing of flutes small particles in the internal surface are driven and forced into a trembling motion so that thus the internal surface is made to vibrate and makes oscillations communicating with the air, but in what fashion this explanation is consistent with the laws of nature and motion they themselves are forced to question; I myself cannot honestly conceive how the differences in the sounds of flutes of different length but of the same loudness can be explained in this way. For I am unable to see why the internal particles if they ever start moving should oscillate in different fashion in tubes of different length. I judge that not even a single experiment on flutes can be explained by this theory.

29) In order that I should obtain a true explanation of this matter, it was first necessary to examine more carefully the structure of flutes and what goes on in them when they are blown. A flute or pipe is a tube to which at the lower end is joined a hollow mouthpiece suitable for receiving air. This mouthpiece ends in a slit along the tube directly opposite to one side of the inside surface of the tube, so that the air blown in through the mouthpiece can force its way through the slit along the length of the tube by creeping over the inside surface. If a whistle or pipe is constructed in this way, when blown it will emit sound, as is easily shown; if a tube without a mouthpiece is also blown so that the air in the tube creeps along the inside surface even it will emit a sound like that of a flute. In any case the internal surface of the tube ought to be both hard and smooth so that the rushing air shall not stop.

30) Let us now see what happens in the tube when it is blown so that the air inside is able to become vibratory or in what fashion the air blown into the tube in the way described is able to render vibratory the air contained in the tube. It is clear that when the air enters the tube the air already contained in it will be compressed along its length. When this air expands again it goes too far and in turn is compressed again by the surrounding atmosphere, so that vibratory motion is thus produced in the tube. This vibration is the cause of the sound. And so the true cause of the sound of flutes is found. And its truth and reality have become more abundantly clear when it has been applied to the explanation of the observed events relating to the sounds of flutes. But earlier the way had to be considered more carefully just how the trembling motion is produced.

31) An air column in a tube oscillates along its length by expanding and contracting in the fashion of strings and hence I shall consider such a column as a little bundle of aerial strings stressed by atmospheric pressure. Though the weight of the string has the tendency to try to pull the string asunder, in the air column case here considered, the contrary prevails, since the air column is pressed together by the atmospheric pressure. Yet the analogy is none the less legitimate. For the atmospheric pressure exerts on the air column the same effect as the stretching force does on the string, if in each case the same thing happens when expansion takes place, i.e., just as the stretching force tends to shorten the string when it is stretched, so the atmospheric pressure tends to oppose the expansion of the air column. Whereas, however, ordinary

strings excited at a single point give out sound, the bundle of aerial strings is unable when excited at a single point to vibrate as a whole because of the discontinuous distribution of its parts. Hence it has to be excited along its entire length. This is what happens in the flute, in which the air introduced from the outside compresses the air in the tube along its entire length.

32) To determine the aerial oscillations in a flute or to determine the frequency of oscillations in any pipe whatever, we consider the air in the tube replaced by a string stressed by atmospheric pressure. Using this analogy we find the frequency by the argument in para. 21. Thus let the length of the tube equal a, let p which was the tension in the string now be the atmospheric pressure or the height of the mercury in the barometer (minimum 2260 scruples and maximum 2460 scruples). Let q which was the mass of the string now become the mass of the air in the tube. Again let the ratio of the specific gravity of the mercury to that of air be $n:1$ and let k be the height of the mercury in the barometer. Then we have $p/q = nk/a$.

33) With the appropriate substitution in the formula for the frequency of a stretched string in para. 20, we get for the fundamental of the open pipe

$$f = \frac{22}{7a} \sqrt{3166nk}$$

[Editor's note: Again π is taken as 22/7. The frequency, as in the case of the string, is taken as double what we take as the frequency nowadays. The pipe must be open at both ends, though Euler does not specifically state this.

In modern terminology

$$f = \text{frequency} = \frac{V}{2a}$$

where V is the velocity of sound in air. But

$$V = \sqrt{p/\rho} = \sqrt{\frac{\rho_{Hg} \cdot h_{Hg} \cdot g}{\rho_{air}}} \qquad \text{(Newtonian formula)}$$

where ρ_{Hg} is the density of mercury, h_{Hg} is the height of the mercury column in the barometer, ρ_{air} = density of air, and g = acceleration of gravity.

Now, in Euler's notation $\rho_{Hg}/\rho_{air} = n$. Also $g = \pi^2 l$, where l is length of the seconds pendulum = 3166 scruples or about 100 cm. Hence

$$f = \frac{1}{2a} \cdot \sqrt{3166 n h_{Hg} \cdot \pi^2} = \frac{\pi}{2a} \sqrt{3166nk}$$

since $h_{Hg} = k$ in Euler's notation. As before we must multiply by 2 to get Euler's result. Note in any case he is bound to be in error because his expression for the velocity of sound in air is Newton's. So far as can be seen he does not use his own expression for the velocity of sound in air, as given in para. 8. This is an interesting point.]

It appears from this that since n and k change with different weather conditions, the frequency must also change. With increasing nk the sound goes up in pitch, with decreasing nk, the pitch goes down. Hence the sounds of flutes will be of highest pitch at maximum temperature and of lowest pitch at lowest temperature. [Euler also adds "heaviest" air to highest heat. This seems to be a slip!] These differences have commonly been observed by musicians and organists. Since, however, this change will be the same in all flutes, the harmony is not altered.

34) In order to express numerically the frequency of pipes, if one takes first the case in which the barometer and the thermometer stand highest, we put $n = 12000$ and $k = 2460$ scruples. If the length a is expressed in scruples, we then get for the frequency

$$f = \frac{22}{7a} \sqrt{(3166)(12000)(2460)} = \frac{960771}{a}$$

In the other extreme for cold weather we set $n = 10000$ and $k = 2260$ scruples. This gives

$$f = \frac{22}{7a} \sqrt{(3166)(10000)(2260)} = \frac{840714}{a}$$

[Editor's note: For a pipe 1 meter long, these give, respectively, about 300 and 280 Hz. Actually in modern notation, these should be 165 and 140 Hz, respectively. If we calculate the frequencies from $f = V/2a$ with $V = 344$ m/sec, we get in place of 165 Hz, the value 172 Hz. The difference is due to the smaller value of V from Newton's formula which Euler used.]

35) This makes clear the reason why pipes emit sounds of frequency inversely proportional to their lengths and why the amplitude of the sound (loudness) has nothing to do with the matter, nor does the material of the pipe either. [Editor's note: Euler neglects the increase in pitch connected with overblowing in which the loudness is greater. Of course the quality or timbre of the pipe, i.e., the composition of the various overtones, varies with the material somewhat.] However, although neither the loudness of the sound nor the material of the pipe contribute to change the pitch of the sound produced, nevertheless these contribute much to the agreeableness of the sound. For the amplitude (or intensity) in the tube corresponds to the loudness of the sound. The greater the amplitude in the tube, the louder is the sound. The amplitude in the case of pipes is analogous to the thickness of strings. And so while any particular string is not adapted to give out any sound whatever, since the thicker string is required for the louder sound, so in the case of pipes the wider ones correspond to the larger amplitude.

36) Since the ratio of sounds differing in frequency from each other by a whole tone is as 8 to 9, the same pipe with changing condition of the air can emit sounds differing at most by a tone. Let the pipe be 4 Rhenish feet long and arranged to emit the sound C. The frequency of the emitted sound has a maximum value of 240 and a minimum of 210. [Editor's note: If we calculate the frequency according to the formula $f = V/2a$, where $V = 344$ m/sec at 20°C and $a = 4(31.38)$ cm $= 125.52$ cm (since the Rhenish foot $= 31.38$ cm), we get $f = 135$ Hz. Since Euler's frequency is always double what we now call frequency, this should correspond to 270 in his

units. This is somewhere near what he gets. His minimum value should correspond to freezing temperature. But for this we get 259. It seems he must have taken more extreme temperature values. Of course his Newtonian velocity values are lower than the true sound velocity. In any case it is hard to know what he means by C. In the modern musical notation (Helmholtz) C means 65.4 Hz, which in Euler's terminology would be 120.8 and this is obviously off. He may have meant c which corresponds to 131 or 262 in his notation. This is probably the situation.] This result [his figures of 240 and 210] agrees well enough with what we found before in dealing with strings. There to be sure the frequency for C was found to be 116 [Editor's note: this more or less agrees with the modern figure, as we have noted above], whence it appears that the sound C in pipes is approximately an octave above the sound C in strings. [Editor's note: This is a strange statement for Euler to make. It was obviously not based on experience!] When two sounds are an octave apart they are often judged to be the same. This is not surprising, since it is difficult to judge of sounds of different quality when they are in unison. I do not find that a single octave difference discredits the confirmation of my theory. [Editor's note: Euler was correct in the use of his formula, but the C should have been c, in modern notation.]

37) What has been said up to now concerning the sound of pipes must be interpreted as referring to open cylindrical tubes, in which an exit is provided for the air blown into the tube. When however the tube is closed at the upper end, the air blown in is not able to get out. Hence it must turn back in order to get out at the lower orifice. So it follows that the air has to go along the tube and back again before it can get out. Hence the air in the tube must be considered like a string of double the actual length. Hence a closed tube will emit sound of the same frequency as an open one of double the length, that is of lower pitch (than the open tube of the same length). [Editor's note: In endeavoring to explain the experimental fact, which he evidently knows very well, Euler departs from the wave theory which he seemed to grasp in the case of a string. His explanation has no validity, though he evidently felt it was plausible. Here he was considerably hampered by his dependence on the string analogy. The latter can correspond to a tube open at both ends or closed at both ends, but not the open–closed case.] I leave to my honorable competitors the examination of the sounds of pipes which do not have the same width at all points, i.e., are either convergent or divergent, as well as the sounds of pipes which are partly closed at the upper end.

12

Investigation of the Curve
Formed by a Vibrating String

D'ALEMBERT

Jean le Rond, called d'Alembert (1717–1783) was one of the most colorful, though at the same time controversial figures of 18th century French mathematics and physics. Left as a foundling on the steps of the Church of Jean le Rond in Paris because his high society parents did not choose to acknowledge him, he was brought up by foster parents. Precocious as a young man, he studied both law and medicine in college, but was largely self-taught in mathematics to which he actually devoted his life. He made notable contributions to the physics of fluids, to cosmology, and to mechanics in general. His treatise "Traité de Dynamique" (1743) was a landmark in the development of mechanics, containing a statement of his famous principle governing the motion of bodies subject to constraints. He was one of the leading contributors to the great French Encyclopedia of the 18th century and a close friend of Diderot. Extremely controversial in his relations with other scientists throughout Europe, his correspondence often was unnecessarily polemical.

So far as is known, d'Alembert was the first to derive and publish the differential equation of wave propagation. This he did in connection with the vibrations of a stretched string. He also derived the general solution of the one-dimensional wave equation. These results are presented in the following paper (a translation of "Recerches sur la courbe que forme une corde tendue mise en vibration") published in 1747. The notation is of course not that of modern times. This is clarified by the accompanying editorial notes. d'Alembert was sloppy in his writing and there are many misprints. A supplement to this paper (not presnted here) considers special cases of harmonic solutions and provided the first indication of the method of solving a partial differential equation by separation of variables.

Translated by R. Bruce Lindsay
from "Recerches sur la courbe que forme une corde tendue mise en
vibration," *Hist. Academy of Sciences, Berlin,* **3**, 214–219 (1747)

I. I propose to make clear in this memoir that there are an infinity of curves
other than the mate of the elongated cycloid, which satisfy the problem in question.
I shall always assume that the excursions or vibrations of the string are very small,
so that the arcs such as AM [Figure 1] of the curve formed in the vibration can
always be considered approximately equal to the corresponding abscissas AP. I assume
in the second place that the string is of uniform thickness throughout its length.
In the third place, I assume that the force of tension F on the string is to the weight
of the string as m is to unity, whence it follows that if p is the weight per unit length
and l is the length of the string, we have $F = pml$. Fourth, if we call AM or its
(approximate equivalent) $AP = s$, and call $PM = y$ and consider ds constant, the
accelerating force on the point P is $-F(ddy/ds^3)$ if the curve is concave toward AC,
and $F(ddy/ds^2)$ if the curve is convex.

Figure 1.

II. Having assumed this, let us imagine [see Figure 2] that Mm and mn are
two consecutive elements of the curve at any arbitrary instant of time and that $Pp =$
$p\pi$, that is, ds is constant. Let t be the time that has elapsed since the string started
to vibrate. It is then certain that the ordinate PM can only be expressed as a function
of the time t and of the abscissa or of the corresponding arc s or AP. Hence let
$PM = \phi(t,s)$, that is, equal to an unknown function of t and s. We then have $d\phi(t,s) =$
$p\,dt + q\,ds$ [Editor's note: do not confuse the coefficient p here with p in Figure 2.
The author is careless in his terminology], where p and q are also unknown functions
of t and s. From Euler's theorem (Vol. VII of *Memoir of St. Petersburg Acad.*) it follows

that the coefficient of dt in the expression for dp is equal to the coefficient of ds in the expression for dq. If we write them $dp = \alpha dt + \nu ds$ we must have $dq = \nu dt + \beta ds$, where α, ν, and β are also functions of t and s.

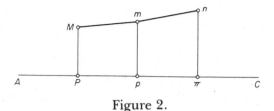

Figure 2.

III. From this it follows that as the arcs Mm, mn belong to the same curve, we shall have $pm - PM$ equal to the difference in $\phi(t,s)$ due to a variation in s only, that is, that $pm - PM = qds = ds \cdot q$; and that the quantity above called ddy, that is, the second difference of PM corresponding to the variations of s alone, becomes $ds \cdot \beta ds$ and hence $F(ddy/ds^2) = F\beta$.

IV. Let us now suppose that the points M, m, n go into M', m', n' [Figure 3]. It is certain that the excess of PM' over PM will be the difference in $\phi(t,s)$ corresponding to a variation in t only, that is $PM' - PM = pdt = dt \cdot p$ and that the second difference in PM taken with respect to a change in t alone, that is, the difference MM', or what is the same thing, the space traversed by the point M by virtue of the accelerating force acting on it will be equal to αdt^2.

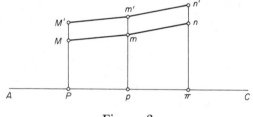

Figure 3.

V. This being assumed, let a be the space through which a heavy body actuated by its weight p per unit mass will move from rest in time θ; it is then evident that we shall have (Lemma XI, Section 1, Book 1, *Principia Mathematica* of Newton)

$$\frac{\alpha dt^2}{2a} = \frac{F\beta dt^2}{p\theta^2}$$

and hence

$$\alpha = \frac{2aF\beta}{p\theta^2} = \frac{2apml\beta}{p\theta^2} = \beta \cdot \frac{2aml}{\theta^2}$$

VI. We now remark that one can represent the given time θ by a constant line of any magnitude one wishes. It is necessary only to take care in expressing the variable and indeterminate parts of the time, to take lines for t which are the line marked θ in the ratio of the variable parts of the time to the constant time θ, during which the heavy body traverses the space a. We can then suppose that θ is such that $\theta^2 = 2aml$, whence $\alpha = \beta$. Then, since $dp = \alpha dt + \nu ds$ it is necessary that dq or $\nu dt + \beta ds$ should equal $\nu dt + \alpha ds$.

VII. To determine from these conditions the quantities α and ν we note that as $dp = \alpha dt + \nu ds$ and $dq = \nu dt + \alpha ds$, we have

$$dp + dq = (\alpha + \nu)(dt + ds)$$

and

$$dp - dq = (\alpha - \nu)(dt - ds)$$

whence it follows that

1) $\alpha + \nu$ is equal to a function of $t + s$ and that $\alpha - \nu$ is equal to a function of $t - s$
2) we have $p = [\Phi(t + s) + \Delta(t - s)]/2$
3) or (by proper choice) more simply $p = \Phi(t + s) + \Delta(t - s)$.

Similarly

$$q = \Phi(t + s) - \Delta(t - s)$$

From this we deduce that PM as the integral of $pdt + qds$ will also be the sum of functions of $t + s$ and $t - s$, which we may call Ψ and Γ. Hence we finally have as the general equation of the curve of the string:

$$y = \Psi(t + s) + \Gamma(t - s)$$

VIII. But it is easy to see that the above equation includes an infinity of curves. To see this, we take here only one special case, namely that for which $y = 0$, when $t = 0$, that is, we assume that the string lies initially along a straight line and that it is forced to depart from its state of rest by the action of some cause. It is evident that we shall then have $\Psi(s) + \Gamma(-s) = 0$ or $\Psi(s) = -\Gamma(-s)$. Furthermore, since the string always passes through the fixed points A and C it is necessary that $y = 0$ when $s = 0$ and $s = l$, no matter what t is. Hence $\Psi(t) + \Gamma(t) = 0$, for all t; or $\Psi(t) = -\Gamma(t)$. From this it follows that $\Psi(t - s) + \Gamma(t - s) = 0$ for all values of the argument. Hence the fundamental equation above becomes $y = \Psi(t + s) - \Psi(t - s)$.

It is further necessary that $\Gamma(s) = -\Psi(-s) = -\Psi(s)$. Hence $\Psi(s)$ must be an even function of s (only even powers of s may enter if Ψ is a polynomial series). Moreover the condition that $y = 0$ for $s = l$, leads to

$$\Psi(t + l) = \Psi(t - l)$$

It is therefore necessary to find a function $\Psi(t + s)$ such that $\Psi(s) - \Psi(-s) = 0$ for all s and $\Psi(t + l) - \Psi(t - l) = 0$ for all t.

IX. To arrive at this, let us imagine the curve toT in Figure 4 with coordinates $TR = u$, $QR = z$ and which are such that $u = \Psi(z)$. Then since $\Psi(s) - \Psi(-s)$ must be equal to zero, it is evident that taking $QR = Qr$, it is essential that $rt = RT$, and that thus the curve toT will have equal and similar parts on either side of o, namely, to and oT. Further, since $\Psi(t + l)$ must equal $\Psi(t - l)$ and the difference between

$t+l$ and $t-l$ is $2l$, it is evident that the curve toT must be such that if we suppose it extended, any two ordinates distant from each other by $2l$ must be equal. Then if we suppose that $QR = l$, we see that the part TK must be equal and similar to the part To, and that the part KX must also be equal and similar to oT, etc. Since the segments to, oT are already similar and equal, it follows that the curve we seek extends to infinity on the two sides of the point o, and that it is made up of segments which are all equal and similar to the segment oTK, whose abscissa $QV = 2l$ and which is divided by its middle point T into two similar and equal parts. But mathematicians know that such a curve can always be generated by means of another curve $TV'SR'T$ (see Figure 5), which is a closed curve and whose two segments $TR'S$ and $TV'S$ are similar and equal. For if through any point L on the axis TS we draw a straight line LH which is equal to a multiple of the arc TR' added to any function of the abscissa TL and of the ordinate LR, or even if we make the line LH equal to an arbitrary function of the abscissa TL and the ordinate LR added to the space TLR divided by an arbitrary constant, it is certain that one will get by this means a curve oTK, the two segments of which are equal and which will extend to infinity, having all its segments similar and equal to oTK like the ordinary cycloid.

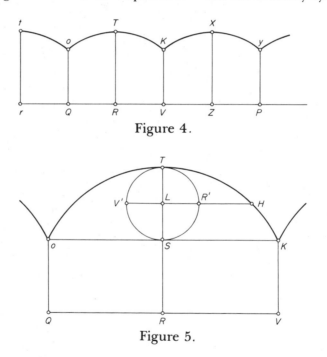

Figure 4.

Figure 5.

X. Having thus described such a curve oTK, it will be easy to determine for any arbitrary time t the curve assumed by the stretched string, for the latter curve is always constructed by taking for the ordinate which corresponds to an arbitrary abscissa s the difference between two ordinates of the curve oTK with respect to any arbitrary axis ZV, and of which the one, $\Psi(t + s)$, is the distance from Z corresponding to $t + s$, whereas the other, $\Psi(t - s)$, is the distance from the same point Z corresponding to the quantity $t - s$.

XI. We have already noted that $\Psi(s)$ must be an even function of s, whence

$\Psi(t + s)$ must likewise be an even function of $t + s$. Hence the difference $\Psi(t + s) -$ $\Psi(t - s)$, taken with t alone varying, that is to say, $dt \cdot [\Delta(t + s) - \Delta(t - s)]$ must be such that $\Delta(t + s)$ and $\Delta(t - s)$ are odd functions of $t + s$ and $t - s$, respectively. But it is easy to see that $\Delta(t + s) - \Delta(t - s)$ or $(PM' - PM)/dt$, expresses in general the velocity of the point M and that $\Delta(s) - \Delta(s)$ expresses the initial velocity of the same point. Hence the expression for the initial velocity impressed on each point of the string when it lies along the straight line and just commences to move ought to be such that when expressed in the form of a series it will contain only odd powers of s. Otherwise if the function of s which expresses the initial velocity were not an odd function of s, the problem would be impossible, that is, one would not be able to assign a function of t and s which represents in general the value of the ordinates of the curve for an arbitrary abscissa s and an arbitrary time t.

There are many other deductions which can be drawn from the general solution we have just given. They will be the subject of a second memoir.

[Editor's note. D'Alembert's notation is so different from the modern differential calculus notation that the reader may find it difficult to convince himself that the author in this paper really accomplished what has been claimed in the introductory editorial statement. However, a translation into modern terminology clarifies matters. We note that if we take y as the displacement of any point of the string then though D'Alembert did not use the partial derivative notation, what he says in Section II is equivalent to the following

$$p = \frac{\partial y}{\partial t}, \; \alpha = \frac{\partial p}{\partial t} = \frac{\partial^2 y}{\partial t^2}$$

Thus α is the acceleration of any point on the string. In Section III it is established that $\beta = \partial^2 y/\partial s^2$. In Section V the equation for α then becomes

$$\frac{\partial^2 y}{\partial t^2} = \frac{2aF}{p\theta^2} \cdot \frac{\partial^2 y}{\partial s^2}$$

But from the physical meaning of the symbols

$$\frac{2a}{p\theta^2} = \frac{g}{p}$$

where g = the acceleration of gravity. Hence the above equation becomes

$$\frac{\partial^2 y}{\partial t^2} = \frac{F}{p/g} \cdot \frac{\partial^2 y}{\partial s^2}$$

This is the one-dimensional wave equation, with the wave velocity

$$c = \sqrt{\frac{F}{p/g}}$$

where p/g is the mass of the string per unit length. This wave equation is solved in general terms by D'Alembert in Section VII. The solution is applied to the special case of the finite string fastened at both ends in Sections VIII to XI.

Further helpful commentary on D'Alembert's paper is found in Clifford Truesdell: "The Rational Mechanics of Flexible or Elastic Bodies, 1638–1788" (introduction to Leonhardi Euleri Opera Omnia, Vols. X and XI Seriei Secundae; Orell Füssli, Zurich, 1960, p. 237ff].

13

Researches on the Nature and Propagation of Sound

JOSEPH LOUIS LAGRANGE

Joseph Louis Lagrange (1736–1813), though born in Turin, Italy, of French ancestry, spent practically all his professional career in Paris and Berlin and is considered one of the most eminent of French mathematicians and theoretical physicists. He succeeded Euler as Director of Mathematics in the Berlin Academy in 1766 but returned to Paris in 1787. Like Euler he turned his attention to a host of problems in pure and applied mathematics. His most famous work is probably his *Mecanique Analytique* (1788), the first treatise to put mechanics on a firm mathematical basis. Lagrange early became interested in acoustics (see the introduction to the papers of Euler in this volume). He solved the problem of the vibrating string by a method which shed new light on the propagation of waves through elastic media. He then applied these ideas to aerial propagation of sound. We present here a translation of the introduction to his great memoir of 1759 ("Recherches sur la nature et la propagation du son"). This explains clearly his method but avoids the mathematical details. Euler's method for handling this same problem is analytically simpler than that of Lagrange (see p. 136), but it is significant that Euler in a number of places expresses his deep indebtedness to the work of Lagrange. d'Alembert severely criticized the work of both, but it is now agreed that his criticism was not wholly justified.

Translated by R. Bruce Lindsay from
"Recherches sur la nature et la propagation du son,"
Miscellanea Taurinensis, **1**, 1 (1759) and *Oeuvres,* **1**, 39 (1867)

Introduction

Although the calculus has in recent years been brought to a very high degree
of perfection, it docs not appear however that there has been much advance in
the application of this science to the phenomena of nature. The theory of fluids
which is assuredly one of the most important in physics is still very imperfect in
its elements in spite of the efforts of the many great men who have tried to examine
it with thoroughncss. It is the same with the problem which I shall undertake to
examine here, and which one can with reason regard as one of the principal parts
of that theory. For sound consists only of certain disturbances impressed on sounding
bodies and communicated to the elastic medium which surrounds them; hence it
is only through the knowledge of the motions of fluids that we can hope to discover
its real nature and determine the laws it must follow in its propagation.

Newton, who was the first to endeavor to submit fluids to calculation, also made
the first analytical researches on sound. He has been able to determine the velocity
of sound by a formula which does not deviate too much from experience. But if
this theory has satisfied physicists, most of whom have adopted it, this is not true
of mathematicians, who in studying the demonstrations on which it is based have
not found in them the degree of solidity and clarity which characterizes the rest
of Newton's work. But as far as I know, no one has ever tried to discover and make
known the principles which make Newton's demonstration insufficient. Still less has

anyone tried to substitute for his principles others that are more trustworthy and rigorous.[1]

The commentators on the *Principia* have indeed tried to patch up this passage on the velocity of sound by a purely analytical method; they have not only envisaged the question from a very special point of view, but their calculations are besides so complicated and so encumbered with infinite series that it does not appear that one is able in any way to agree with the conclusions they have been forced to draw.

I have then thought it necessary to take up the whole subject again from the beginning and to treat it as an entirely new subject without borrowing anything from those who have worked on it up to the present.

Such is the object which I have proposed for myself in the following researches. To make my object better understood, I begin by giving an idea of Newton's theory and the difficulties to which it is subject.

The whole of Newton's theory is found in Sections VIII of Book II of the *Principia*. The author first considers the propagation of motion in elastic fluids, and considers that this consists of successive dilatations and compressions which are also like pulsations and which spread in every direction throughout the fluid. He then passes on to consider how these pulsations can be produced by the tremblings of the parts of a sounding body. He imagines that a particle of fluid, impelled by the vibrations of the surrounding body, condenses through a certain distance the following particles, until the condensation having become greatest, the same particles begin to dilate one from another. According to Newton, it is as if there were an infinity of sound-carrying fibers, which all spread out from the same point as from a common center. He wishes, moreover, each of these first fibers to create another equal one at its extremity, when it has made a complete oscillation. This in turn is to create a third, and so on successively, so that these form outside the sounding body, so to speak, several spherical domes, which forever go on enlarging, just as one observes in the waves which are excited on the surface of quiet water, by agitation through any foreign object.

According to the illustrious Newton this is the nature of the motions of the particles of air which produce and propagate sound. But Newton went still further. He has calculated all the particular motions which make up each of the pulsations. To do this, he looks upon the elastic fibers of air as composed of an infinity of physical points lying along a straight line and spaced at equal distances from each other. The method he employs to determine the oscillations of these points consists

---

[1]Here is how one of the most celebrated mathematicians of our time speaks about the matter in his *Traite des Fluides* (art. 219). "This would be the place to give the methods for determining the velocity of sound, but I confess that I have not been able to find anything on the subject which satisfies me. I know at present only two authors who have given formulas for the velocity of sound, namely Newton in his *Principia* and Euler in his *Dissertation on Fire*, which won the Academy prize in 1738. The formula given by Euler without demonstration is very different from that of Newton and I do not know the road that led to it. With respect to Newton's formula, he has given a demonstration of it in his *Principia* but that passage is perhaps the most obscure and difficult of the whole work. John Bernoulli, the younger, in his *Essay on Light*, which won the Academy award in 1736 said he did not dare to flatter himself that he understood this part of the *Principia*.

in supposing them from the very first to be isochronous and the same for all. Newton proves then that this hypothesis is in complete agreement with the mechanical laws which govern the mutual action which the points exert by virtue of their elasticity. From this he concludes that the motions are in fact such as he has assumed them to be; and since for each oscillation there should be created a new fiber equal to the first, he finds the space through which sound travels in a given time, by calculating only the period of a single oscillation.

John Bernoulli, the younger, in his excellent *Essay on Light* has also used the same hypothesis to determine the velocity of sound. His procedure differs, however, from that of Newton in that he assumes from the beginning that the vibrations of the particles are perfectly isochronous, a result the great mathematician had proposed to demonstrate. It is not surprising that the two authors arrived at the same formula for the velocity of sound and the apparent agreement of their calculations cannot be taken as a proof of the theory they have used.[2]

With respect to the first propositions on the formation of elastic fibers, and especially on this analogy to waves, I believe it is useless to stop to examine them. For several authors have already made plain the lack of solidity of the scheme and its insufficiency for the explanation of the phenomena of sound.[3] Moreover the manner in which the method is presented in the *Principia* makes it clear that the author adopted it as a simple hypothesis for simplifying his rather complicated problem. But even if this hypothesis were true would it not be right to demand a demonstration of it? But this demonstration would necessarily depend on the general solution of the proposed problem. It would be necessary to admit that the theory of Newton, even in this respect would be far removed from being able to satisfy its object. Even

---

[2]Bernoulli has proved conclusively that every body held in equilibrium by two equal and oppositely directed forces, if displaced slightly from equilibrium, will return to its equilibrium state through simple and regular oscillations. But this theory is applicable to the one case in which there is only one movable body. To appreciate this, suppose with Bernoulli that the body is subjected to the two forces P and Q acting in opposite directions. It is clear that these forces must be functions only of the distance of the body to some fixed point. Then if the body is displaced through a very small distance ds, the sum of the forces will be expressible as pds, which will provide the accelerating force tending to make the body reassume its equilibrium position. Since only very small displacements are contemplated p may be treated as constant, whereupon the net force will be proportional to the displacement and oscillations will follow in accordance with the known laws of isochronism. But the situation will not be the same if there are several bodies which are maintained in equilibrium by forces P, Q, R, ... acting along the same straight line. In this case the resulting force corresponding to very small displacements will be of the form $p_1 ds_1 + p_2 ds_2 + \ldots$ in which p_1, p_2, ... can be regarded as constants. In this case it can be seen that the bodies will not in general oscillate in isochronism. And this is what happens with the particles in the elastic fibers of air. Hence for this reason the commentary on Newton's method is still insufficient even when it does not include approximations. d'Alembert has made this difficulty clear for the case of a vibrating string loaded with several small weights. See *Memoirs of the Berlin Academy* for the year 1750, p. 359.

[3]See for example, the memoir of M. de Mairan, in Memoirs of the Academy of Paris, year 1737. Also Perrault's *Physics* and others.

worse, the theorem on which he bases the laws of oscillations of the particles is founded on principles which are insufficient and even faulty.

The celebrated Euler appears to have recognized this from the year 1727 as we see in his *Dissertation on Sound*, submitted at Basel in that year. It was, however, Mr. Cramer, who I believe was the first to give a solid and convincing proof. He made it clear that the procedure of Newton could equally be applied to demonstrate this other proposition, to wit: that the elastic particles follow in their motion the laws of a heavy body which rises and falls freely. This is completely incompatible with the isochronism of the oscillations which the celebrated English author has sought to establish. This comment would in itself be sufficient to demolish completely Newton's theory. However, since great men ought to be judged only after the most careful and most rigorous examination, one would be wrong to reject their ideas before having demonstrated their insufficiency in a manner that leaves nothing to be desired.

There is the first step which I have thought ought to be made at the beginning of the researches which I have proposed to do on the nature of the propagation of sound.

I have started then by studying with all the attention of which I have been capable the propositions of Newton which are in question, and I have found in fact that they are founded on suppositions which are incompatible among themselves and which necessarily lead to faulty results. This is what I have tried to make clear in two different ways in the first chapter of the following dissertation. Having achieved this object, I have applied myself to develop the direct and general methods for the solution of the proposed problem without employing any other principles than those immediately associated with the well-known laws of dynamics.

To give my investigations the greatest possible generality, and at the same time to make them applicable to what really goes on in nature, I have first viewed the question from the same point of view with which all mathematicians and physicists have looked at it up to now, and I doubt that one will ever be able to reduce the problem of the motions of the air which produce sound to a simpler enunciation than the following:

Being given an indefinite number of elastic particles distributed on a straight line, which are maintained in equilibrium by their mutual repulsive forces, to determine the motions which these particles must follow when they are disturbed in any way whatever, without leaving the straight line.

To facilitate the solution, I assume only that the particles are all of the same size and all subject to the same elastic force, and furthermore that their motions are always infinitely small, conditions which I do not believe impose the least restrictions on the nature of the problem from the physical point of view.

In examining the equations found after using these single assumptions I have perceived that they differ in no way from those that pertain to the problem of vibrating strings; provided that one supposes that the same particles are disposed in the same manner in the one case as in the other. From this it follows that in increasing their number to infinity and diminishing their masses in the same ratio the motion of a sonorous fiber whose elastic particles are in mutual contact should be analogous to that of a corresponding vibrating string.[4]

This has then led me to speak of the theories which the great mathematicians, Brook Taylor, d'Alembert, and Euler have provided for the vibrating string problem. I make clear in a few words the points on which they differ and the objections which Daniel Bernoulli has made to the latter two, and after having most carefully examined the reasoning of all of them, I have concluded that the calculations made up to the present time cannot decide such questions and that it is necessary for the general solution that we should have in mind what is necessary to review.

I try then to give that solution the analysis of which appears to me to be both new and interesting, since there are an indefinite number of equations to solve at the same time. Fortunately the method which I have followed has led me to formulas which are not too complicated, having regard to the large number of operations I have had to go through. I consider first the formulas for the case in which the number of movable bodies is finite, and from this I easily deduce the whole theory of mixtures of simple and regular vibrations which Daniel Bernoulli has found only in special and indirect ways. I pass then to the case of an infinite number of moving particles, and having then demonstrated the insufficiency of the earlier theory for this case, I deduce from my formulas the same construction of the problem of vibrating strings given by Euler, which has been so strongly contested by d'Alembert. Moreover, I give to this construction all the generality of which it is capable, and by the application which I make to musical strings, I obtain a general and rigorous demonstration of that important truth of experience, namely, that no matter what initial configuration one gives to the string, the period of its oscillations nevertheless remains the same.[5]

At this point I develop the general theory of the harmonic sounds which result from the same string, as well as those of wind instruments. Although these two theories have already been proposed, the one by Sauveur and the other by Euler, I believe I am the first who has derived them directly from analysis.

I turn now to the principal object of my researches, to find the laws for the propagation of sound. I assume that a particle of air receives an arbitrary impulse from the sounding body. I then find from the application of my formulas that this is communicated from one particle to the next by a motion which is not only instantane-

---

[4]We owe it to d'Alembert to admit here that he had already found the analogy between the two problems in article XLVI of his first Memoir on vibrating strings in the Memoirs of the Berlin Academy, but so far as I know it does not appear that he ever made any use of it.

[5]The scholar d'Alembert, cited above, in Article III of his Appendix to his Memoir on vibrating strings, printed in the volume of Memoirs of the Berlin Academy, makes the following remark concerning this proposition: "It is probable that in general, no matter what configuration the string starts with, the period of vibration will always be the same, and this is apparently confirmed by experience; but it would be difficult, perhaps impossible, to demonstrate this rigorously by calculations." I repeat these words of such a great mathematician only to give an idea of the difficulty of the problem I have solved.

ous but does not depend at all on the force of the first impulse. The velocity with which the communication takes place turns out to be by the same formula which Newton has already given for the velocity of sound, the results of which are sufficiently in agreement with experience. The calculation leads me here to provide a treatment of simple and compound echoes, and the theory which I establish is not subject to any of the difficulties which are met in the explanations given by physicists up to the present time. These researches are followed by an examination of mixtures of sounds and of the manner in which they can spread through the same space without interfering with each other in any way. Finally I deduce from my formulas a rigorous and incontrovertible explanation of resonance and the natural vibration of a string one of whose harmonics agrees with the frequency of the sound to which it is exposed: a phenomenon known for a long time, and for the explanation of which many systems have been invented, without however being able to give a satisfactory reason.

Here are then the principal subjects I have treated in the present dissertation, which the lack of time and certain other obstacles have prevented me from explaining with greater order and clarity. I am far from believing that they contain a complete theory of the nature and propagation of sound; but it will at least have contributed to the advancement of the physical and mathematical sciences to have demonstrated by calculation some truths of nature which have up to now appeared inexplicable. The agreement of my results with experience will perhaps serve to destroy the prejudices of those who seem to despair that mathematics can ever bring true light into physics. This is one of the principal goals which I have now set for myself.

14

Letter of Leonhard Euler to Joseph Louis Lagrange

Translated by R. Bruce Lindsay
from *Oeuvres de Lagrange*, **14**

Berlin, October 23, 1759

Sir:

Having received the excellent present which you have had the kindness to send to me [Editor's note: a copy of a book containing Lagrange's "Researches on the nature and propagation of sound"], I have at once looked into it with the greatest interest and I cannot sufficiently admire the skill with which you handle the most difficult equations for the determination of the motion of strings and the propagation of sound. I am infinitely obliged to you for having defended my solution against all caviling, and it is as a result of your profound calculations that everyone should now recognize the use of irregular and discontinuous functions in the solution of problems of this kind. For the thing now seems to me to be so clear that there cannot remain the least doubt. Let us suppose that one is to find a function z of the two variables t and x such that $\partial z/\partial t = \partial z/\partial x$; it is then evident that every function of $t + x$, whether regular or irregular can be taken for z. For example, having traced any curve whatever, such as AM in Figure 1, if one takes the abscissa AP as $t + x$, the ordinate PM will furnish a value for z. It is the same with the problem of the string. I have then observed that my solution is not sufficiently general, for in giving to the string initially any shape whatever, such as AMB in Figure 2, my original solution demanded that in this state there must be no motion. But now I can solve the problem not only when the cord has initially any arbitrary shape, but also that in this condition an arbitrary velocity is imposed at every point, e.g., Mm at point m. I see that you have treated this case when the string is stretched initially along the straight line AB, but I do not know whether your solution can be extended also to the case in which the string in addition to its motion has an arbitrary configuration.

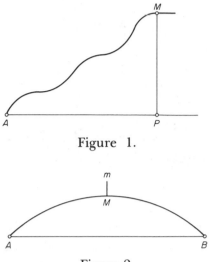

Figure 1.

Figure 2.

I pass now to the propagation of sound, about which I have never been able to come to a conclusion in spite of efforts I have made in this direction, for the attention I gave it in my youth was based on some illusory notion in the endeavor to make the theory of the velocity of sound agree with experiment. I have therefore read your memoir on this subject with the most lively satisfaction and I cannot sufficiently admire your sagacity in surmounting all obstacles. I now see that one should be able to deduce the same solution of the equation

$$\frac{\partial^2 z}{\partial t^2} = \alpha \frac{\partial^2 z}{\partial x^2}$$

by using discontinuous functions. But then M. d'Alembert would raise the same objections as against the motion of strings. It is only after your investigations that I would be able to make this method valid. I have solved in this way the case where one assumes initially not only an arbitrary displacement of the molecules of air but gives also an arbitrary motion to each molecule, just as in the case of the string, in this process restricting attention to a single line of air or rather a narrow straight tube, full of air, as you have done. This generalization appeared to me to be more useful in that it reveals to us more clearly the motion to which all the particles of air are successively excited. It enables us also to answer a misgiving which has long bothered me. This is, that a displacement excited at A [see Figure 3] spreads itself equally on the two sides of A, but having arrived at X, it spreads only toward E, leading us then to ask what difference there is between an initial displacement at A and one derived from it at X, which makes the former spread out toward both D and E, whereas the latter spreads only toward E. This misgiving is dissipated by the aforesaid general solution, through which one will see that the initial displacement of the particles at A, through the motion impressed on each one could be such that the propagation can take place only in the direction of E, and one will then perceive in turn that this circumstance will always hold for the derived displacements. It is indeed remarkable that the propagation of sound actually takes place faster than calculation would indicate. I have now renounced the idea I had previously, that the succeeding displacements would be able to accelerate the propagation of the earlier ones, so that the shriller the sound, the faster it would travel, as you have perhaps seen in our latest memoirs. It has also occurred to me to wonder whether the magnitude of the displacements could not cause some acceleration, since in calculations one assumes them to be infinitely small; it is evident that including the size would alter the calculations and make them intractable. However, so far as I can see, it seems to me that such a circumstance would decrease the velocity rather than increase it.

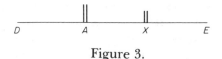

Figure 3.

It is indeed a loss that this same problem cannot be solved by giving the air three dimensions, or even only two, for then one may well doubt that the propagation would be the same. At least it is certain that in these cases the displacements would become weaker the farther they take place from their origin. I have indeed found the fundamental equations for the case in which the air extends to two dimensions or where it is confined between two planes. Let Y in Figure 4 be a particle of air in a state of equilibrium. After agitation let it be transported to y.

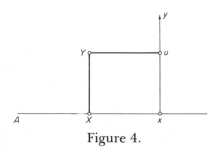

Figure 4.

Take $AX = X$, $XY = Y$, $Xx = Yu = x$, and $uy = y$. This being supposed, x as well as y will be certain functions of XY and of the time t and hence of three variables. For their determination I find the two equations:

$$\frac{\partial^2 x}{\partial t^2} = \alpha \frac{\partial^2 x}{\partial X^2} + \alpha \frac{\partial^2 y}{\partial X \partial Y}$$

$$\frac{\partial^2 y}{\partial t^2} = \alpha \frac{\partial^2 y}{\partial Y^2} + \alpha \frac{\partial^2 x}{\partial X \partial Y}$$

From this, if I suppose that the initial displacement takes place at A [see Figure 5] and that it spreads out from there in the form of circular waves in such a fashion that an arc ZV (in the equilibrium state) has been transported to zv, placing $AZ = Z$ and $Zz = z$, the quantity z will be a certain function of the two variables t and Z, for the determination of which I find this equation:

$$\frac{\partial^2 z}{\partial t^2} = \alpha \frac{\partial^2 z}{\partial Z^2} + \frac{\alpha}{Z} \frac{\partial z}{\partial Z} - \frac{\alpha z}{Z^2}$$

If we reject the last two terms there remains the same equation that applies to the case in which the air extends only along the straight line AE. But from this equation it does not appear that the propagation takes place with the same velocity in the two cases. One would therefore strongly suspect that the analysis stops at the point of being able to solve equations of this kind, and I hope that this glory is reserved for you. What you have to say about the echo is as important in analysis as in physics. And everyone ought to agree that this first volume of your work is a real masterpiece and contains much more profundity than so many other volumes of established academies, and never has a particular society deserved more the support of its sovereign.

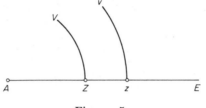

Figure 5.

As regards the sounds of music I am completely of your opinion that the harmonic sounds which M. Rameau believes he hears from the same string actually came from other vibrating bodies. And I do not see why this phenomenon should be regarded as the fundamental principle of music in place of the true relations which are its foundation. I still believe that I have accurately determined the degree of agreement with which we hear two given sounds, and that two sounds in the frequency ratio 8:9 are perceived more easily than if they are in the ratio 7:8. But here I think one has to take into account a presumption by which one imagines in advance the sound ratios, and then a deviation from this is intolerable. So far as the violin is concerned, if two strings are found to be in the interval of a sixth, he considers them out of tune, since he claims that their interval should be a fifth. For the interval 7:8, it will be very difficult to take it for what it is. One would imagine always that it ought to be 8:9, being badly tuned. It is only necessary to anticipate this presumption in order to produce the interval 7:8, but for that one also needs particular rules of agreement.

I have just finished the third volume of my Mechanics, which treats the motions of rigid bodies. I have discovered some new principles of the greatest importance. In order that such a body should rotate freely about an axis, it is not only necessary that this axis should pass through the center of gravity (or rather the center of mass [inertia]) of the body, but it is also necessary that all the centrifugal forces should be done away with. It is, of course, clear that in all bodies not all lines which pass through the center of mass have this property. But I have shown that for all such bodies, no matter how irregular they may be, there are always three such mutually perpendicular lines, which I have named the three principal axes of the body with respect to which I have determined *the moments of inertia,* and this consideration has enabled me to solve a host of problems which had previously appeared insoluble. For example, having impressed on any body whatever any motion whatever, it is desired to determine the continuation of this motion, leaving aside all forces which would act on the body.

I hope you have safely rceived my last letter, which I entrusted to the hand of a friend, in Geneva, M. Bertrand, who has applied himself to mathematics with great success.

I have the honor to be, Sir,

Your very humble and obedient servant,

L. Euler

To: M. de Lagrange, Professor of Mathematics and member of the Royal Academy of Science and Belles-Lettres of Prussia in Turin.

15

On the Propagation of Sound

LEONHARD EULER

Translated by R. Bruce Lindsay from
Memoirs de l'Academie des Sciences de Berlin (1766).
Also in *Opera Omnia III*, i, 428–451

1) Physicists as well as mathematicians have devoted much effort to explain how sound is transmitted through air. But it must be admitted that the theory has up to now been highly incomplete. What the great Newton has contributed to this problem is more a feat of ingenuity than a sufficient treatment, since he based his reasoning on purely arbitrary hypotheses. M. de Lagrange, a learned mathematician in Turin has just remarked very judiciously in the first volume of *Miscellanea Physico-Mathematica* published in Turin in 1759 that no matter what other hypotheses Newton had adopted he would have reached the same conclusion. This should suffice to assure us of the correctness of the conclusion so far as regards the velocity with which sound is transmitted through the air. But the actual motion with which the particles of air are displaced remains unknown to us. And we shall not be able to boast of an understanding of the propagation of sound unless we are in a state of being able to explain clearly how these displacements are transmitted through air.

2) All who have treated this problem after Newton have either fallen into the same difficulty or, wishing to examine more profoundly the real motion of the air, have encountered intractable calculations from which they have been able to draw absolutely no conclusions. And I ought to admit that the one or the other fate has befallen me after every time I have entered upon such investigations. I was then agreeably surprised when I saw in the excellent book which I have just cited that M. de Lagrange has fortunately surmounted all these difficulties and that by calculations that would appear to be completely intricate. Without contradiction this is one of the most important discoveries made in mathematics in a long time and one which will doubtless lead us to others.

3) In examining these prodigious calculations, I have given some thought to the possibility of attaining the same end but by a simpler route and by some effort I have achieved this. I shall then have the honor to explain here the method which seems to me the most proper for this research. But no matter how simple it may

136

be made to appear I must admit that it would never have come into my mind if I had not already seen the ingenious analysis of M. de Lagrange. There is indeed a circumstance which would stop us short, namely, if the analysis were applicable only to continuous quantities whose nature could be represented only by a regular curve or contained in a certain equation. It is only the expertness of introducing discontinuous quantities in the calculation that can lead us to the solution we seek. This is done in a manner similar to that by which I have determined the motion of a string which has been assigned an arbitrary initial configuration, inexpressible in terms of any equation.

4) One has really only to envisage the propagation of sound as it actually takes place. The air being abruptly agitated at some one place, the particles of air which are far removed from the place at first experience nothing; it is only after a certain time that they are displaced. And thereafter they are restored to equilibrium. Let us then think of an arbitrary particle at distance x from the origin of the disturbance, and after time T it is excited during an interval of time we call θ. Calling its velocity v, if we consider the state of the particle, the velocity ought to depend on x and on the time t in such a way that if $t < T$, $v = 0$ and that the velocity shall have a finite (nonvanishing) value while t lies between T and $T + \theta$, but that for $t > T + \theta$, the velocity reverts to zero. We see that this situation cannot be represented by a regular (continuous) function of the time t.

5) It is not proper to think that a function similar to those representing curves bounded in a closed space should be proper to represent the state of the particles of air in the propagation of sound. Such a function of t which would have real values only so long as $T < t < T + \theta$, is not at all suitable in our case to represent the value of v, since for $t < T$ and $t > T + \theta$, it would give imaginary values, in place of the fact that v is not imaginary under those conditions, but strictly zero. One could not even say that the values of v are there extremely small, but joined continuously with the finite values of v while t is in the interval T, $T + \theta$, since before the excitation arrives at the particle and after it has passed on it is completely at rest, as if it had never been disturbed. This is without doubt one of the principal reasons why the propagation of sound has refused to yield to calculation.

6) M. de Lagrange has happily avoided this stumbling block, since he has considered the particles of air as isolated, without forming a complete continuum. On this view he has assigned finite size to them, so that the number of particles distributed through any arbitrary interval remains finite. He has used the same method by which he has determined, in the same memoir, the vibrations of a string loaded with a finite number of weights. By this method he has been able to show, by the solution of the equations, that a single particle may be in motion while all the others are at rest. But at the end it is seen that the number of particles does not really enter into the problem and that the situation should not change if the number of particles of air filling a certain space is infinite. Everything comes back then to the fact that one should introduce discontinuous functions into the analysis which is to solve the problem. This would appear to be a great paradox.

7) As a matter of fact when I gave my general solution for the vibration of a string, which also involved the case in which the string was initially given an irregular configuration not expressible in terms of any equation, my solution was the subject of suspicion on the part of some eminent mathematicians. M. d'Alembert liked best

to maintain that in this case it was absolutely impossible to determine the motion of a string, which admitted my solution, even if it differed in no respect from his own in the other case. It was not even sufficient to make clear, as I had done, that my construction would satisfy completely the differential equation of the second degree which contains without contradiction the true solution. The discontinuity appeared to him always incompatible with the laws of the calculus. But now that M. de Lagrange, has completely justified my solution in an incontestable fashion, I do not doubt that the world will shortly recognize the necessity of using discontinuous functions in analysis, especially, when it is seen that this provides the only means of explaining the propagation of sound.

8) The paradox will appear even greater when I say that there is a very large part of the integral calculus where one is obliged to admit such discontinuous functions, even though one admits arbitrarily constants in the ordinary integrations. For the integral calculus is a method of finding functions of one or more variables when one knows some relation between their differentials of the first or higher order. Wherever it is a question of function of two or more variables, arbitrary functions are allowable, not excepting the discontinuous variety. This is true for the same reason that functions of a single variable, which are found by integration, have to be supplemented by an additive arbitrary constant which must be determined ultimately by the essential conditions governing each problem.

9) To put this in clear fashion, let us seek a function z of the two variables x and t such that

$$\left(\frac{dz}{dt}\right) = a\left(\frac{dz}{dx}\right)$$

where we already know that (dz/dt) means the fraction dz/dt (i.e., the ratio of the differentials), supposing only that t is variable, and similarly for (dz/dx). This condition is similar to that which governs the motion of vibrating strings, namely

$$\left(\frac{ddz}{dt^2}\right) = a\left(\frac{ddz}{dx^2}\right)$$

which differs from the former case only in the fact that here we are dealing with differentials of the second degree, so that the same circumstances hold in both cases. But it is evident that one can satisfy the condition

$$\left(\frac{dz}{dt}\right) = a\left(\frac{dz}{dx}\right)$$

by taking for z any function of $x + at$, without excluding discontinuous functions. For if we conceive any wave whatever drawn by free hand and not following any equation, if one takes $x + at$ as the abscissa, the ordinate will give a value of z which will satisfy the above equation and since there is no other condition, nothing obliges us to believe that a regular and continuous curve is any more necessary to satisfy the given conditions than an irregular and discontinuous curve, still less that the latter should be excluded.

10) Let us suppose the question is the motion of a string and that the conditions are such that after any arbitrary time t there corresponds to the abscissa x the ordinate

z such that

$$\left(\frac{dz}{dt}\right) = a\left(\frac{dz}{dx}\right)$$

Then I say that if we take z as any arbitrary function of the quantity $x + at$, that is

$$z = \Phi(x + at)$$

we shall have provided a general solution of the problem, no matter what function, regular or irregular, is denoted by Φ. The exact nature of the function is determined by the nature of the problem, which is given if the configuration of the string is known at some moment, say $t = 0$. Then $z = \Phi(x)$ will represent the equation of the initial configuration of the string, whatever it may happen to be, regular or irregular. Then, knowing this configuration, one can determine easily the configuration for any arbitrary time t, since for any arbitrary abscissa x there will correspond the same ordinate which corresponded in the initial configuration to the abscissa $x + at$.

11) My solution of the problem of the vibrating string is based on reasoning similar to that just presented, reasoning which is now safe from all objection. It is on the same foundation that I shall establish the solution of the problem of the propagation of sound and which enables me to dispense with the embarrassing calculations which M. de Lagrange has been obliged to develop. I look at the problem from the same point of view as that able mathematician in considering only those particles of air which are situated on a single straight line along which the propagation of the sound takes place. For, although the sound spreads equally in all directions, it seems certain that the propagation along each straight line is not disturbed by the motions of the neighboring particles. However it is much to be wished that this matter could be resolved by determining the disturbance throughout the whole atmosphere. But here we encounter difficulties which appear well-nigh insurmountable. I shall then content myself, like M. de Lagrange, with motion confined to a straight line.

Analysis of the Propagation of Sound Along a Straight Line

12) I consider then only the air confined to the straight line AE [Figure 1], very much like the air confined in a tube of very narrow bore, which I shall suppose to be terminated at the ends A and E. This makes the situation to which we are to apply the calculus perfectly definite. Let the length of the tube AE be a and its area of cross section, which is supposed uniform and almost infinitely small, be ee, so that the volume of the air in the tube is aee. Let the air at first be in equilibrium and of the same density throughout its length, so that its elasticity also remains the same. Let the height h be the measure of the elasticity in this equilibrium state,

by which is meant that the elasticity is balanced by the weight of a column of air of height h, or one can also say that each particle of air is subject to the pressure of a similar column of air whose volume is *hee*.

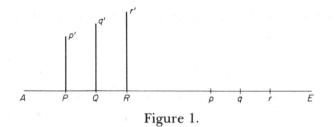

Figure 1.

13) It is now supposed that the air in the tube is subjected to a disturbance so that its equilibrium is disturbed. To represent the effect, we consider three points very close together $P, Q, R,$ separated by equal and very small intervals, i.e., $PQ = QR = \omega$, and such that by the disturbance the particles of air between $P,Q,$ and R are found after time t between the points $p,q,$ and r, and are thereby more or less condensed, according as the intervals pq and qr are smaller or greater than the original intervals PQ and QR. We suppose that the elasticity is changed in the same ratio. To express this change, we set for the equilibrium state

$$AP = x, \qquad AQ = x' = x + \omega, \qquad AR = x'' = x + 2\omega$$

and for the disturbed state

$$Pp = y, \qquad Qq = y', \qquad Rr = y''$$

from which we shall have the intervals

$$pq = \omega + y' - y, \text{ and } qr = \omega + y'' - y'$$

and the masses of the particles of air which are contained therein will be the same as the masses contained in the equilibrium state in the intervals PQ and $QR = \omega$ whose volume equals *eew*.

14) We observe here that the quantities x relate to the equilibrium state and express the distance of each particle of air from the fixed point A, but that the quantities y mark the displacement of each particle (caused by the disturbance) suffered in the time t. Thus the particle of air which in the state of equilibrium was distance x from the fixed end A will after time t find the interval $x + y$ open to it; hence the air will not be in equilibrium unless all the y vanish. If we erect perpendiculars to AE at the points P,Q, R with the ordinates Pp', Qq', Rr' equal respectively to the intervals Pp, Qq, Rr, the curved line which passes through the points p', q', r' will represent the disturbed state of the air in the tube at the time t. It is clear that the first ordinate at A and the last ordinate at E must vanish. For since the tube is assumed to be closed at the two ends the particles of air at A and E are unable to be displaced from their positions.

140

15) The elasticity which in the equilibrium state was represented by the height h will in the interval pq be represented by a height $= (h \cdot QR)/qr$. Then, since $PQ = QR = \omega$, we have

$$pq = \omega + y' - y, \qquad qr = \omega + y'' - y'$$

and therefore the height which measures the elasticity in the interval pq will be $= h\omega/(\omega + y' - y)$ and that for the interval qr will be $h\omega/(\omega + y'' - y')$. But it is the difference between these two heights which accounts for the acceleration or deceleration of the motion of the particle at q. Having divided the whole interval AE into very small subintervals all equal to ω, each containing the volume of air $ee\omega$ in the equilibrium state, let us now conceive of these particles as united at the points P, Q, R so as to have now at q a volume of air $= ee\omega$, which will be pushed from the rear toward A by the force

$$= \frac{eeh\omega}{\omega + y' - y}$$

and pushed from the other side toward E by the force

$$= \frac{eeh\omega}{\omega + y'' - y'}$$

16) Combining these two forces we find that the particle of air at q will be impelled toward E by the force

$$= \frac{eeh\omega\,(y'' - 2y' + y)}{(\omega + y' - y)(\omega + y'' - y')}$$

in which the distance from the fixed point A is $Aq = x' + y'$, and in which the part x' remains invariant with respect to the time, the other part y' being the only one which suffers the effect of the force. We assume that dy' corresponds to the interval dt and hence, using the principles of mechanics, we arrive at the following equation

$$\frac{ddy'}{dt^2} = \frac{2gh\,(y'' - 2y' + y)}{(\omega + y' - y)(\omega + y'' - y')}$$

in which g corresponds to the distance of free fall of a body under gravity in one second [Editor's note: g here is one half the normal acceleration of gravity] if t is expressed in seconds. The problem then is to find for each abscissa x and each time t the value of the interval y.

17) We now consider x as a variable and it is then clear that y will be a function of the two variables x and t. Since in the formula ddy/dt^2 we supposed x constant, we must write (ddy/dt^2) to avoid ambiguity. Placing $\omega = dx$, we get

$$y' - y = dx\left(\frac{dy}{dx}\right) \text{ and } y'' - 2y' + y = dx^2\left(\frac{ddy}{dx^2}\right)$$

whence the force equation takes the form

$$\left(\frac{ddy}{dt^2}\right) = \frac{2gh(ddy/dx^2)}{1 + (dy/dx)^2}$$

141

This indeed would be very difficult to solve. But if we further assume that the disturbance is very small we can neglect $(dy/dx)^2$ compared with unity and have for the propagation of sound for very small disturbances the equation

$$\left(\frac{ddy}{dt^2}\right) = 2gh\left(\frac{ddy}{dx^2}\right)$$

18) The situation here is the same as with the vibrations of a string in which we suppose that the displacements are also very small and not in opposition to the assumptions of the mathematicians who claim to have explained the motions of the string. So also in our case I am investigating the phenomena of motions which are extremely small so that the curve passing through the points p', q', r' is displaced only infinitely little from the axis AE, in the same way as the curve of the string is envisaged. The analogy goes even further since the same equation which expresses the propagation of sound also determines the vibrations of a string terminated at A and E. We therefore have the same integral for the solution throughout the tube, i.e.,

$$y = \Phi(x + t\sqrt{2gh}) + \Psi(x - t\sqrt{2gh})$$

This integral is complete, since it involves two arbitrary functions, as demanded by the order of the dynamical equation above.

19) To determine the nature of the two functions, it is necessary to apply the conditions prescribed in the problem. It is clear at first that, putting $t = 0$, the equation $y = \Phi(x) + \Psi(z)$ expresses the state of the air in the tube when it is first disturbed. Now let us suppose that the air in the tube is disturbed so that its state is represented by the curve AZE in plane of AE [see Figure 2]. This means that the arbitrary point X in the air column has been displaced in the direction towards E by the amount XZ. If we now call $AX = x$ and $XZ = z$ we shall have $z = \Phi(x) + \Psi(x)$. Let z be a function of x in the form $z = \theta(x)$. Hence

$$\Phi(x) + \Psi(x) = \theta(x)$$

This determines the form of one of the two arbitrary functions Φ and Ψ. But it must be remarked that the curve AZE must be infinitely close to AE and moreover it should join AE at the two extremities A and E. This means z must be very small and indeed equal to zero when $x = 0$ and $x = a$.

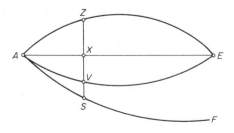

Figure 2.

142

20) The determination of the other function rests on knowledge of the initial motion of the air. Let us conceive therefore another curve AVE such that the ordinate $XV = v$, expressing the velocity which has been impressed on the particle of air at X in the direction XE. Hence v will also be some given function of x. Thus whatever the disturbance is its initial effect will always be determined by the two curves AZE and AVE, the first of which shows the initial space through which the particle of air at X is moved and the other shows the velocity impressed in this motion. If one wishes to assume that the particles of air are initially displaced but have *no* initial velocity, then the curve AVE will coincide with AE and everywhere in the air column $v = 0$. In any case since the tube is closed at both ends $v = 0$ always at A and E.

21) To take advantage of this condition, let us find the general expression for the velocity, namely (dy/dt). It is necessary to differentiate our functions. We employ the symbolism

$$d\Phi(u) = du \cdot \Phi'(u)$$
$$d\Psi(u) = du \cdot \Psi'(u)$$

whence from

$$y = \Phi(x + \sqrt{2gh}\ t) + \Psi(x - \sqrt{2gh}\ t)$$

we get

$$\left(\frac{dy}{dt}\right) = \sqrt{2gh}\ [\Phi'(x + \sqrt{2gh}\ t) - \Psi'(x - \sqrt{2gh}\ t)]$$

and consequently, since initially the velocity $= v$, we have

$$\frac{v}{\sqrt{2gh}} = \Phi'(x) - \Psi'(x)$$

We multiply by dx and integrate to get

$$\int \frac{vdx}{\sqrt{2gh}} = \Phi(x) - \Psi(x)$$

Now $\int vdx$ is the area included by AXV and is therefore also a function of x. Let us set $\int vdx = \sum(x)$ and get

$$\Phi(x) + \Psi(x) = \frac{\sum(x)}{\sqrt{2gh}}$$

22) This equation joined with

$$\Phi(x) + \Psi(x) = \theta(x)$$

serves to determine both Φ and Ψ in terms of the two given functions $\theta(x)$ and $\sum(x)$, so that

$$\Phi(x) = \frac{1}{2}\theta(x) + \frac{1}{2}\frac{\Sigma(x)}{\sqrt{2gh}}$$

$$\Psi(x) = \frac{1}{2}\theta(x) - \frac{1}{2}\frac{\Sigma(x)}{\sqrt{2gh}}$$

Hence our general equation which gives the position of the particle X at time t becomes

$$y = \frac{\theta(x + \sqrt{2gh}\ t) + \theta(x - \sqrt{2gh}\ t)}{2}$$

$$+ \frac{\Sigma(x + \sqrt{2gh}\ t) - \Sigma(x - \sqrt{2gh}\ t}{2\sqrt{2gh}}$$

and the velocity of the same particle is given by

$$\left(\frac{dy}{dt}\right) = \frac{[\theta'(x + \sqrt{2gh}\ t) - \theta'(x - \sqrt{2gh}\ t)]\ \sqrt{2gh}}{2}$$

$$+ \frac{[\Sigma'(x + \sqrt{2gh}\ t) + \Sigma'(x - \sqrt{2gh}\ t)]}{2}$$

where we remind ourselves that

$$\Sigma'(x) = v, \text{ since } \Sigma(x) = \int v\,dx$$

23) Now the whole solution will be determined if the two extremities A and E are removed to infinity. For, if we now draw another curve ASF so that the ordinate XS gives the area AXS with now $XS = \Sigma(x)$, one would be able to take in the two curves ASF and AZE (with $XZ = \theta(x)$), the ordinates which correspond to all the abscissas $x + \sqrt{2gh}\ t$ and $x - \sqrt{2gh}\ t$, and from this one will have for every time t the quantities y which correspond to each particle of air X. But since the thread of air AE is terminated by the points A and E beyond which the disturbance cannot be communicated, the curves formed on the initial state of the air will no longer furnish the ordinates corresponding to the abscissas $x + \sqrt{2gh}\ t$ when they are greater than $AE = a$, nor to the abscissas $x - \sqrt{2gh}\ t$, when the latter are negative. There is no question here of the natural continuation of these curves. This does not enter into consideration since the given curves AZE and ASF can even be discontinuous.

24) We then have need of some accessory facts to provide us with the true ordinates of our two curves of condition, when we take the abscissa either greater than a or negative. For this purpose we have only to look at the conditions mentioned above, namely that where $x = 0$ and $x = a$, the ordinates y shall always vanish. From this we deduce

$$\theta(a + \sqrt{2gh}\ t) + \theta(a - \sqrt{2gh}\ t) + \frac{\Sigma(a + \sqrt{2gh}\ t) - \Sigma(a - \sqrt{2gh}\ t)}{\sqrt{2gh}} = 0$$

and

$$\theta(\sqrt{2gh}\ t) + \theta(-\sqrt{2gh}\ t) + \frac{\Sigma(\sqrt{2gh}\ t) - \Sigma(-\sqrt{2gh}\ t)}{\sqrt{2gh}} = 0$$

If we now consider an abscissa either greater than a, say $a + u$, or negative, say $-u$, we shall have

$$\theta(a + u) + \frac{\Sigma(a + u)}{\sqrt{2gh}} = -\theta(a - u) + \frac{\Sigma(a - u)}{\sqrt{2gh}}$$

$$\theta(-u) - \frac{\Sigma(-u)}{\sqrt{2gh}} = -\theta(u) - \frac{\Sigma(u)}{\sqrt{2gh}}$$

From these we can always assign ordinates for those values of x that lie outside the limits of A and E.

The Propagation of Sound

25) We now proceed further with the explanation of the propagation of sound along the line AE [Figure 3]. We suppose that by some force a small particle of air mn has been displaced and put into the state represented by the small curve mon. While this displacement subsists, it is assumed that the rest of the line Am and nE remains still in equilibrium. Let us see how the displacement communicates itself with the other particles of air. On the hypothesis just stated $\Sigma = 0$ and there remains only the function θ which gives the ordinates of the curve mon whose abscissas lie in the region mn. But since initially, when $t = 0$, the particles of air save for those in mn are in equilibrium, the whole line representing the initial state consists of the straight line Am, the curve mon, and the straight line nE, and consequently a line of mixed character $AmonE$, for which, taking u as abscissa the ordinates will be given by the function $\theta(u)$. Furthermore, since $\theta(-u) = -\theta(u)$ [Editor's note: How does he know that θ is an *odd* function?], it follows that the continuation of the line to the left of A, namely $AA' = a$ will mirror the part to the right of A, that is (see the figure), $A\mu\omega\nu A'$ will reverse the situation in $AmonE$. Moreover, since

$$\theta(a + u) = -\theta(a - u)$$

it is necessary on the continuation EE' to establish the same line in reverse and so in turn with the other intervals $= u$ taken one after another on this line.

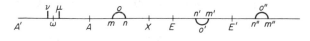

Figure 3.

145

26) From the above we see that the ordinate $\theta(u)$ will always be zero unless the abscissa u measured from the point A to the right falls between the limits

$$\left\{\begin{matrix} Am \\ An \end{matrix}\right\} \quad \text{or between} \quad \left\{\begin{matrix} An' \\ Am' \end{matrix}\right\} \quad \text{or} \quad \left\{\begin{matrix} Am'' \\ An'' \end{matrix}\right\} \quad \text{etc.}$$

$$\text{or between} \quad \left\{\begin{matrix} -A\mu \\ -Av \end{matrix}\right\} \quad \text{or} \quad \left\{\begin{matrix} -Av' \\ -A\mu' \end{matrix}\right\} \quad \text{etc.}$$

Then if we take $Am = m$ and $An = n$, the limits outside of which the ordinate $\theta(u)$ is everywhere zero are

$$\left\{\begin{matrix} m \\ n \end{matrix}\right\} \quad \text{or} \quad \left\{\begin{matrix} 2a-n \\ 2a-m \end{matrix}\right\} \quad \text{or} \quad \left\{\begin{matrix} 2a+m \\ 2a+n \end{matrix}\right\}$$

$$\text{or} \quad \left\{\begin{matrix} 4a-n \\ 4a-m \end{matrix}\right\} \quad \text{or} \quad \left\{\begin{matrix} 4a+m \\ 4a+n \end{matrix}\right\} \quad \text{etc.}$$

or

$$\left\{\begin{matrix} -m \\ -n \end{matrix}\right\} \quad \text{or} \quad \left\{\begin{matrix} -2a+n \\ -2a+m \end{matrix}\right\} \quad \text{or} \quad \left\{\begin{matrix} -2a-m \\ -2a-n \end{matrix}\right\}$$

$$\text{or} \quad \left\{\begin{matrix} -4a+n \\ -4a+m \end{matrix}\right\} \quad \text{or} \quad \left\{\begin{matrix} -4a-m \\ -4a-n \end{matrix}\right\} \quad \text{etc.}$$

In general these limits will have the form

$$\left(\begin{matrix} \pm 2i\,a \pm m \\ \pm 2i\,a \pm n \end{matrix}\right)$$

where i is any integer. Unless the abscissa u falls between two such limits the ordinate of u will always equal zero.

27) Let us now take any point X on the straight line AE, letting $AX = x$, and let us seek the disturbances it will undergo, given by the quantity y whose value at time t is

$$y = \tfrac{1}{2}\,\theta\,(x + \sqrt{2gh}\ t) + \tfrac{1}{2}\,\theta\,(x - \sqrt{2gh}\ t)$$

At first we see that the first term on the right vanishes unless $x + \sqrt{2gh}\ t$ lies between the limits

$$\left\{\begin{matrix} m \\ n \end{matrix}\right\} \quad \text{or} \quad \left\{\begin{matrix} 2a-n \\ 2a-m \end{matrix}\right\} \quad \text{or} \quad \left\{\begin{matrix} 2a+m \\ 2a+n \end{matrix}\right\} \quad \text{etc.}$$

The second term on the right also vanishes identically unless the quantity $x - \sqrt{2gh}\ t$ falls between the limits

$$\left\{\begin{matrix} m \\ n \end{matrix}\right\}$$

or its negative $\sqrt{2gh}\ t - x$ falls between

$$\begin{Bmatrix} m \\ n \end{Bmatrix} \quad \text{or} \quad \begin{Bmatrix} 2a - n \\ 2a - m \end{Bmatrix} \quad \text{or} \quad \begin{Bmatrix} 2a + m \\ 2a + n \end{Bmatrix} \quad \text{etc.}$$

Thus, if we suppose that $x > n$, this particle will remain at rest until we have

$$x - \sqrt{2gh}\ t = n \quad \text{or} \quad t = \frac{x - n}{\sqrt{2gh}}$$

It is then only after this time t that the particle at X will begin to move (under the disturbance). After that its rest will be re-established after the expiration of time $t = (x-m)/\sqrt{2gh}$ so that the motion lasts for the time interval $(n-m)/\sqrt{2gh}$. From this we see that each particle of air is excited only during a very short interval of time in accordance with the extent of the original disturbance mn and it is then that the sound is perceived.

28) It then takes the time $t = (x-n)/\sqrt{2gh}$ for the sound to travel from n to X or to travel through the space $nX = x - n$. From this we see that the time is proportional to the space, just as we know from experience. Expressing the time in seconds, we have also agreed to let h be the distance a body falls freely under gravity in one second. Hence in one second, sound will travel through a space equal to $\sqrt{2gh}$. But we know that $g = 15\frac{5}{8}$ Rhenish feet. In the elasticity of the air is represented by a column of water 32 feet high, and if we assume that water is 800 times as dense as air, we have

$$h = (32)(800) \text{ feet}$$

Hence

$$\sqrt{2gh} = (31\frac{1}{4})(32)\ 800 = 400\sqrt{5} = 894 \text{ feet}$$

Hence on this basis the velocity of sound in air is 894 feet per second.

But we know from experience that sound is transmitted in air at the velocity of 1100 feet per second. No one has yet discovered the reason for this excess of the experimental over the theoretical value.

29) But after the particle of air at X has been displaced by the sound disturbance the first time it will then be disturbed again and again and indeed an infinite number of times for which t is included in the following limits

$$t\sqrt{2gh} = \begin{Bmatrix} x + m, & 2a - n - x, & 2a - n + x, & 2a + m - x \\ x + n, & 2a - m - x, & 2a - m + x, & 2a + n - x \end{Bmatrix} \quad \text{etc.}$$

If the line AE is not terminated finitely at all, the particle at X will be excited only once. If the line is terminated only at the extremity A, the distance $AE = a$ being infinite, the particle will still receive one other excitation and this will happen after the time $(x+m)/\sqrt{2gh}$. This is the explanation of the simple echo. But if the line is terminated at both A and E, the disturbance arrives at X time after time, serving to explain the existence of multiple echoes. For this to hold, however, it is essential

147

that the particles of air at A and E are not at all movable, for this is the necessary condition for the production of echoes.

30) Since we have found that

$$y = \tfrac{1}{2}\theta (x + \sqrt{2gh}\ t) + \tfrac{1}{2}\theta (x - \sqrt{2gh}\ t)$$

we must note again that the disturbance of the particle X is only one-half that of the particle originally at mn. For the quantity y gets its magnitude when the one or the other member falls into the interval mn and since both do not fall there at the same time, the quantity y will be equal to only the half of the ordinate in the interval mn; whence it follows that the disturbances of the particle X are twice as feeble as the primitive disturbance at the particle mn. This is also a necessary consequence of the principle that the effect cannot be greater than the cause. For, since the original disturbance in mn is communicated equally towards A and E, at every instant there will be two particles at equal distance from mn on either side which will be displaced, whose motions taken together ought to be equal to the primitive motion in mn so that each can be only one-half of the latter. This diminution will be even greater when the disturbance from mn spreads out in every direction. From this we see that the sound transmitted through a tube should be the strongest.

Explanation of a Paradox

31) A doubt presents itself here which is not easy to remove. It seems that the disturbance which is now at X ought to be regarded like the original disturbance at mn and like that should be transmitted backwards as well as forwards. But this does not happen, for we have seen that the disturbance at any given moment at X is transmitted forward towards E and not at all backwards toward A. It is the same with the disturbance starting from mn in the direction towards A which is always transmitted in the same direction without creating new disturbances in the opposite direction. I now remove the limits A and E or I consider them as transferred to infinity, since I introduced them into the calculation anyway only to explain echoes. It will then be demanded, with reason: What is the difference between the initial disturbance at mn and that which is created later at X? For if everything is at rest save the particles around X, which find temselves displaced from their natural equilibrium positions, it seems as if this disturbance ought to be communicated just as well toward A as toward E. However, this would be quite contrary to experience, and we know that there is a vast difference between the place where the sound is created and the places where it is perceived.

32) It is necessary then that there should be an essential difference between the disturbance communicated to the particles of air at X and the primitive disturbance at mn. We must therefore expose this difference. But, having introduced into the calculation the primitive disturbance at mn, I have supposed a restriction in neglecting the functions symbolized by the sign $\sum$, which contains the condition that the particles in the space mn, having been displaced from their natural positions are not found in motion at all and that from this state of displacement they are suddenly released.

It is necessary to conclude from this that if the particles at X, after having been displaced, again find themselves at rest, there should result the same effect as the primitive disturbance in mn. But although each particle, having arrived at its greatest displacement, should then be reduced to rest, this does not happen to all the particles around X at the same instant. Consequently it is here, without doubt, that we must look for the solution of our difficulty.

33) From this we understand that the propagation depends not only on the displacement of the particles at mn but also on the motions impressed on them at the beginning, which influences the propagation insofar as in a certain case it takes place in one direction. It is therefore very important to treat the subject in all its elaborateness without neglecting the functions $\sum$. For this effect I shall not limit myself to a line or tube with finite termination, and I shall therefore suppose the line infinite, since there is no longer a question here of echoes. In the beginning it is assumed that the particles of air contained in the space mn [Figure 4] have been displaced so that the point x has been displaced toward E by the amount xz, so that z is on the curve mzn. It is also assumed that at the same point x there has been impressed a velocity $= xv$ directed also toward E. By quadrature (integration) of this lower curve mvn one forms a new curve $ms\xi$, so that the ordinate $xs = mxv/\sqrt{2gh}$ and since the velocity line mvn coincides on either side of the space mn with the same axis, i.e., mA and nE, the continuation of the curve $ms\xi$ will be toward A along the axis mA and toward E along the straight line ξe parallel to the axis nE.

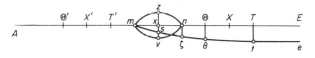

Figure 4.

34) This being assumed, if we take any point X and place $AX = x$, after time t, the point will be displaced toward E by a distance y, where

$$y = \tfrac{1}{2}\,\Theta(x + t\sqrt{2gh}\,) + \tfrac{1}{2}\,\Theta(x - t\sqrt{2gh}\,)$$

$$+ \tfrac{1}{2}\,\Sigma(x + t\sqrt{2gh}\,) - \tfrac{1}{2}\,\Sigma(x - t\sqrt{2gh}\,)$$

since the denominator $\sqrt{2gh}$ which turned up in para. 22 is already included in the function $\sum$. But here Θ denotes the ordinates of the curve mxn which on either side coincides with the axis AE, so that $\Theta(u)$ is always zero unless u is included between the limits Am and An. Here A is a fixed point chosen at will, as the origin of the axis of abscissas. It has no reference to the end of line or tube, which is no longer considered closed. Similarly $\sum$ denotes the ordinates of the curve, $Ams\zeta e$, so that the value of $\sum(u)$ is zero when $u < Am$ and equal to $n\zeta = Ee$, where $u > An$. But if u finds itself between these two limits, i.e., if $u = Ax$, then $\sum(u) = xs$. It is scarcely necessary to remark that if any ordinate occurs in a sense contrary to that shown in the figure, it is necessary to consider it negative.

149

35) Let us consider at first a point X farther from the fixed point A than the interval mn and since $AX = x$, let us take on either side the intervals $XT = X\Theta = t\sqrt{2gh}$ in order to have $AT = x + t\sqrt{2gh}$ and $A\Theta = x - t\sqrt{2gh}$. It is then clear that as long as $X < Xn$, we shall have

$$y = \tfrac{1}{2}Tt - \tfrac{1}{2}\Theta\theta = 0$$

since

$$\Theta(AT) = 0, \qquad \Theta(A\Theta) = 0$$
$$\textstyle\sum(AT) = Tt \qquad \text{and} \qquad \sum(A\Theta) = \Theta\theta$$

But when the point X falls in the space mn where $X = t\sqrt{2gh} = Xx$, we shall have

$$\Theta(AT) = 0, \qquad \Theta(Ax) = xz$$
$$\textstyle\sum(AT) = Tt = n\zeta, \qquad \sum(Ax) = xs$$

and hence

$$y = \tfrac{1}{2}(xz + n\zeta - xs)$$

which is the space through which the point X will be moved toward E from its equilibrium position after time $t = Xx/\sqrt{2gh}$. But after the time $t = Xm/\sqrt{2gh}$, we shall have

$$y = \tfrac{1}{2}n\zeta$$

which will also remain as the value of y when $t > Xm/\sqrt{2gh}$ so that after this time it will be at rest, even though it is at a distance from its natural place equal to $\tfrac{1}{2}n\zeta$, the disturbance having lasted from time $t = Xn/\sqrt{2gh}$ to time $t = Xm/\sqrt{2gh}$.

36) Let us now consider any arbitrarily chosen point X' on the other side of the interval mn, so that $AX' = x$ and taking on either side equal intervals $X'T' = X'\Theta' = t\sqrt{2gh}$, we see that as long as $X'T' < X'm$ or $t < X'm/\sqrt{2gh}$ the point X' will remain at rest. But if T' increases, so that $t = X'x/\sqrt{2gh}$ whereby

$$\Theta(Ax) = xz$$
$$\Theta(A\Theta') = 0$$
$$\Sigma(Ax) = xs$$
$$\Sigma(A\Theta') = 0$$

we shall have

$$y = \tfrac{1}{2}xz + \tfrac{1}{2}xs = \tfrac{1}{2}(xz + xs)$$

and after the time $t = Xn/\sqrt{2gh}$, there results

$$y = \tfrac{1}{2}n\zeta$$

which will remain as the value of y, so that the particle X', after having been displaced, will find itself displaced toward E from its original position by the amount $\tfrac{1}{2}n\zeta$. Hence after all the displacements have taken place, all the line of air AE will have advanced in the direction AE by the amount $\tfrac{1}{2}n\zeta$.

37) From this we see that the displacements of the particles X and X' of which one is on one side and the other on the other side of the initial disturbance mn, are completely different, since of X the greatest displacement is $\tfrac{1}{2}(xz - xs + n\zeta)$ whereas that of X' is $\tfrac{1}{2}(xz + xs)$, and therefore in this case the sound is transmitted quite differently forward from what it is backward, whereas in the preceding case in which the initial velocities xv vanish, the propagation is the same in both directions. But we further see that it would be possible for the propagation to take place in one direction only. This would happen if throughout the whole interval mn, we were to have $xz - xs + n\zeta = 0$. For this to happen, since xz and xs vanish at m, it is necessary that $n\zeta = 0$ and $xz = xs$. If we then place

$$xz = z, \quad xv = v, \quad \text{and} \quad xs = \frac{\int v\, dx}{\sqrt{2gh}}$$

this condition requires that

$$x\sqrt{2gh} = \int v\, dx \text{ and } \int v = \frac{dz\sqrt{2gh}}{dx}$$

In this case the curve $ms\zeta$ will be equal and will coincide with the other curve mzn and will rejoin the axis at n, so that $n\zeta = 0$. Then the particle at X situated in the side of the disturbed region mn towards E will not be disturbed or displaced at all and the propagation can take place only on the other side toward A.

38) It is precisely the case of displacements which are produced by an arbitrary primitive disturbance, which are always such that even if they themselves were primitive would transmit themselves only in one direction. To assure oneself of this one has only to give to z the value of y found previously and to v the value of (dy/dt). One then gets

$$z = \tfrac{1}{2}\Theta(x + t\sqrt{2gh}) + \tfrac{1}{2}\Theta(x - t\sqrt{2gh})$$

$$+ \tfrac{1}{2}\,\Sigma(x + t\sqrt{2gh}) - \tfrac{1}{2}\,\Sigma(x - t\sqrt{2gh})$$

$$\frac{v}{\sqrt{2gh}} = \tfrac{1}{2}\Theta'(x + t\sqrt{2gh}) - \tfrac{1}{2}\Theta'(x - t\sqrt{2gh})$$

$$+ \tfrac{1}{2}\,\Sigma'(x + t\sqrt{2gh}) + \tfrac{1}{2}\,\Sigma'(x - t\sqrt{2gh})$$

Taking the differential of z, assuming that x alone is varied,

$$\left(\frac{dz}{dx}\right) = \tfrac{1}{2}\,\Theta'\,(x + t\sqrt{2gh}\,) + \tfrac{1}{2}\,\Theta'\,(x - t\sqrt{2gh}\,)$$

$$+\tfrac{1}{2}\,\Sigma'\,(x + t\sqrt{2gh}\,) - \tfrac{1}{2}\,\Sigma'\,(x - t\sqrt{2gh}\,)$$

But, as we have seen above, it is always only one of the two abscissas $x + t\sqrt{2gh}$ or $x - t\sqrt{2gh}$ for which there is a finite ordinate. Therefore, if it is the first, there will evidently be

$$\left(\frac{dz}{dx}\right) = \frac{v}{\sqrt{2gh}}$$

and therefore such a disturbance will be propagated in only one direction. This then is the real explanation of the paradox under discussion.

Why Several Sounds Do Not Disturb Each Other

39) From the above we can clearly understand the reason why several sounds do not disturb or interfere with each other. This is a problem which has for a long time tormented physicists. The theory of the great Newton, although fundamentally accurate, was not sufficient to explain this phenomenon, since it did not at all determine the true nature of the disturbances to which the particles of air are subjected. M. de Mairan[1] has imagined that each sound, according as it is of low or high pitch, is transmitted only by certain particles of air whose elasticity is appropriate to it. But besides the fact that the state of equilibrium absolutely demands that all the particles of air should be endowed with the same degree of elasticity this explanation is contradicted by the first principles on which our theory is founded and whose certainty cannot be challenged by doubt. For propagation is accounted for by only a single displacement excited in the air and it does not matter whether this is followed by others or not. Still less does the propagation depend on the order of the succession of displacements, from which one judges sounds of low and high pitch.

40) In order to clear up entirely the preceding point one has only to assume several primitive disturbances a, b, c, d, α, β on the straight line AE [Figure 5]. Considering an arbitrary particle of air at P, we see by what I have just explained that the disturbance at α will be communicated to it after the time $P\alpha/\sqrt{2gh}$. It follows that P will receive the disturbance a after the time $Pa/\sqrt{2gh}$ and so for the others, so that each disturbance is transmitted to the particle P in a definite time and an ear placed at P will perceive all the disturbances, without any one of them being troubled by the others. It could also happen that two displacements arrive at the same instant at the same particle, such as O, the distances αO and aO being

---

[1]J. J. Dortous de Mairan (1678–1771), Discourse on the propagation of sound, Memoirs of the Academy of Sciences of Paris, 1737.

equal. But then this particle will be disturbed quite otherwise than if it were to receive a simple displacement and it will consequently transmit its displacement as much forward as backward. But this is precisely the case in which one should consider the disturbances as disturbing each other, something that does not happen at O or any other point P.

Figure 5.

Reflections on the Preceding Theory

41) It is necessary to comment at the outset on the fact that I have here considered propagation along a straight line only, just as if the air were confined in a very narrow cylindrical tube. From this one would be inclined to think that in free air the propagation ought to follow altogether different laws. At any rate it is clear that the disturbances, spreading out in every direction ought to decrease considerably faster than in the case of a tube. But as far as the nature of the sound disturbance is concerned and in particular the velocity with which it is transmitted to distant points it appears certain that the latter will be the same in free air as in air enclosed in a tube. For since sound like light travels in straight lines, which we can call sound rays, the transmission along each of these straight lines ought to follow the same rules which I have just disclosed, with the single difference that the disturbance becomes more feeble the greater the distance of travel. However it is much to be wished that one were in a condition to solve the problem of transmission in free air.

42) In the second place there always remains the great difficulty that sound travels effectively a greater distance in a given time than the theory indicates. I now recognize that the later displacements cannot be the cause, as I had previously thought. But it is really essential to compare the case of experience with that to which the theory is restricted. Without pretending that propagation in free air can cause the difference in question it is necessary to remember than our calculation assumes that the disturbances are infinitely small, which would produce sounds too weak for one to be able to observe the distance of their propagation in one second. Hence, since the sounds which are employed in experiments are produced by very powerful disturbances, it is very probable that in the principal equation in para. 17, which is

$$[1 + \left(\frac{dy}{dx}\right)^2] \frac{ddy}{dt^2} = 2gh \left(\frac{ddy}{dx^2}\right)$$

it is not permitted to neglect that term $(dy/dx)^2$ as I have done in the preceding calculations. Perhaps it is here that one must look for the solution of the difficulty.

153

43) So, although this important discovery is really owing to M. de Lagrange, I flatter myself that this Memoir is not lacking in very interesting research results. For, leaving aside the fact that my analysis is very simple, I have introduced the free use of discontinuous functions, objected to by many great mathematicians, but a usage which is absolutely necessary every time it is a question of finding by integration functions of two or more variables and when one requires a general solution. After that, although the solution is analogous to that of the vibrating string, which I have given previously, I have here determined with more exactness the arbitrary functions by conditions appropriate to the nature of the problem. Moreover, my solution when applied to strings, is more general, since in the initial state one can give to the string not only any arbitrary configuration but also give to every element of it an initial arbitrary motion. This I had not done in my earlier memoir, nor have others who have treated the same subject. Finally I believe that the explanation of the paradox, that the displacements caused by the propagation of the sound are in nature altogether different from the initial disturbances, furnishes very considerable clarification of this intricate problem.

16

Discoveries in the Theory of Sound

ERNST FLORENS FRIEDRICH CHLADNI

Ernst Florens Friedrich Chladni (1756–1827), a German musician and scientist, was first trained in law at the University of Leipzig, but soon turned to physics. Through his interest in music his attention was turned to the need for understanding the nature of the vibration of solid bodies which give rise to sound. He made many experiments on the vibrations of plates and showed how these could be made visible by sprinkling sand over the vibrating surface. The sand tends to gather along the nodal lines where there is no motion. The figures so formed have long been known as Chladni figures. We present here a translation of extracts from his famous book on discoveries in the theory of sound (Entdeckungen über die Theorie des Klanges) published in 1787, one of the benchmarks of acoustics.

16

Discoveries in the Theory of Sound

ERNST FLORENS FRIEDRICH CHLADNI

Translated by R. Bruce Lindsay from
Entdeckungen über die Theorie des Klanges (Wittenberg) 1787

Introduction

The elastic vibrations of strings and rods in which we must pay attention to individual curved lines only, have been calculated by so many investigators so accurately that there does not seem to be much left to say about them. On the other hand, the real nature of the sounds of bodies in which simultaneous elastic bending of whole surfaces in several dimensions must be taken into consideration, is still concealed in utter darkness, since neither calculations which agree with experiment nor correct observations on these matters are available. I have succeeded in discovering a means of rendering every possible kind of sound from such bodies not only audible but also visible. Hence I hope through the publication of my observations to provide correct hypotheses for a more precise investigation of this still much undercultivated part of mechanics. I am confident that the incompleteness of my remarks will be forgiven by all who know from their own experience how many difficulties confront the person who wants to embark on the hitherto untrod paths of nature on every step of the way.

Every sounding body can give off different tones and for each of these assumes a different kind of vibrating motion. Because of the bending caused thereby the surface or shape of the body is either intersected by one, two, three, or more lines, or perhaps in some cases by none at all. The places where the snakelike oscillation lines intersect the surface were called by Sauveur, de la Hire, and others vibration *nodes*. They remain at rest, whereas the remaining parts of the sounding body move, and one can touch the body at such places or otherwise apply damping there without thereby stopping the sound. The latter happens quickly, however, when one touches a part of the body between vibration nodes or otherwise applies damping to such a region.

It is well known that in the case of the fundamental of a string there are no nodes [Editor's note: save at the fixed ends]. For the higher tones, which we call harmonics, however, there are 1, 2, 3, ... nodes. If we refer to the fundamental as 1 (first harmonic), the higher tones are given the numbers 2, 3, . . . (second, third, . . . harmonics).[1] Every arbitrary harmonic tone of a string can be readily produced if one touches a nodal point and at the same time bows a vibrating part with a violin bow or uses some other means to set it in motion. Such tones are excellently executed on the marine trumpet. I have heard cello players produce them with good effect in many cases. For example on the "d" and "a" strings on the cello many pure harmonic tones can be produced with unusual intensity and indeed just as successfully as when produced in the usual fashion, so that many tones in the octave still sound very pleasant from the same string.[2]

In elastic rods and metal plates and also for rings, disks, and bells and similar sounding bodies the relations of the different tones of the same body to each other as well as the corresponding tones of different bodies differ very much from those taking place in strings. For most such bodies it is not possible to elicit by striking them all tones with sufficient precision. For the investigation of these it is better to strike the body at a right angle with a violin bow which has been rubbed with resin of some kind. In this way definite and controlled tones can be produced. Even wooden rods, caskets, and shells can be made to give off long continued sounds.[3] All places on the sounding body where the surface is intersected by the snakelike curved lines, can be made visible, if the surface is flat, when one before or during the stroking sprinkles some sand on it. The sand is thrown off from the vibrating portions of the surface, often with considerable violence, whereas it remains at rest on the places where there is no motion.

---

[1]It is also well known that horns, trumpets, and open pipes give the same series of harmonics. For a closed pipe, however, the corresponding numbers are the odd integers.

[2]In Sulzer's theory of the fine arts in the section on sound it is maintained incorrectly that the fourth harmonic of e is the highest usable tone. The author of this article bases his statement on Euler's *Tentamen Novae Theoriae Musicae*, Chap. 1, para. 13. However, he either did not read the context or he did not at all understand it. For Euler did not himself characterize the fourth harmonic of e, but rather the fifth harmonic of e as the audible tone of highest pitch. Following investigations of sounding bodies, I am inclined to believe that the middle of the fourth harmonic of e is the limit of usable sound, whereas approximately the fifth harmonic of e or f is the limit of the tones that can be discriminated.

[3]The stroking of bars, bells, and such like sounding bodies with the violin bow is no discovery of mine. For the so-called violin harmonica is an instrument that has been known for a long time, and knowledge of which has been provided several years ago by Schröter, Senal, and others. Also the Abbot Mazzocchi constructed an instrument similar to the harmonica, consisting of metal bells which were stroked with a violin bow. A reference to this is to be found in the third volume of Forkel's critical library of music, p. 321 and in his musical almanac of 1782 (p. 33). So far as I know, however, no one has used the violin bow for the investigation of such sounding bodies, an investigation which cannot be conveniently carried out in any other way.

Since in the sound emission of bars and flat metal plates the vibrating motion is far simpler than in the case of bells and curved disks, it is appropriate for the greater clarity of what follows to say something about it at first. There are six cases in which a rod can produce definite sequences of tones and in each of these cases can describe curved lines:

1) If one end is rigidly fastened and the other free.
2) If one end is lightly fixed and the other is free.
3) If both ends are free.
4) If both ends are lightly fixed.
5) If both ends are rigidly fixed.
6) If one end is rigidly fixed and the other lightly fixed.

What I call rigidly fixed here is what Euler called *infixus*, assuming thereby that such an end is fastened into a wall so that it cannot move at all. For greater convenience in experimental investigation one can equally well tighten one end of the bar in a bench vise.[4] What I here term lightly fixed was called by Euler *simpliciter fixus*. By this he meant that the end of the rod is fixed to an immovable pivot in such a way that it is free to rotate about the latter.

The progressions of tones which are produced by the same rod or plate in all these six cases and the patterns of the curved lines formed thereby have been calculated by L. Euler in the Acts of the Imperial Academy of Sciences in St. Petersburg for 1779.[5] He also worked out the first, third, fourth, and sixth cases in his *Methodus inveniendi curvas . . . .*[6] However, in the last mentioned work there are some erroneous assertions, which were competently disproved by Count Giordano Riccati, who moreover calculated the oscillations of freely vibrating cylinders with the greatest precision.[7] The first one who investigated with success the oscillations of elastic plates and rods was Daniel Bernoulli. Many writing by him concerning this subject appeared in the Commentaries and New Commentaries of the Petersburg Academy of Sciences.

In the first case, in which one end of the rod is immovable and the other end is free, in the lowest and simplest mode of sound, the line of the rod *ab* [Figure 1] is never intersected by the curved vibration line *ac*, but touches it only at the fixed end *a*. This vibration mode can be made to appear very readily in a rod or nail which is rigidly fastened to a wall or in a vise, if one strokes or strikes it at

---

[4]It must be remarked here that if the direction of the stroking is parallel to the mouth of the vise the tones are deeper than when one strokes the bar at right angles to the mouth of the vise. The reason is that with the stroking parallel to the mouth the lower end of the rod does not stand fast enough against the vise. Here the difference amounts to a whole tone in the case of the simpler sounds, but for the more complicated sounds it is less.

[5]*Investigatio motuum, quibus laminae et virgae elasticae contremiseunt*, autore L. Eulero in Actis Acad. Scient. Imp. Petrop. pro anno 1779, P.I. p. 103 sequ.

[6]*Methodus inveniendi curvas maximi minimive proprietate gaudentes, solutio problematis isoperimetrici, latissimo sensu accepti*, autore L. Eulero, additam I, de curvis elasticis, p. 282, sequ.

[7]In the article: *Delle vibrazioni sonore dei cilindri*, which appeared in the first volume of the *Memorie di matematica e fisica della societa Italiana*, Verona, 1782.

a point not too close to the fixed end. This sound or vibration mode is utilized in the so-called harmonica with iron nails or the violine harmonica which consists of iron brads, which are fastened to the bridge of a resonance back and are stroked with a violin bow. In addition to the first mentioned vibration mode, several others can be produced in the same rod. In these the curved vibration line intersects the equilibrium position of the rod *ab* in 1, 2, 3, or more points. In order to produce any particular one of these modes, hold the rod at one of the intersection points and stroke or strike it in the middle of a vibrating segment. For example, the mode in which the curved vibration line *ace* intersects *ab* in a single point *d* [Figure 2] may be produced by stroking or striking the rod around *c* while keeping *d* from moving. Since the greater the number of intersections that the curved line makes with the horizontal line, the more closely the outermost vibration node approaches the end of the rod, one can hardly fail to bring out any one of the whole series of tones produced by the same rod in this way, by bringing the finger which touches the rod even closer to one end and stroking the rod even closer to the fixed end. At the same time the number of vibration nodes gets even greater. For the mode of vibration in which one node exists [Figure 2] the tone will be nearly two octaves and a fifth above that for the simplest mode [Figure 1]. If there are two nodes, the pitch is approximately an octave plus a fifth above the previous one; if there are three nodes, the difference between that and the one next lower is about an octave. If there are four nodes, the tone is about a large sixth higher, etc. Hence if one takes the contra C as the tone corresponding to the simplest mode, the higher notes appear at follows:

No. of nodes	0	1	2	3	4	5	6
Tones	C	gs−	$\overline{\overline{\text{d}}}$	$\overline{\overline{\text{d}}}$−	$\overline{\overline{\text{h}}}$−	$\overline{\overline{\text{f}}}$	$\overline{\overline{\text{h}}}$

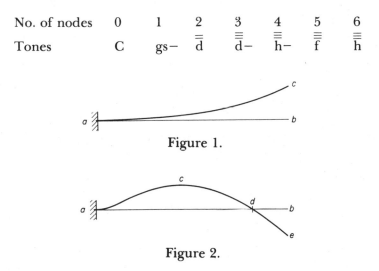

Figure 1.

Figure 2.

With the exception of the fundamental tone, the pitches of these vary approximately as the squares of 3, 5, 7, etc.

Exactly the same state of affairs prevails for the vibration modes of forks and other bars in which both ends are free, and in which the middle is kept motionless in a handle or fastened in a vise. Without regarding how both halves share their vibrations, we can treat each half as a rod in itself with one end motionless: the same series of tones results.

In the second case in which one end is lightly fastened with the other end free, the vibration nodes will be found at almost the same places as in the first cases, but the shape of the curved lines is somewhat different. This can be seen by comparing Figure 3 (the second case) with Figure 2. For the lowest tone in this case there is still one node. For the next higher mode there are two nodes, as shown in Figure 4. In order to produce the sounds corresponding to these modes, hold the rod between two fingers at a node, rest one end not too rigidly on a table or a resonating platform and stroke the rod in the middle of an oscillating segment. If the fundamental of the rod in the first case is still taken as C, the upper tones for the second cases under discussion will be as follows (their pitches varying approximately as the squares of 5, 9, 13, 17, . . .).

No. of nodes	1	2	3	4	5	6
Tones	d	h̄−	h̄+	g̿s	d̿s+	a̿

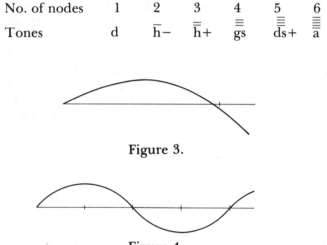

Figure 3.

Figure 4.

In the third case, in which both ends are free, the fundamental has two nodes, as shown in Figure 5. The next higher mode has three, as shown in Figure 6, and so on.

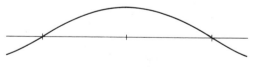

Figure 5.

Figure 6.

In his *Methodus inveniendi curvas . . .*, Euler denied the possibility of the existence of modes for which there are an odd number of nodes. He assumed that in such cases a rotational motion would take place (*motus rotatorius*) about the middle of the rod. After more careful investigations, however, he finally admitted the possibility of such an odd number of nodes (in the *Actis* of 1779). Moreover Count Giordano Riccati, in the article already mentioned, *Delle vibrazioni sonore dei cilindri*, showed clearly that the modes with an odd number of nodes are just as possible as those with an even number. This I had myself already established before I was acquainted with the work of Euler and Riccati on this subject. Actually Riccati has, in his work on the modes of the freely suspended cylinder as well as of the cylinder in general, added much new material to Euler's results.

In order to obtain every possible mode of a rod free at both ends, lay the rod at two vibration nodes lightly on some surface or on a piece of cloth, which can be moved at will, and stroke or strike the rod between two nodes.

[Editor's note: From this point on Chladni proceeds to discuss in turn and in detail the other cases of rods clamped or free as indicated in the preceding list. It is unnecessary to follow him here as the results are only detailed variations of what he has already said. We therefore proceed to what he has to say about the vibrations of plates.]

In all the cases of vibration modes so far discussed, only elastic curvatures of single lines entered into consideration. The working out of such modes for the case in which whole surfaces in several dimensions are subjected to elastic curvature is fraught with much greater difficulty. For such cases the detailed calculations have not been made, nor have usable attempts been made in this direction. The few things that have been said about such vibrations do not for the great part agree with experience. Since in my own investigation here I have had to enter regions hitherto untrod, I deserve hopefully some consideration if I err.

For the distinct production of such vibration modes it is essential to touch with the finger or otherwise hold those places on the vibrating body which are to stay at rest in the particular mode required and then to stroke the edge, at a suitable place with a violin bow held at a right angle to the edge. Moreover, if it is desired to make the division of the body into its various vibrating parts visible, one should strew sand on the horizontal surfaces of the body. This will be thrown off from the vigorously vibrating points and will collect along the nodal lines. By merely striking the body it is not in general possible to hear many such individual modal tones without the intermixture of others. Moreover it is not possible in general in this way to make the individual vibrations visible by the use of sand. For the latter purpose it is essential to employ stroking with a violin bow. If the edge is too sharp it must be made more blunt by means of a file. Otherwise the hairs of the violin bow will suffer damage. It proves most effective to hold the vibrating body at a place where two nodal lines intersect. In this way the excitation of the neighboring parts of the body is less hindered. In this way not only is the emitted sound clearer, but the visible pattern will be more distinct. Many such varieties of vibrating motion can be obtained in this way without much trouble. For others, however, much patience is required and continual practice in this sort of investigation is necessary. People must therefore not ascribe to me false assertions if the desired kinds of vibrations are not obtained at once. By repeated experimental attempts one ultimately reaches the goal: often a mode very difficult to achieve will show up when one is looking for another one.

Through the elastic surface curvature the original shape of the vibrating body is intersected by certain *lines*, just as happens for certain points in the case of vibrating strings and bars. Two portions which are separated from each other, as *anb* and *bod* in Figure 7 vibrate simultaneously in opposite directions, that is, the portion on one side of the nodal line is momentarily displaced *above* the equilibrium position and the portion on the other side is at the same time displaced *below* its equilibrium position and vice versa. Two portions at opposite angles from each other like *anb* and *cmd* in Figure 7 vibrate simultaneously in the same direction. In most modal patterns certain nodal lines assume a snakelike form; their number is fixed for each individual pattern. In the case of such nodal lines running beside each other, the position if the curved displaced parts is so arranged that either two nodal lines come very close to each other and then separate or in a few cases two snakelike nodal lines separated by a straight line come close to each other and then separate. At every point of closest approach the lines can combine so that they cross each other, so that the curved nodal lines in Figure 8 can eventually take the form of the corresponding straight lines in Figure 7. In the same way two straight nodal lines as in Figure 7 can become two curved nodal lines as in Figure 8. Occasionally the nodal lines, both straight and curved will encounter the edge of the plate or come very close to it, yielding patterns like Figures 9 and 10, where *gf* is the edge of the plate. Many patterns are at times so changed that without proper effort it becomes very difficult to judge their real form. In any case in very few cases do the patterns appear in the regular form suggested by the figures just given. At times the patterns are so confused that corresponding modes must be evaluated in terms of their relation to other clearer patterns. Even if the pattern may not be recognizably clear, the sound emitted will usually be the same as if the pattern were regular. The reason for such irregularities in patterns is to be found in part in the form of the vibrating body, but also in part in nonuniform thickness and elasticity of the vibrating object. It often happens that irregular patterns are produced when one does not touch or hold the body at precisely the places through which nodal lines should pass. Sometimes patterns can be changed perceptibly by slight displacements of the finger without changing the sound emitted.

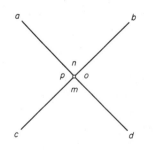

Figure 7.

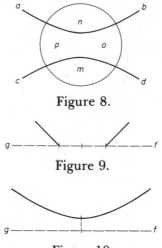

Figure 8.

Figure 9.

Figure 10.

In the case of metal or glass strips in rectangular form besides the vibration modes already calculated by others, there are still an infinite number of possible ones which have never been investigated by anyone.

[Editor's note: In the footnote following this paragraph the author calls attention to the vibration of a square plate as shown in Figure 11, where nodal straight lines like *b* are possible. He points out that under certain circumstances these nodal lines become curved as in Figure 12 or Figure 13.]

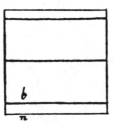

Figure 11.

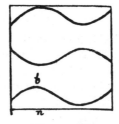

Figure 12.

163

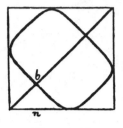

Figure 13.

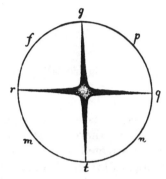

Figure 14.

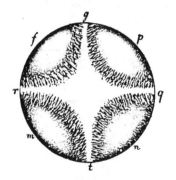

Figure 15.

[*Editor's note:* The author proceeds with the detailed examination of the vibration patterns of rectangular plates. The figures shown above are a sufficient indication of what may be expected in this case. We pass therefore to his discussion of circular plates.]

In the fundamental mode of a bell or circular disk, which is the only one that is customarily used, the disk divides into four segments if one (as in Figure 14) bows or strikes it at p, n, f, or m. Here the nodal lines are gt and rq. One can touch or otherwise damp the disk anywhere along these nodal lines without disturbing the oscillating motion. As a matter of fact the sound of the fundamental comes out more clearly when one does keep the disk motionless along the nodal lines. Such damping along the nodal lines often damps down the unpleasant discordances of the higher modes. The division of the disk into parts as shown in Figure 14 can be made visible with sand. When water is placed in a round drinking glass or similar vessel and the sides are bowed at place corresponding to p, n, m, f in Figure 14, the water is disturbed in striking fashion and is indeed driven toward the center of the surface so that its appearance is something like that shown in Figure 15. Here the segments gq, qt, tr, and rg tend to stay at rest and the disturbance is strongest toward the center. The result of the investigation will appear even better if one places the vibrating vessel in a much larger vessel and flows water into both to the same height. In this case from the four vibrating segments streams of water will be forced out and in. It is understood of course that one must fasten or press firmly the inner vessel against the floor of the outer vessel, so that it will not be moved as a whole in the bowing process.

17

Invention of the Stethoscope

RENE THEOPHILE HYACINTHE LAENNEC

René Théophile Hyacinthe Laennec (1781–1826) was a French physician of the Napoleonic era and the founder of chest medicine. In connection with his study of diseases of the chest and heart he was led to the invention of the stethoscope and thus introduced an application of acoustics to medicine of far reaching significance. It was a forerunner of much modern research on the use of acoustics both for medical diagnosis and therapy.

The following extract is taken from Laennec's celebrated treatise "Traité de la auscultation médiate et les maladies des pommons et du coeur" (1804) in English translation by Dr. John Forbes, M.D.

In these cases some physicians have attempted to gain further information by the application of the ear to the precordial region; and, doubtless, such a proceeding will encrease the certainty of the diagnosis. Even this, however, is very insufficient; and there are, besides, many reasons why it cannot be followed, as a general guide, in practice. Nevertheless, I had been in the habit of using this method for a long time, in obscure cases, and where it was practicable; and it was the employment of it which led me to the discovery of one much better.

In 1816, I was consulted by a young woman labouring under general symptoms of diseased heart, and in whose case percussion and the application of the hand were of little avail on account of the great degree of fatness. The other method just mentioned being rendered inadmissible by the age and sex of the patient, I happened to recollect a simple and well-known fact in acoustics, and fancied, at the same time, that it might be turned to some use on the present occasion. The fact I allude to is the augmented impression of sound when conveyed through certain solid bodies,—as when we hear the scratch of a pin at one end of a piece of wood, on applying our ear to the other. Immediately, on this suggestion, I rolled a quire of paper into a sort of

cylinder and applied one end of it to the region of
the heart and the other to my ear, and was not a
little surprised and pleased, to find that I could
thereby perceive the action of the heart in a manner
much more clear and distinct than I had ever been
able to do by the immediate application of the ear.
From this moment I imagined that the circumstance
might furnish means for enabling us to ascertain the
character, not only of the action of the heart, but of
every species of sound produced by the motion of
all the thoracic viscera. With this conviction, I
forthwith commenced at the Hospital Necker a
series of observations, which has been continued to
the present time. The result has been, that I have
been enabled to discover a set of new signs of diseases of the chest, for the most part certain, simple,
and prominent, and calculated, perhaps, to render
the diagnosis of the diseases of the lungs, heart and
pleura, as decided and circumstantial, as the indications furnished to the surgeon by the introduction
of the finger or sound, in the complaints wherein
these are used.

In prosecuting my enquiries I made trial of instruments of various composition and construction.—
The general result has been that bodies of a moderate density, such as paper, wood, or indian cane,
are best suited for the conveyance of the sound, and
consequently for my purpose. This result is 'perhaps contrary to a law of physics;—it has, nevertheless, appeared to me one which is invariable.

I shall now describe the instrument which I use
at present, and which has appeared to me preferable

to all others. It consists simply of a cylinder of wood, perforated in its centre longitudinally, by a bore three lines in diameter, and formed so as to come apart in the middle, for the benefit of being more easily carried. One extremity of the cylinder is hollowed out into the form of a funnel to the depth of an inch and half, which cavity can be obliterated at pleasure by a piece of wood so constructed as to fit it exactly, with the exception of the central bore which is continued through it, so as to render the instrument in all cases, a pervious tube. The complete instrument,—that is, with the funnel-shaped plug infixed,—is used in exploring the signs obtained through the medium of the voice and the action of the heart; the other modification, or with the stopper removed, is for examining the sounds communicated by respiration. (See Plate VIII.) This instrument I commonly designate simply the *Cylinder*, sometimes the *Stethoscope*.

In speaking of the different modes of exploration I shall notice the particular positions of the patient, and also of the physician, most favourable to correct observation. At present I shall only observe that, on all occasions, the cylinder should be held in the manner of a pen, and that the hand of the observer should be placed very close to the body of the patient to insure the correct application of the instrument.

The end of the instrument which is applied to the patient,—that, namely, which contains the stopper or plug,—ought to be slightly concave to insure its greater stability in application; and when there is much emaciation, it is sometimes necessary to insert between the ribs a piece of lint or cotton, or a leaf of paper, on which the instrument is to be placed, as, otherwise, the results might be affected by the imperfect application of the cylinder. The same precaution is necessary in the examination of the circulation in cases where the sternum, at its lower extremity, is drawn backwards, as frequently happens with shoemakers, and some other artisans.

Some of the indications afforded by the stethoscope, or *mediate auscultation,* are very easily acquired, so that it is sufficient to have heard them once to recognise them ever after: such are those which denote ulcers in the lungs, hypertrophia of the heart when existing in a great degree, fistulous communication between the bronchia and cavity of the pleura, &c. There are others, however, which require much study and practice for their effectual acquisition

The employment of this new method must not make us forget that of Avenbrugger; on the contrary, the latter acquires quite a fresh degree of value through the simultaneous employment of the former, and becomes applicable in many cases, wherein its solitary employment is either useless or hurtful. It is by this combination of the two methods that we obtain certain indications of emphysema of the lungs, pneumo-thorax, and of the existence of liquid extravasations in the cavity of the pleura. The same remark may be extended to some other means, of more partial application, such, for example, as the *Hippocratic succussion,* the *mensuration* of the thorax, and *immediate* auscultation; all of which methods, often useless in themselves, become of great value when combined with the results procured through the medium of the stethoscope.

18

Memoir on the Theory of Sound

SIMEON DENIS POISSON

Siméon Denis Poisson (1781–1840), French mathematician, whose name will forever be associated with the well-known equation in potential theory, was interested in the application of mathematics to almost every branch of physics. As a relatively young man he turned his attention to the theory of sound and showed how to solve the wave equation under conditions that had baffled earlier workers. The following extract constitutes the introduction to his memoir of 1807 and indicates his appreciation of the outstanding problems of acoustics as well as his ideas as to appropriate methods for solving them. His discussion of the question of the velocity of sound in air is of particular interest.

Translated by R. Bruce Lindsay from
Journal de l'Ecole Polytechnique, **7**, 319 (1808)

We have known for a long time about the vibrations of air set in motion by a sounding body. But M. Lagrange is the first who has subjected these motions to mathematical analysis, and who has deduced the principal results of the theory of sound. His beautiful researches on this subject are known so well to all mathematicians that it would be useless to repeat them here. At the time when they were published, the calculus of partial differentials, on which the solution of questions of this kind depends, was scarcely known. We were not in agreement on the matter of the use of discontinuous functions, which however is indispensable to employ to represent the state of the air at the beginning of its motion. Fortunately progress in analysis has caused these difficulties to disappear. We are far from thinking that we have surmounted all difficulties. Our aim in this memoir is to demonstrate several theorems which in our opinion ought to be of equal interest to physicists and mathematicians.

These theorems are independent of the particular movements of the air molecules and of the cause which has produced the sound. They relate to the propagation and reflection of sound, which we have considered from a more general point of view than has hitherto been the case. In discussing air in three dimensions, Lagrange and Euler have assumed that the velocity of the air molecules depends only on the distance from the origin of the motion, or in other words that the intensity of the sound is the same at all points of the sound wave, where by sound wave they understand the part of the air in motion at each instant. [Editor's note: The author here seems to refer to what later came to be called the *wave front.*] In this particular case the equation which includes the theory of sound can be integrated in closed form, and the problem has a solution which leaves nothing to be desired. But in the general case in which the intensity of the sound varies in an arbitrary manner from one point to the next on the wave front, the equation for sound propagation is not integrable in closed form. However, we shall demonstrate by simple means and without recourse to series expansion that in the general case the velocity of sound is still the same along the different sound rays, which amounts to saying that the wave front always preserves a spherical form with center at the point of original disturbance. The law in accordance with which the sound intensity varies from point to point on the same wave front depends on the cause which produces the sound. If it has been produced by compressing or dilating a portion of the mass of air, then reestablishing at once communication between this portion of air and the entire mass, it is easy to prove that the sound intensity will be the same at all points of the wave front, at any rate when the wave has traveled a long way from the point of original disturbance. For, let us decompose the portion of air that is compressed or dilated into an infinite number of infinitely small parts. Each of these particles if it alone were compressed or dilated would produce a wave of which this particle would be the center and which would spread itself equally in every direction. But in accordance with the principle of Daniel Bernoulli on the coexistence of small oscillations, all these waves can exist at the same time without interfering with each other in any way. Besides, it is clear that when the rays of all the waves have become very great in comparison with the dimensions of the portion of air which has been compressed or dilated, all the waves will have combined into a single one. From this one can conclude that whatever the shape of the portion of air which has been compressed may be or whatever may be the law of density distribution in the interior of this portion of air, the intensity of sound will be the same at all points of the wave front. But when the portion of air which is compressed or dilated receives at the same time velocities which can be different in size and direction for all the molecules composing it, one can no longer say how a disturbance of this kind will spread through the mass of fluid. Thus, for example, the portion of air at the open end of an instrument like a speaking trumpet is at the same time compressed and set in motion by the vibrations of the interior air. Experience proves that the intensity of the sound is greatest in front of the open end of the trumpet along the direction of the axis; but what the ratio is between this intensity and that in another direction at the same distance from the source of the disturbance, is a problem we have not yet solved. It is not indeed the same with the intensity of sound on the same sound ray; that is to say, on any given straight line passing through the center of the original disturbance. Whatever this disturbance may be, the velocities of the air molecules situated on the same ray are in inverse ratio to the distances from the center of motion, when such distances are very large with respect to the dimensions of the portion of air initially disturbed. If one then considers the sound due to the impact of a fluid against a fixed obstacle, in this case the organ of hearing, and if we suppose this impact is proportional to the square of the fluid velocity, the result will be that the sound intensity along a given sound ray should decrease proportionally to the square of the distance from the sounding body, at least when this distance has become sufficiently great. The angle which the direction of the velocity of the molecule of air makes with the sound ray which passes through this molecule continually decreases in proportion with the length of the ray, so that eventually this angle may be regarded as zero. But, it is natural for us to think that it is the direction of the velocity of the molecules of air which enables us to determine the direction of the sound, since it is the magnitude of the velocity of the molecules which determines the sound intensity. This then explains why the direction of the sound indicates to us the place where it originated.

We have obtained these results relative to the direction and to the attenuation of the velocity of the air molecules in expanding the integral of the sound equation into a series of decreasing powers of the distance from the center of the disturbance. After having found this series, we have transformed it into a definite integral, which led us to the discovery of a very extensive class of particular integrals in finite form, satisfying the sound equation. Each of these integrals corresponds to a particular hypothesis as to the nature of the original disturbance. This has the advantage of showing how this disturbance, produced in a very small portion of air can expand into the entire mass of air in such a way that the velocity of the molecules will not be the same in every direction. One can verify for each of these that in spite of this difference in molecular velocity, the velocity of sound is always the same along all sound rays, as we have demonstrated in general.

When the mass of air is terminated by a surface whose position is rigidly fixed, experience indicates that the sound on striking the surface is reflected by it. Lagrange has considered this reflection of sound in an infinitely narrow cylindrical channel. But no one yet has determined the reflection in a satisfactory manner when all three dimensions of the air are considered and when the reflecting surface is struck at all possible angles by the sound rays. When this surface is a plane of unlimited extent we shall show that the reflection of sound takes place like that of light on a plane mirror; that is to say the reflection is the same in direction and intensity as if the original disturbance were in back of the plane at a distance equal to that of the actual disturbance in front of the plane and in such a way that the straight line joining the two centers is perpendicular to the plane. It results from this that each particle of air is set in motion at two different times. Each time it describes a short straight line, the first time on the direct ray and the second time on the reflected ray. These two motions are separated by an interval of time which depends on the distance from the molecule to the fixed plane. When the molecule is very close to the plane the two motions coincide in part. One commences about at the end of the other. Consequently this molecule describes a small curve, the nature of which depends on that of the initial disturbance. We can suppose the disturbance such that the curves of which we are speaking have several points of inflection. It is the case in which the molecules of air make several comings and goings on the direct and reflected rays. But whatever these curves may be, the molecules which are not in contact with the plane are able to approach indefinitely close to it, but never quite to reach it and the lines which they describe will always be terminated by two straight lines, the one directed along the direct ray and the other along the reflected ray. On the other hand, the molecules which are in contact with the plane never leave it and their motion always takes place in the plane itself.

In the theory of the emission of light it suffices to know how the light is reflected from a plane in order to obtain the reflection from any arbitrary surface, because one can consider each light ray in isolation from the rest and substitute for the reflecting surface its tangent plane at the point where the direct ray is incident. It is thus only a simple matter of geometry to determine the intensity of the reflected light at any given point, or, what amounts to the same thing, the number of reflected rays which pass through this point. But when one wishes to know how the waves produced in an elastic fluid or in an incompressible fluid are reflected by a surface which bounds the fluid, it is necessary to consider the reflection of the entire wave and to determine the shape and intensity of the reflected wave. It is this which makes the problem difficult and the problem can be solved only for each individual reflecting surface by itself.

If the sound originates from one of the foci of an ellipsoid of revolution and is to be reflected from its surface, we demonstrate rigorously that the reflected sound forms a wave whose center is at the other focus of the ellipsoid.

The direct and reflected sound rays which come together at the same point of the reflecting surface make equal angles with the normal at this point. Thus one can say that the angle of incidence is equal to the angle of reflection, just as in the reflection of light, though the latter does not have the same cause as the former. [Editor's note: Poisson was here evidently adhering to the corpuscular theory of light.] Furthermore, we shall deduce the intensity of the reflected sound in terms of that of the incident sound. Calculation shows that the ratio of these intensities for two sound rays which coincide at the same point on the surface is the same as that which would be observed in the case of light under similar circumstances. On a given sound ray the intensity of the reflected sound increases as one approaches the second focus of the ellipsoid in such a way that for points close to the focus the intensity of the reflected sound is much greater than that of the direct sound. This result is confirmed by experience. For we know that if one speaks in a low voice at one focus of an elliptical vault, the voice can be heard distinctly at the other focus, whereas it disappears at every other point. As for the velocity of the reflected sound, it is the same as that of the incident sound. It follows from this that sound arrives at a given point by a broken line in the same time that the direct sound would take to traverse a straight line equal in length to the broken line.

That which is demonstrated in an ellipsoid of revolution with arbitrary distance between the foci, holds equally for a paraboloid, if we assume that the distance in question becomes infinite. The sound produced at the focus of a paraboloid of revolution will then be reflected by the surface of the paraboloid parallel to its axis and conversely.

We have also considered the sound originating at one of the foci of a hyperboloid of revolution and reflected by either the concave or convex surface. This reflection is found to possess properties analogous to those found in the reflection from an ellipsoidal surface. Finally, if one reduces the air to two dimensions, one finds that the sound produced at one focus of any conic section is reflected to the other focus.

There exists an appreciable difference between the velocity of sound, calculated by theory, and that found in experience. All physicists who have measured this velocity have found a value greater than that calculated. The members of the Academy of Sciences have found that sound in air travels 337 meters per second. But if one uses the given values of the elasticity and density of air, resulting from the most recent measurements, one finds that the velocity of sound, calculated according to theory is equal only to 282 meters per second, a quantity too small by a sixth part.

We shall not recall here all the hypotheses which have been advanced to produce agreement between theory and observation. A single one of these alone merits the attention of physicists.

Lagrange has already remarked (Memoirs of the Turin Academy, volume 2) that we could deduce by analysis the observed value of the velocity of sound by assuming that in the compression of air, its elasticity increases more rapidly than its density. But he himself observes that this hypothesis is not really tenable, since Mariotte and all those who have repeated his experiments found that the density of air increases in the same proportion as the weight which compresses it as long as the temperature remains constant. The reconciliation of this law of Mariotte [Editor's note: what is called Mariotte's law on the continent of Europe is usually referred to as Boyle's law in English speaking countries] with the increase in elasticity demanded by the velocity of sound, has been carried out by Laplace, who takes account of the heat accompanying the compression of the air. It is now well known, indeed, that when one compresses a given volume of any gas one releases a quantity of heat sufficient to ignite a combustible body but which dissipates itself rapidly in the form of radiation or otherwise. But with density held constant the elasticity of the air increases or diminishes with the increase or decrease in temperature. The law that density is proportional to pressure supposes then that one has left the compressed air time enough to lose the increase in heat produced by the compression and that at the very beginning the elasticity ought to increase faster than the density. We are lacking direct experiments to determine the quantity of heat released in the compression of air. Besides, in the production of sound each layer of air is compressed between two other layers of air, and it could well happen that in compression of this kind the heat released would not be the same as that released when we compress a volume of air in a closed vessel. It is therefore impossible to determine *a priori* the increase in the elasticity due to the development of heat which accompanies the propagation of sound. In any case this increase is nonetheless incontestable and one is permitted to doubt only what influence it has on the velocity of sound. Is the difference between calculation and observation owing to this cause alone? It seems to us that this conclusion will be justified if one can find no other objection to the analytical method by which the velocity of sound is calculated, and if at the same time one pays attention in the calculation to all the physical circumstances which could have an influence on the result. By comparing then the calculated and observed values of the velocity of sound, one can determine the quantity of heat released in the production of the sound and used to increase the elasticity of the air. The result of the comparison of the two velocities will be found in the course of this memoir.

In the calculation leading to the deduction of the velocity of sound, we consider the velocity of the air molecules as very small and neglect the square and higher powers of this velocity. Euler thought that the difference between the theoretical value of the sound velocity and the observed value was due to this circumstance. According to him the velocity of the air molecules producing the sound is not small enough so that one is justified in neglecting its square without making an error, especially when the sound intensity is very great, as is the case when one determines the velocity by observation. To overcome this difficulty we have considered the propagation of sound along a straight line of air and we have not neglected any term in the equation of motion. Then this equation is no longer linear in form. It is no longer integrable in closed form. But it admits a particular integral which includes an arbitrary function and which suffices to determine the motion of the air when we assume that the velocity of the molecules, while not very small, is only smaller than that of sound. With the use of this integral one can rigorously demonstrate that the velocity of sound is independent of the velocity of the air molecules and hence of the cause which has produced the sound. Sound, strong or weak, is then propagated with the same velocity, in agreement with experience.

Everything we have just said presupposes that the temperature and density remain constant throughout the mass of air. But since the sound velocity is equal to the square root of the ratio of the elasticity of the air to its density (as long as one neglects the correction due to the development of heat) it follows that as long as this ratio does not change, neither will the velocity of sound. From this we are able to conclude that in supposing all the layers of the atmosphere to be at the same temperature, the velocity of sound should be the same as if the density of air did not vary in passing from one layer to the next, for in passing across the layers, the elasticity varies in the same ratio as the density, whatever the law of gravity may be. It will not be the same if the temperature varies at the same time as the density. In this case the sound will no longer be transmitted by a uniform motion and furthermore its velocity will not be the same for all sound rays. In this case the sound wave front will not have a spherical form, as in the case of constant temperature. Suppose, for example, that the temperature decreases in proportion to the vertical height above the surface of the earth, the naturally prevailing situation, as is shown by the theory of light refraction compared with experience (see Book 12 of Laplace's *Mecanique Celeste*). In this case if we imagine a sound ray starting from a point higher in the atmosphere and hitting the surface of the earth, the temperature and consequently the ratio of the elasticity to the density of the air, will increase along the ray in proportion to its length multiplied by the cosine of the angle which it makes with the vertical. We conclude from this that the motion of sound along each ray will be of the same nature as that of a heavy body which falls sliding along the ray as along an inclined plane and which leaves the origin of the ray with some given velocity. Accordingly the velocity of sound will be the greater the less it deviates from the vertical, and this velocity along a given ray will accelerate in proportion to the time that has elapsed from the origin of the motion.

The partial differential equation which is the expression for the theory of sound changes its form when one takes into account the weight of the air and the variation in temperature. In order to deduce directly the sound propagation velocity, it is necessary to use the integral of this equation, expressed by means of a definite integral. In this way one is led to the expression for the velocity which has been indicated in the preceding considerations. This same integral also furnishes a means for determining the intensity of the sound produced at different heights in the atmosphere. When the temperature is assumed to be constant, we arrive at this remarkable result: the intensity of sound depends only on the distance it has traveled. [Editor's note: presumably inversely, though the author does not say so] and on the density of the layer of atmosphere where it started. Hence this intensity is the same, in every case, as if the atmosphere were homogeneous and with density equal to that of this layer. It follows from this that people who go up in a balloon ought to hear a noise whose origin is at the surface of the earth as well as if they were resting on the surface. At the same time the noise that they produce in an elevated layer of the atmosphere is also heard on the surface of the earth as well as it would be in the same layer at a distance equal to the distances to the earth. One can readily assure oneself that a variation in temperature from one layer to another of the atmosphere will not appreciably alter this result, which indeed appears to agree with experience.

In general it is easy to obtain definite integrals which satisfy a given partial differential equation, but these integrals are often only a simple game of analysis, which teach us nothing as to the nature of the function we wish to determine and hardly advance the solution of the problem proposed for solution. Those of which Laplace has shown the utility in the solution of partial differential equations of the second order in three variables (Memoirs of the Academy for 1779) are not in the same class. On the contrary it is integrals of this form which have given us the velocity and intensity of sound when the density and the temperature of the air are variable. Laplace has already used them to determine the velocity of sound in the case in which one dimension of the air is cut out and where one supposes that the intensity depends only on the distance to the center of the original disturbance. Many other questions, like the transmission of motion in a heavy chain, the propagation of sound in a tube whose diameter is not constant, the vibrations of strings of nonuniform thickness, all may be solved with the use of definite integrals. To give an example, we have solved at the end of this memoir the problem of the heavy chain, homogeneous and of equal thickness throughout.

[Editor's note: The above constitutes the introduction to the Memoir of 1807. It provides in clear fashion the program the author has set for himself. The remainder of the Memoir (some 60 pages) is devoted to the appropriate mathematical analysis to support the author's claims in the introduction. Thus Poisson derives the wave equation for sound waves in a fluid by introducing the velocity potential and the condensation, writing the equation of motion in Newtonian fashion as well as the equation of continuity for fluids. Elimination of the condensation between these two equations yields the familiar wave equation. All this is essentially what the modern theoretical acoustical scientist does, and it is of interest to realize how old the method was even when Rayleigh repeated it in his famous *Theory of Sound* in 1877. One can only marvel at the mathematical genius of Poisson as he goes on to secure special solutions for different kinds of wave fronts and also tackles the problem of finite amplitude waves. Poisson was much exercised over the difficult problem posed by the velocity of sound. Though he was clearly intrigued by the suggestion of Laplace to resolve the difficulty he was still some nine years away from Laplace's final breakthrough in terms of the specific heats of gases.]

19

Velocity of Sound in Air and Water

PIERRE SIMON LAPLACE

Pierre Simon Laplace (1749–1827), the French mathematician of the late 18th century, is principally known for his contributions to celestial mechanics. But he also devoted attention to physics, with special reference to hydrodynamics, capillarity, and heat. His one important contribution to acoustics came as a result of his ingenious assumption with respect to the nature of sound disturbances in fluids, showing that it is more plausible to treat them as adiabatic rather than isothermal. On this assumption he was able to correct Newton's derivation of the expression for the velocity of sound. Though much criticized when it was proposed, his theory was finally accepted and became the basis for the experimental evaluation of the ratio of the two specific heats of gases. It is only in very recent times that Laplace's theory has been shown to be inadequate for very high frequency sound passing through a gas at very low pressure.

Laplace's paper is presented here in a translation by the editor of this volume.

Translated by R. Bruce Lindsay from
Annales de Chimie et de Physique, III (1816)

In the second book of his *Principia Mathematica* Newton has given the expression for the velocity of sound. The manner in which he obtained it is one of the most remarkable exhibitions of his genius. The sound velocity in air resulting from this expression is smaller by about one-sixth than that resulting from experimental measurements made with great care by the members of the French Academy in 1738. Newton, already aware of the discrepancy through the measurements made in his own time, tried to explain it. But the modern discoveries on the nature of atmospheric air have destroyed this explanation and all others which various mathematicians have proposed. Fortunately these discoveries present us with a phenomenon which appears to me to be the true cause of the excess of the observed velocity of sound in air over the calculated value. This phenomenon is the heat which the air develops through its compression. When its temperature is raised, the pressure remaining constant, only a part of the caloric which it receives is employed in producing this rise; the rest, which becomes latent, serves to increase its volume. It is this heat which is released when by compression one reduces the expanded air to its original volume. The heat disengaged by the close approach of two neighboring molecules of a vibrating aerial fiber then increases their temperature, and this diffuses itself little by little through the air to the surrounding bodies. But since this diffusion takes place very slowly relative to the velocity of the vibrations, we may suppose without sensible error that during the period of a single vibration the quantity of heat remains the same between two neighboring molecules. Thus these molecules in approaching each other repel each other the more, in the first place because, since their temperature is supposed to stay constant, their mutual repulsion increases in the inverse ratio of their separation, and in the second place because the latent heat which is developed raises their temperature. Newton took account of only the first of these two causes of repulsion, but it is clear that the second cause should increase the velocity of sound since it increases the elasticity of the air. In including this in the calculation I arrive at the following theorem:

The real velocity of sound is equal to the product of the velocity given by the

Newtonian formula by the square root of the ratio of the specific heat of air at constant pressure and varying temperatures to the specific heat at constant volume. [Editor's note: See the *Collected Works of Laplace*, Vol. 5, pp. 109, 134, 157.]

If we assume, as several physicists have done, that the heat contained in a mass of air submitted to constant pressure and at different temperatures is proportional to its volume (a result which cannot be far from the truth) the square root just mentioned becomes the square root of the ratio of the difference between two pressures to the difference in the quantities of heat developed by two equal volumes of atmospheric air subjected to these pressures respectively in dropping from a given temperature to the same lower temperature, it being understood that the smaller of the heat quantities and the smaller of the pressures are taken as unity.

Being desirous of comparing the above theorem with experiment, I have fortunately found the necessary observational results among the numerous results of the interesting work of LaRoche and Berard on the specific heats of gases. [Editor's note: See *The Collected Works of Laplace*, vol. 5, p. 143.] These able physicists have measured the quantities of heat which two equal volumes of atmospheric air give up in a drop in temperature of about 80°, the one being compressed by the weight of the atmosphere while the other is compressed by the same weight increased by 0.36. They have found that the heat released under the greater pressure was 1.24 times the heat released under the smaller pressure, the latter heat being taken as unity. According to the theorem mentioned above it is therefore necessary in order to find the real velocity of sound to multiply the velocity found by Newton by the square root of the ratio of 0.36 to 0.24 or by the square root of 1.5. At the temperature of 6° Newton's formula gives 282.42 meters/sec. Multiplication by $\sqrt{1.5}$ yields 345.9 meters/sec. The observations of the members of the French Academy give 337.18 meters/sec. The difference between these two results can be attributed to the uncertainty in experimental measurements. But the smallness of the difference establishes incontestably that the excess of the observed velocity of sound over that calculated from the Newtonian formula is due to the latent heat developed by the compression of the air.

We conclude from the above that if at constant pressure one expands a given volume of air by increasing its temperature and then reduces it by compression to its original volume, it will give up through this compression a third of the quantity needed by the original expansion. It is desirable that physicists should determine by direct measurement the ratio of the specific heat of air at constant pressure to that at constant volume, a ratio which we have found above to be 1.5. The velocity of sound determined by the members of the French Academy gives 1.4254 for this quantity. In the light of the difficulty associated with the direct measurement of specific heats, the velocity of sound is the more precise way of obtaining the ratio.

I have calculated [Editor's note: See *The Collected Works of Laplace*, vol. 14, p. 293.] the velocity of sound in rain water and in sea water as, respectively, 2642.8 meters/sec and 2807.4 meters/sec. These figures are based on the measurements of Cauton on the compressibility of these liquids, paying attention only to the linear decrease in the compressed volume. I recognize now that it is necessary to consider the total decrease in volume. Hence the numbers given above should be divided by $\sqrt{3}$. They are then reduced to 1525.8 meters/sec and 1620.9 meters/sec, respectively. This means that the velocity of sound in fresh water is about 4.5 times the velocity in air.

20

New Experiments on Sound

CHARLES WHEATSTONE

Charles Wheatstone (1802–1875), British physicist, is perhaps best known to students of physics for his connection with the so-called Wheatstone bridge for the measurement of electrical resistance. However, early in his career he became interested in acoustics, probably because of his work in the making of musical instruments. As professor of experimental philosophy in King's College, London, he turned his attention to a wide variety of problems in sound, light, and electricity. He is usually credited with the introduction of the term *microphone* to denote a sensitive receiver of sound. The following extract is taken from an account of some of Wheatstone's early experiments in sound, in which he studied the vibrations of surfaces in somewhat different fashion from the method employed by Chladni and foreshadowed Faraday's later experiments (see the article by Faraday in this volume).

ANNALS

OF

PHILOSOPHY.

20

AUGUST, 1823.

ARTICLE I.

New Experiments on Sound. By Mr. C. Wheatstone.

(To the Editor of the *Annals of Philosophy.*)

On the Phonic Molecular Vibrations.

SIR,

BEFORE I enter on the immediate subject of this article, it may be necessary to exhibit a general view of those bodies, which, being properly excited, make those sensible oscillations, which have been thought to be the proximate causes of all the phenomena of sound. These bodies, to avoid many circumlocutions otherwise inevitable, I have termed Phonics.

Linear Phonics.

Transversal,	*Longitudinal,*
Making their oscillations at right angles to their axis.	Making their oscillations in the direction of their axis.
1. Capable of tension, or variable rigidity: chords, or wires.	1. Columns of aeriform fluids or liquids: cylindric and prismatic rods.
2. Permanently rigid: rods, forks, rings, &c.	

Superficial Phonics.

1. Capable of tension: extended membranes.
2. Permanently rigid: laminæ, bells, vases, &c.

Solid Phonics.

1. Volumes of aeriform fluids.

The sensation of sound can be excited by any of these bodies when they oscillate sufficiently rapidly, either entire, or divided into any number of parts in equilibrium with each other. The laws of these subdivisions differ in the various phonics according to their form and mode of connection or insulation; and the velocities of the oscillations, or degrees of tune, depend on the form, dimensions, mode of connection, mode of division, and elasticity of the body employed. The points of division in linear phonics are called nodes, and the boundaries of the vibrating parts of elastic surfaces are termed nodal lines. The parts at which the oscillatory portions have their greatest excursions are named centres of vibration; these are always at the greatest mean distances from the nodal points or lines.

These mechanical oscillations are not, however, themselves the immediate causes of sound; they are but the agents in producing in the bodies themselves, and in other contiguous substances, isochronous vibrations of certain particles varying in magnitude according to the degree of tune. I convinced myself of this important fact by the following simple experiments: I took a plate of glass capable of vibrating in several different modes, and covered it with a layer of water; on causing it to vibrate by the action of a bow, a beautiful reticulated surface of vibrating particles commenced at the centres of the vibrating parts, and increased in dimensions as the excursions were made larger. When a more acute sound was produced, the centres consequently became more numerous, and the number of coexisting vibrating particles likewise increased, but their magnitudes proportionably diminished. The sounds of elastic laminæ are generally supposed to be owing to the entire oscillations of the simple parts as shown by Chladni, when, by strewing sand over the sonorous plates, he observed the particles repulsed by the vibrating parts, accumulate on the nodal lines, and indicate the bounds of the sensible oscillations. Did no other motions exist in the plate but these entire oscillations, the water laid on its surface would, on account of its cohesion to the glass, show no peculiar phenomena, but the appearances above described clearly demonstrate that the oscillating parts consist of a number of vibrating particles of equal magnitudes, the excursions of which are greatest at the centres of vibration, and gradually become less as they recede further from it, until they become almost null at the nodal lines.

To multiply these surfaces, and to observe whether the magnitudes of these particles vary in different media, in a glass vessel of a cylindric form, I superposed three immiscible fluids of different densities; namely, mercury, water, and oil. On producing the sounds corresponding with each mode of division, I observed a number of vibrating parts, agreeing with the sound, and showing similar appearances to the plate, formed on the surfaces of each of the fluids; not the least agitation appeared in the uniform parts. I afterwards inserted this glass in another vessel of water in order to observe the vibrations of the external surface, and found the same results as in the interior, though the levels of the surfaces were different.

The most accurate method to observe these phenomena is by employing a metallic plate of small dimensions, which must be fixed horizontally in a vice at one end, and covered on its upper side with a surface of water : on causing it to oscillate entirely by means of a bow, a regular succession of these vibrating corpuscles will appear arranged parallel to the two directions of the plate, and if the action of the bow be rendered continuous, their absolute number might be counted with the aid of a micrometer. Diminishing the oscillating part of the plate to one half of its length, the double octave to the preceding was heard, agreeably to the established rule, that the velocities of the oscillations are inversely as the squares of the lengths ; four vibrating corpuscles then occupied the space before occupied by one, and the absolute number was double to that in the former instance ; but the absolute number of these corpuscles have no influence whatever on the degree of tune, which entirely depends on their relative magnitude in the same substance ; theory shows us that in plates of this description alteration of breadth does not affect the degree of tune ; let us, therefore, reduce this half of the plate to half its breadth, and we shall find the note remain the same, but the absolute number of the corpuscles will in this case be equal to that in the entire plate. Let us now take two plates of equal lengths and breadths, but one double in thickness to the other; the rule is, that the velocities of the oscillations are as the thicknesses of the plates ; we shall, therefore, in the thicker plate see a double number of particles to that of the other, occupying the same extent of surface. The last circumstance in which two plates may differ is their specific rigidity, and in this respect it will be found that two plates of exactly equal dimensions, and covered with the same number of vibrating corpuscles of equal magnitudes, but of different substances, differ in sound ; therefore, the absolute magnitudes of the particles cannot be assumed as a standard of tune, unless regulated by the specific rigidity.

Unassisted by any means of actual admeasurement, the above are but the proximate results sensible to the eye; more extended and accurate experiments are necessary to confirm the results with mathematical certainty. As the absolute magnitudes of these particles will, I imagine, be hereafter a most useful element for calculation, I will here indicate the most effectual way I am acquainted with to arrive at this knowledge. A thick metallic slip of considerable length and breadth, bent similarly to a tuning fork, and fixed at its curved part in a vice, is very easily excited by friction, and a more considerable surface of regularly arranged vibrating particles is seen than in most other superficies; any description of common exciter may be employed,

G 2

When this bent plate is excited by percussion, the particles, before their disappearance, will assume an apparent rotatory motion, on account of the force exerted, and its susceptibility of continuing the vibrations. Employing a parallelopedal rod, the appearances of the higher modes of subdivisions are particularly neat; the entire vibrating parts between the nodes form ellipses, and the semi-part at the free end, a regular half of the same figure. It is important to remark, that the crispations of the water only appear on the sides in the plane of oscillation; the other two sides, on one of which the exciter must be applied, do not show similar appearances.

I have also rendered the phonic molecular vibrations visible, when produced by the longitudinal oscillations of a column of air; the following were the means employed: I placed the open end of the head of a flute or flagiolet on the surface of a vessel of water, and on blowing to produce the sound, I observed similar crispations to those described above, forming a circle round the end of the tube, and afterwards appearing to radiate in right lines; on the harmonics of the tube being sounded, the crispations were correspondently diminished in magnitude. These phenomena will be more evident if the tube be raised a little from the surface of the liquid and a thin connecting film be left surrounding it; the vibrating particles will then occupy a greater space, and be more sensible.

The existence of the molecular vibrations being now completely established, it becomes a critical question, in what manner the sensible oscillations induce these vibrating particles. I do not know whether what I am now going to adduce will be admitted as the right explanation, but it is certainly analogous, so far as the superficial and transversal linear oscillations are concerned. A flexible surface, covered with a coat of resinous varnish, being made to assume any curve, the cohesion of the varnish will be destroyed in certain parts, and a number of cracks will be observed more regularly disposed as the force inducing the curve has been more regularly applied; when the original position of the surface is restored, the cracks will be imperceptible, but will again appear at every subsequent motion. Be this as it may, these particles are invariable concomitants of the sensible oscillations, and there is no reason to suppose otherwise than that their vibrations are isochronous with them. To avoid confusion, I have restricted the word vibrations to the motions of the more minute parts, and the term oscillations to those of the sensible divisions. We may reasonably suppose that the molecular vibrations pervade the entire substance of a phonic; their excursions, however, are not the same in all parts, and they can only be rendered visible, when these excursions are large; they may be so few in number as to be entirely inaudible, as in their transmission through linear conductors; but however few, when they are properly directed, they induce the mechanical divisions of sonorous bodies, each of which will give birth to numerous vibrating corpuscles whose excursions are greater, and the sound will be rendered audible. Dr. Savart has well investigated the modes of division in surfaces put in motion by communicated vibrations. All those phonics whose limited superficies preclude them from exciting in themselves a sufficient number of vibrating corpuscles, when insolated, produce scarcely any perceptible sound, as extended chords, tuning forks, &c. but those whose superficies or solidities are more extended, as bells, elastic laminæ, columns of air, &c. produce sufficient volume of sound without accessory means.

Loudness of sound is dependent on the excursions of the vibrations; volume, or fulness of sound, on the number of co-existing particles put in motion. Thus the tones of the Æolian harp, on account of the number of subdivisions of the strings, are remarkably beautiful and rich, without possessing much power; and the sounds of an Harmonica glass, in which a greater number of particles are excited than by any other means, are extraordinarily so united, according to the method of excitation, with considerable intensity; their pervading nature is one of the greatest peculiarities of these sounds.

The following is a recapitulation of the various properties of sound, which are attributable to modifications of the vibrating corpuscles :

The tune — velocities of the vibrations.
The time — continuance of the vibrations.
The intensity — excursions of the vibrations.
The richness, or volume — number of co-existing vibrations.
The quantity (timbre) — magnitudes of the vibrating corpuscles.

(Depends on the)

It has often been thought necessary to admit the existence of more minute motions than the sensible oscillations, in order to account for many phenomena in the production of sound. Perrault in his " Essai du Bruit," insisted on their necessity more than any other author I have read : he imagined, that the vibrations have a much greater velocity than the oscillations which cause them, but the experiment he adduced to prove this is far from conclusive ; he mistook for these vibrations the oscillations of the subdivisions of the long string he employed. Other distinguished philosophers have had ideas of a similar nature, and Chladni thinks their existence necessary to account for the varieties of quality. I, however, conceived I was the first who had indicated these phenomena by experiment, until a few days ago repeating them, together with the others which form the subject of this paper, in the presence of Prof. Oersted, of Copenhagen, he acquainted me with some similar experiments of his own. Substituting a very fine powder, Lycopodion, instead of the sand used by Chladni, for showing the oscillations of elastic plates, this eminent philosopher found the particles not only repulsed to the nodal lines, but at the same time accumulated in small parcels, on and near the centres of vibration; these appearances he presumed to indicate more minute vibrations, which were the causes of the quality of the sound: subsequently he confirmed his opinion, by observing the crispations of water, or alcohol, on similar plates, and showed that the same minute vibrations must take place in the transmitting medium, as they were equally produced in a surface of water, when the sounding plate was dipped into a mass of this fluid. These experiments were inserted in Lieber's History of Natural Philosophy, 1813.

Rectilineal Transmission of Sound.

As the laws of the communication of the phonic vibrations are more evident in linear conductors, I shall confine the present article to a summary of their principal phenomena.

In my first experiments on this subject, I placed a tuning fork, or a chord extended on a bow, on the extremity of a glass, or metallic rod, five feet in length, communicating with a sounding board; the sound was heard as instantaneously as when the fork was in immediate contact, and it immediately ceased when the rod was removed from the sounding board, or the fork from the rod. From this it is evident that the vibrations, inaudible in their transmission, being multiplied by meeting with a sonorous body, become very sensibly heard. Pursuing my investigations on this subject, I have discovered means for transmitting, through rods of much greater lengths and of very inconsiderable thicknesses, the sounds of all musical instruments dependant on the vibrations of solid bodies, and of many descriptions of wind instruments. It is astonishing how all the varieties of tune, quality, and audibility, and all the combinations of harmony, are thus transmitted unimpaired, and again rendered audible by communication with an appropriate receiver. One of the practical applications of this discovery has been exhibited in London for about two years under the appellation of " The Enchanted Lyre." So perfect was the illusion in this instance from the intense vibratory state of the reciprocating instrument, and from the interception of the sounds of the distant exciting one, that it was universally imagined to be one of the highest efforts of ingenuity in musical mechanism. The details of the extensive modifications of which this invention is susceptible, I shall reserve for a future communication; the external appearance and effects of the individual application above-mentioned have been described in the principal periodical journals.

The transmission of the vibrations through any communicating medium as well as through linear conductors is attended by peculiar phenomena; pulses are formed similar to those in longitudinal phonics, and consequently the centres of vibration and the nodes are reproduced periodically at equal distances; in this we observe an analogous disposition with regard to light. I had intended to include in this paper all the analogical facts I have observed illustratory of the identity of the causes of these two principal objects of sensation, but want of time, and the danger of delay, now the subject is occupying so much the attention of the scientific world, has induced me hastily to collect the present experiments, and to defer the others for a future opportunity.

The thicknesses of conductors materially influence the power of transmission, and there is a limit of thickness, differing for the different degrees of tune, beyond which the vibrations will not be transmitted. The vibrations of acute sounds can be transmitted through smaller wires than those of grave sounds: a proof of this is easy; attach a tuning fork to one end of a very small wire, and apply the other end to the ear, or a sounding board; on striking the fork rather hard, two co-existing sounds will be produced, that which is more acute will be distinctly heard, but the other will not be transmitted. If the vibrations of a tuning fork be conducted through a piece of brass wire of the size and thickness of a large needle, the sound, imperfectly transmitted, will become more audible by the pressure of the fingers on the conducting wire; but if a steel wire of the same length and thickness be employed, the sound will be unaltered by any pressure, because steel has a greater specific elasticity than brass.

Polarization of Sound.

Hitherto I have only considered the vibrations in their rectilineal transmission; I shall now demonstrate, that they are peculiarly affected, when they pass through conductors bent in different angles. I connected a tuning fork with one extremity of a straight conducting rod, the other end of which communicated with a sounding board; on causing the tuning fork to sound, the vibrations were powerfully transmitted, as might be expected from what has already been explained; but on gradually bending the rod, the sound progressively decreased, and was scarcely perceptible when the angle became a right one; as the angle was made more acute, the phenomena were produced in an inverted order; the intensity gradually increased as it had before diminished, and when the two parts were nearly parallel, it became as powerful as in the rectilineal transmission. By multiplying the right angles in a rod, the transmission of the vibrations may be completely stopped.

To produce these phenomena, however, it is necessary that the axis of the oscillations of the tuning fork should be perpendicular to the plane of the moveable angle, for if they be parallel with it, they will be still considerably transmitted. The following experiment will prove this: I placed a tuning fork perpendicularly on the side of a rectilinear rod; the vibrations were, therefore, communicated at right angles; when the axis of the oscillations of the fork coincided with the rod, the intensity of the transmitted vibrations was at its maximum; in proportion as the axis deviated from parallelium, the intensity of the transmitted vibrations diminished; and, lastly, when it became perpendicular, the intensity was at its minimum. In the second quadrant, the order of the phenomena was inverted as in the former experiment, and a second maximum of intensity took place when the axis of the oscillations had described a semicircumference, and had again become parallel, but in an opposite direction. When the revolution was continued, the intensity of the transmitted vibrations was varied in a similar manner, it progressively diminished as the axis of the oscillations deviated from being parallel with the rod, became the least possible when it arrived at the perpendicular, and again augmented until it remained at its first maximum, which completed its entire revolution.

The phenomena of polarization may be observed in many corded instruments: the cords of the harp are attached at one extremity to a conductor which has the same direction as the sounding board; if any cord be altered from its quiescent position, so that its axis of oscillation shall be parallel with the bridge, or conductor, its tone will be full; but if the oscillations be excited so that their axis shall be at right angles with the conductor, its tone will be feeble. By tuning two adjacent strings of the harp-unisons with each other, the differences of force will be sensible to the eye in the oscillations of the reciprocating string according to the direction in which the other is excited.

It now remains to explain the nature of the vibrations which produce the phenomena, the existence of which has been proved by the preceding experiments. The vibrations generally assume the same direction as the oscillations which induce them; in a longitudinal phonic the vibrations are parallel to its axis; in a transversal phonic, they are perpendicular to this direction; a circular or an elliptic form can be also given to the vibrations by causing the oscillations to assume the same forms. Any vibrating corpuscle can induce isochronous vibrations of similar contiguous corpuscles *in the same plane* either parallel with, or perpendicular to, the direction of the original vibrations, and the polarization of the vibrations consists in the similarity of their directions, by which they propagate themselves equally in the same plane; therefore the vibrations being transmitted through linear conductors, it is the plane in which the vibrations are made that determines their transmission, or non-transmission, when the direction is altered. A longitudinal or a transversal vibration may be transmitted two ways to a conductor bent at right angles; their axis may be in that direction, as to be in the same plane with the right angle, in which case the former will be transversally, or the latter longitudinally transmitted in the new direction; or their axis may be perpendicular to the plane of this new direction, under which circumstances neither can be communicated.* In explaining the polarization of light, there is no necessity to suppose that the reflecting surfaces act on the luminous vibrations by any actual attracting or repulsing force, causing them to change their axes of vibrations; the directions of the vibrations in different planes, as I have proved exist in the communication of sound, is sufficient to explain every phenomenon relative to the polarization of light.

Let us suppose a number of tuning forks oscillating in different planes, and communicating with one conducting rod; if the rod be rectilinear, all the vibrations will be transmitted, but if it be bent at right angles, they will undergo only a partial transmission; those vibrations whose planes are perpendicular, or nearly so, to the plane of the new direction, will be destroyed. The vibrations are thus completely polarized in one direction, while passing through the new path, and on meeting with a new right angle, they will be transmitted or not, accordingly as the plane of the angle is parallel with, or perpendicular to, the axes of the vibrations. In this point of view, the circumstances attending the phenomena are precisely the same as in the elementary experiment of Malus on the polarization of light.

Double refraction is a consequence of the laws of polarization, by which a combination of vibrations having their axes in different planes, after travelling in the same direction, are separated into two other directions, each polarized in one plane only. That this well-known property of light has a correspondent in the communication of phonic vibrations, I shall now demonstrate. When two tuning forks, sounding different notes by a constant exciter, and making their oscillations perpendicularly to each other, have their vibrations transmitted at the same time through one rod, at the opposite extremity of which two other conductors are attached at right angles, and when each of these conductors is parallel with one of the axes of the oscillations of the forks, on connecting a sounding board with either conductor, those vibrations only will be transmitted through it which are polarized in the same plane with the angle made by the two rods through which the vibrations pass; either sound may be thus

---

* I have just seen a paper by M. Fresnel, entitled " Considerations Mécaniques sur la Polarization de la Lumiere," in which this eminent philosopher had previously arrived at the same conclusions with respect to light, as I have proved in this communication respecting sound. The important discoveries of Dr. Thomas Young, followed by those of M. Fresnel, have recently re-established the vibratory theory of light, and new facts are every day augmenting its probability. The new views in acoustical science, which I have opened in this paper, will, I presume, give additional confirmation to the opinions of these eminent philosophers; and I hope, when I resume the subject, to be enabled to account for the principal phenomena of coloration, with regard to their acoustic analogies, in a way calculated to establish the permanent validity of the theory.

separately heard, or they may both be heard in combination by connecting both the conductors with sounding boards.

The phenomena of diffraction regarding only the form of the surfaces, or the superficies over which the vibrations extend, are by the conformation of the organs of hearing, not of any consequence to the perception of sound, though the same phenomena when the chromatic vibrations are concerned, are very evident to the eye. They, however, undoubtedly take place equally in both instances, and may be well explained by the theory already laid down. Each separate vibration propagating itself in the plane of its vibrating axis, a number of vibrations in different planes, after passing through an aperture, naturally expand themselves transversely as well as rectilineally, and thereby occupy a greater space than they would, were they only longitudinally transmitted.

I have still to indicate a new property of the phonic vibrations, but whether it is analogous to any of the observed phenomena of light, I am yet ignorant. When the source of the vibrations is in progressive motion, the vibrations emanating from it are transmitted, when the conductor is rectilineal and parallel with the original direction, and they are destroyed when the conductor is perpendicular to the direction, though the axis of vibration and the conductor, being in both instances *in the same place,* would transmit the vibrations were the phonic stationary. These circumstances are proved by the following experiments: When a tuning fork placed perpendicularly to a rod, communicating at one or both extremities with sounding boards, and caused to oscillate with its vibrating axis parallel with the rod, moves along the rod, preserving at the same time its perpendicularity and parallelism, the vibrations will not be transmitted while the movement continues, but the transmission will take place immediately after it has remained motionless. When the tuning fork moves on the upper edge of a plane perpendicular to a sounding board, the vibrations rectilineally transmitted will not be influenced by the progressive motion.

21

Experiments on the Velocity
of Sound in Water

JEAN-DANIEL COLLADON

Jean-Daniel Colladon (1802–1893) was a Swiss physicist and engineer who had close relations with French scientists and engineers and taught both in Paris and Geneva. During most of his professional career he was concerned with a wide variety of problems in power engineering. But as a young scientist in Paris he became interested in the measurement of the compressibility of liquids, particularly water. He was aware of the theoretical connection between the compressibility and the velocity of sound. The relatively low compressibility of water meant a correspondingly large sound velocity. Since the velocity of sound in fresh water had never been measured (this was in 1826) Colladon decided to measure it in Lake Geneva. The result appeared in a joint memoir with the French mathematician Charles Sturm in *Annales de Chimie et Physique*, Paris 1827. (See the reference in "The Story of Acoustics" in this volume for more details.) But the whole story of Colladon's experiences in his research on Lake Geneva is delightfully told in the following extract from his autobiography.

Translated by R. Bruce Lindsay from
"Souvenirs et Memoires—Autobiographie de
Jean-Daniel Colladon" (Geneva) 1893

I then traveled by myself with the principal purpose of occupying myself with the measurement of the velocity of sound in water, following which I would bring back my instruments to Paris and repeat the measurements there. This was at the end of September [1826]. I went at first to spend eight days at Avully, where my father, mother, and sister were living and had not seen me for ten months. It was a great pleasure to see them again.

I was looking for an inn which was appropriately situated for my measurements, when M. de Candolle offered to put me up at his country seat "La Perriere" situated two kilometers from Geneva on the right bank of the lake. He offered me the greatest hospitality and said: "You will have the assistance of my son Aphonse and my gardener, whom I place at your disposal; besides that you will have two boats and a small dock where you may keep them."

I selected a bell, which while suspended in water could be struck by a clapper. The arsenal still possessed the old bell (with chains) which had been dismantled and weighed 65 kilograms. It was put at my disposal and I was able to keep it at M. de Candolle's residence. For the day of our first experiment M. de Candolle invited some friends to help me and I made my first measurement at a distance of about 1000 meters. While the gardener and M. de Candolle were located in one boat with the bell, I was in another boat to listen to the sound, and I had a watch which had been lent to me by M. Tavan and which indicated quarter seconds. At the distance mentioned I was able to hear the striking of the bell, having for this purpose plunged my head into the water at the moment of the signal accompanying the striking. M. de Candolle, who held the watch, started the second hand at the instant the clapper hit the bell and stopped it when I gave him a hand signal that the sound had arrived at my ear. There was a light breeze on that particular day, rocking the boat, and I got rather wet. On returning we had tea, our friends left us, and M. de Candolle and I went to bed.

I could not go to sleep right away, for I was trying to think of a way of hearing sound without immersing my head in the water. I then thought that a metallic vase, closed at its base and immersed by means of a weight might perhaps serve to transmit the sound from the water to the air in the vase and that one could then hear it outside. To try this I was able to take a watering pot ballasted so as to sink in the water. I was so impatient to see what result I would obtain that at 5 o'clock in the morning I aroused Alphonse de Candolle and told him of my idea. We dressed in a hurry; we asked the gardener to remain at the dock with the bell, while with de Candolle, I went across the lake, which in this part is about 1500 meters wide. Having arrived at our destination, I immersed the watering pot and gave the signal to strike the bell. Without putting my head in the water I immediately heard in the watering pot the sound from the bell. This produced in me immense pleasure, for henceforth the experimental measurement was much easier. I could now see the signal given by the strike at the bell on the first beat and myself read on a watch the time for the arrival of the sound.

The same day I commissioned the construction of an apparatus composed of a long tube terminated at the end by a kind of spoon closed by a flat plate with an area of about 20 square decimeters. The apparatus was ballasted in such a way that from 60 to 76 centimeters of the tube rested below the water line. When my apparatus was finished I went with M. de Candolle and made some measurements along the lake shore, gradually increasing the distance between the two boats to 6 kilometers. Since these experiments were eminently successful I proposed to repeat them in the body of the lake and I chose the distance from Rolle to Thonon, which is the longest dimension of the lake.

The average depth of the lake between these two points is about 150 meters and the banks are accessible throughout. The distance from Rolle to the bell at Thonon is about 14,237 meters.

I requested my father to station himself on the Rolle side and to use rockets and fireworks for the signals. As the sphericity of the earth prevented one from seeing from Thonon what went on on the other side of the lake, I told my father to take a boat with a mast at least 12 meters high and to install at its top a lamp with an opaque cover which he could raise at the moment of striking the bell.

Thonon at that time was the site of the Savoy custom house and I forsaw that I would have great difficulties to overcome when I arrived there. I knew M. de Magny, who was the Sardinian consul at Geneva. I told him of my discovery and my desire to verify it in the place where the lake is widest. M. de Magny equipped me with letters for the custom house people.

I left with an aide, carrying with me the rockets for the various signals and a stop watch measuring quarters of a second. When we arrived at the customs barrier, we got out in order to permit the inspection of the carriage. I gave the officer in charge the letter of M. de Magny. I waited. In a moment I was asked to proceed to the office, where I saw the local official who was holding the letter of M. de Magny open in his hand. He wanted to know what the rockets I was bringing in were made of. I replied that they were made of powder. He then told me it was impossible to let me take them in, for there was an absolute prohibition on the importation of powder. I replied that it would be impossible for me to carry out my experiments, since the purpose of the rockets was to provide the signal that the bell was being struck.

The man was torn between the desire to let me pass with the rockets and the fear of losing his job if he were to let me go through. He repeated over and over again, "The powder cannot pass." And he showed me at the same time his orders. "You say these rockets are made with powder; consequently I am not able to let you pass under any consideration without encountering the risk of losing my job." I then understood his objection and said "Yes, they are made with powder, but the work of constructing them destroyed the powder and now one can no longer use them as ordinary powder." The man was evidently satisfied by this explanation, and said: "Ah, if that is so, I can then let you pass with them," and he gave me back my rockets.

Having arrived at Thonon, I discharged the carriage and dispatched a man to hire a boat while I took a hurried dinner and measured off a 200 meter cord. Then I got into the boat and took up my station 200 meters from the shore. I then fired a rocket to indicate that the bell should be struck. I instructed them to strike three times in succession and that only the first would be counted. I heard the three strokes perfectly distinctly, but I was not able to see any light signal at Rolle. The experiment was a failure because of this, but I acquired thereby the assurance that the sound would reach me at this distance.

The distance between the bell at Thonon and the tower at Rolle is 14,237 meters. The Thonon bell is 350 meters inland from the shore. The distance between the two banks is then 13,887 meters. In taking out another 400 meters to allow for the distance of the two boats from the shore, we find 13,487 meters as the distance between the two boats.

At this distance the sound is just as clear and sharp as it was at 100 meters. The sound made by a key striking a hard object gives a sufficiently accurate idea of what it is like.

I put off operation to a later time because I had been unable to see the light signals. My father was not able to find a boat with a mast 12 meters in height and had settled for one with a mast only 7 or 8 meters high.

I then changed the striking signal. Near the place where the bell was suspended I placed a metallic plate on which I put a quarter or a half a pound of powder. I assured myself that this powder when burning would produce a light several degrees above the horizon.

With this system I was able to perform various experiments, the three most exact of which were carried out on the 7th, the 15th, and the 18th of November, 1826. The time which the sound took to traverse the distance of 13,487 meters was 9¼ seconds. This yields for the real velocity of sound in water at 8°C, the value of 1437 meters per second.

The formula for the velocity of sound is

$$a = \sqrt{\frac{PK}{D\epsilon}}$$

in which D is the density, in this case equal to 1 [Editor's note: Colladon means here specific gravity]; P is the pressure of one atmosphere, that is, 0.76 meters of mercury, $K = 1$ million, and ϵ is the quantity by which water contracts, i.e., 48. If we insert these values and carry out the calculation we arrive at

$$a = 1437.8 \text{ meters/sec}$$

[Editor's note: One must be careful not to substitute Colladon's numbers as he gives them without regard to proper dimensions. In modern terminology PK/ϵ is the adiabatic bulk modulus of water in dynes/cm², while D is the density in grams/cm³. Colladon's value for this modulus is a bit short of $10^{12}/48$ since he chooses the normal atmosphere as somewhat less than 10^6 dynes/cm². The accepted modern value for the adiabatic compressibility of water is 2.06×10^{10} dynes/cm² at 8°C, leading to the value 1438.8 meters/sec. as the velocity of sound in water at that temperature. Colladon's value is in rather good agreement with this, considering the nature of his method.]

This is the theoretical value of the velocity of sound in water, if we assume that no heat is evolved during the rapid and successive compressions of the liquid molecules in the transmission of the sound. Our experimental result for the velocity is

$$a = 1435 \text{ meters/sec.}$$

This confirms that no heat is evolved in the compressions.

[Editor's note: Colladon uses effectively the isothermal bulk modulus in calculating the velocity. Actually the adiabatic bulk modulus of water is so close to the isothermal that he could hardly expect the difference to show up in his experimental result. Hence his conclusion that "no heat is evolved" etc. does not strictly follow. In any case there is an inconsistency between the two experimental results he quotes.]

We know, however, that this situation is not the same for sound in air. Newton derived the formula for the velocity of sound in air. When this velocity was measured it was found to be greater than the calculated value, the excess amounting to a sixth of the observed value.

M. Laplace has succeeded in providing the explanation of this difference, by attributing it to the increase in the elasticity of the air molecules due to heating. Taking account of this M. Poisson has showed that if the compression and dilation are 1/4460, the temperature should rise or fall by 1/100 of a degree and M. Laplace has found that the velocity of sound is equal to that given by Newton's formula multiplied by the square root of the ratio of the specific heat at constant pressure to that at constant volume. The formula of Newton so corrected then agrees very closely with the experimental value of the velocity of sound. [Editor's note: There is something missing in Colladon's treatment here if taken literally. The translator has added what is necessary to make it correct. At this place in his narrative, Colladon introduces some values of compressibilities of liquids presumably made in the laboratory.]

I now take up again the measurements of the velocity of sound in water.

In the final experiments the station at Rolle, occupied by my father, was arranged as follows. A boat carried a beam, which projected out over the water. The bell

Bateau expéditeur du son.

Figure 1

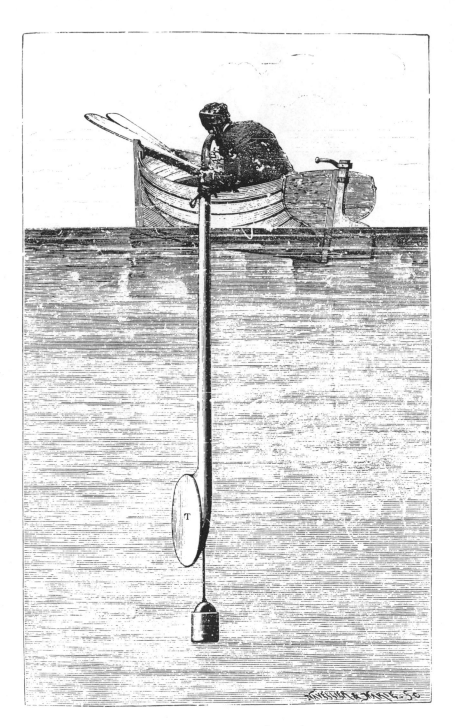

Bateau récepteur du son.

Figure 2

was suspended about two meters below the water surface. A lever bent at right angles dipped into the water near the bell and served as a striker [see Figure 1]. The other and shorter arm of the lever was manipulated by the hand. A small metallic plate was placed near the center of suspension of the bell. On this was placed the powder which a fuse ignited at the moment of striking the bell.

I had my station at Thonon, my ear attached to the extremity of an acoustic tube [see Figure 2]. The boat was oriented so that my face was turned in the direction of Rolle. I was thus able to see the light accompanying the striking of the bell and to hold the watch which served to measure the time taken by the sound to reach me.

Comments on Sound in Water

I have already said that at all distances the sounds were very sharp, like two knife blades or two keys striking each other. This sharpness of the underwater sound makes it easy to identify, in the same way as in air the noise produced by a whistle can be distinguished in a storm.

It should be remarked that in underwater sound transmission the influence of baffles is more marked than in the case of air. In experiments which I made with M. A. de Candolle, near his father's estate, operating on one side of the lake there was a wall which intervened 12 or 15 meters between the two stations and when the line joining the stations encountered the wall there was a marked decrease in the intensity of the signal.

But the most important point to stress is that concerning the enormous elasticity of the water transmitting the sound. It takes only a second to strike the bell. The work done in this action by a single man can hardly exceed 12 kilogram meters. But these 12 kilogram meters suffice to excite a mass of water of the order of 50 billion kilograms. From this we deduce the extreme tenuity of the layers of water that transmit the sound.

One can cite no more striking example of the enormous elasticity of water and moreover of the conservation of *vis viva*.

22

On the Sensitivity of the Ear

FELIX SAVART

Felix Savart (1791–1841), French physicist, is best known for the enunciation of the Biot and Savart law in electrodynamics. He began his professional career in medicine, but soon shifted to physics and became a professor of this subject in Paris. He early became interested in hearing and endeavored to establish with precision the frequency limits of audibility of the human ear. This necessitated the invention of some means of producing pure sounds whose frequency could be varied at will and precisely measured. Savart succeeded in doing this by means of his famous toothed wheel. In the days before electroacoustics, this marked a distinct advance. His 1830 paper is presented here in its entirety in English translation.

Translated by R. Bruce Lindsay from
Poggendorf's *Annalen der Physik und Chemie*,
20, 290 (1830)

Several outstanding physicists have sought to determine to what height and depth in pitch sounds are still perceptible to the ear of man. There is general agreement that the lower limit corresponds to a tone with frequency about 30 simple oscillations per second. [Editor's note: What Savart calls a simple or single oscillation is one half a complete cycle in modern terminology. Thus his 30 simple oscillations per second are equivalent to 15 cycles per second or 15 Hz in modern notation.] This problem, as we shall see a bit further on, cannot be considered as completely solved, yet we are permitted to believe that we are not too far from the truth. As far as the upper limit is concerned, with whose determination I shall concern myself in this article, physicists are by no means in accord. Chladni assumes that tones corresponding to about 12,000 simple oscillations per second are still audible. Biot puts the upper limit at the sound emitted by an open pipe of length 18 lines to which he attributes the frequency 8192 simple oscillations per second. Wollaston maintains that he could never hear tones higher than that produced by a pipe of length one-quarter of an inch. But since he does not say whether the pipe was open or closed nor what its diameter was, we have no way of knowing what frequency was associated with it. In another place this celebrated physicist says that the highest tones which are perceptible have frequencies from six to seven hundred times as great as the lowest audible tones. If we assume that the latter correspond to about 30 simple vibrations per second, it follows that according to Wollaston the upper audible limit lies between 18,000 and 21,000 simple vibrations per second. In a word, if one relies on the textbooks of physics which have appeared to date, one is permitted to conclude that no really exact investigation of this matter has so far been made, and that in this respect acoustics has made no real progress since the time of Sauveur.

The requirements for the solution of this problem reduce to two. The first obviously consists in the precise determination of the freqency of vibration of the body which serves as the source of sound. The second consists in the production of sounds which though unusually high in pitch are yet strong enough to be audible. One might hope to satisfy these conditions by the longitudinal vibrations of cylindrical rods free at both ends. For since on the one hand, the laws governing motions of this kind are well known it would always be easy to determine precisely the number of vibrations of a rod of arbitrary length. If, on the other hand, we use in the research bodies like glass or steel, in which the velocity of sound is large, one might believe it possible to select rods of sufficient length so that they might be readily set in vibration, and hence one would be able to secure from them sounds of sufficient intensity.

With this in mind I made several investigations along this line and found that most persons could still hear distinctly the sound emitted by a glass cylinder with diameter 3 millimeters and length 159 millimeters, though these tones had a frequency of about 31,000 single vibrations per second. When I took rods of smaller diameter and tried to see what would happen if I shortened them further, I found that if the rod had a length of about 150 millimeters, so that the corresponding frequency was about 33,000 single vibrations per second, I sometime heard the sound and then again did not, as if my ear were more sensitive at one moment than at another, which certainly might be possible, or perhaps I did not always set the rod in vibration with the same degree of success.

I endeavored to obtain the same results by using transverse vibrations of smaller rods. Since the rod had a different and smaller mass and was harder to excite in vibration, the sounds had a lower intensity. However, even here I observed that I could hear sounds in the frequency range from 30,000 to 32,000 single vibrations per second. Since, however, these small rods had one end fastened in a vise and hence their lengths could not be precisely determined, these values can be considered only as an approximation, even if somewhere near the truth.

It is even more difficult to estimate the frequency when one employs small air columns, like small organ pipes.

The only way to overcome this difficulty would be to apply pipes of similar form by making use of the laws, in accordance with which for air masses of this kind the frequencies vary as the homologous dimensions. Since, however, the mouth openings are also proportional to these dimensions we see that in the light of the smallness of the pipes this condition is only very partially fulfilled. Moreover, it has not been possible for me to get by this method with any assurance frequencies of over 20,000 single vibrations per second, since above this figure while the sounds are audible they are comparable only with great difficulty.

From these first investigations the conclusion therefore appears to be that the human ear can hear no sounds of frequency above about 32,000 single vibrations per second. If, however, one considers that in order to reach this goal one has to use bodies of very insignificant dimensions and therefore oscillations of unusually small amplitudes, the question naturally arises, whether the human ear has actually reached its limit at this point, or whether even here the sound is imperceptible because of lack of intensity.

In order to overcome this difficulty, one must seek to produce the sounds by a method such that one can at will increase the amplitude of the oscillations or more generally increase its intensity as compared with the surrounding noise, while at the same time determine the frequency with ease and great precision. It seemed to me that one could readily attain this end by means of a more or less rapidly rotating wheel equipped around its periphery with an appropriate number of teeth which can strike one after another on some rigidly fastened thin body like a card or a wedge-shaped sheet of light wood. One is naturally led to believe that in this case tones will be produced whose frequencies, like those of the siren of Cagniard Latour, depend on the number of impacts per unit time. Since with this arrangement one can increase the intensity of the impacts at will by keeping the number of teeth constant while increasing the diameter of the wheel, it is clear that in this way by employing suitably arranged wheels one can produce the highest tones of the musical scale without decrease in intensity.

My first investigations were made with a wheel of brass, 24 centimeters in diameter and with 360 teeth on its periphery. The tones produced went up and down in pitch as the rotational velocity became greater or smaller. Though this first apparatus carried no numbers, it was easy to ascertain with the help of a chronometer that the tones obtained had frequencies proportional to the speeds of rotation of the wheel. If, for example, the speed doubled, the tone went up by an octave. If one compared the tone produced by the wheel with that produced by a monochord it was found that the number of impacts of the teeth per second was just as great as the number of double vibrations per second of the string. [Editor's note: A double vibration per second is the same as the modern cycle per second or Hertz (Hz).]

The tones emitted by this apparatus were very pure so long as the number of impacts per second did not exceed three or four thousand per second, corresponding to six or eight thousand single vibrations per second, since here as in the siren the impact of the tooth and the silence succeeding it must be considered as a double oscillation. Above this limit the sound became weak and lost significantly in its purity. It therefore became clear that in order to go further it would be necessary to take a wheel of larger diameter but without increasing the number of teeth, so that while maintaining the original rotational speed, the impacts would be separated more from each other because of the larger spaces between the teeth. Therefore in place of the first wheel I took a new one, also made of brass, but 48 centimeters in diameter and carrying 400 teeth on its periphery. In this way it is possible to produce very pure tones even when the rotational speed was increased to the point at which ten thousand impacts per second occurred. Above this point the sound lost its intensity and from 12,000 to 15,000 impacts per second it ceased to be perceptible.

Since the latter number still failed to exceed that which I had obtained from longitudinally vibrating rods, I constructed another apparatus with a toothed wheel 82 centimeters in diameter and with 720 teeth on its periphery. With this one could produce sounds corresponding to 24,000 impacts per second or a frequency of 48,000 single vibrations per second. Although the intensity of the sound was very great for 12,000 to 15,000 impacts per second, and thereafter began to decrease perceptibly, I cannot say at what point the sound became completely imperceptible, since the wheel I was using to set the toothed wheel in rotation was not large enough to enable me to increase the speed of rotation further.

It deserves to be remarked that I was not the only one that could hear tones of so high a pitch; they were heard by all persons who helped me in my research. It is therefore not correct to say, as Wollaston did, that the limits above which high notes are audible are different for different people. On the other hand the facts observed by this well-known physicist can be correct in all strictness, only they must be otherwise interpreted. A tone of high frequency with a definite intensity will be audible to some people and not to others, but this will be owing not to the level of its frequency, but to the level of its intensity.

From these investigations we seem able to draw the conclusions that if one continues to increase the diameter of the toothed wheels and the rotational speed without altering the number of teeth, tones can be produced which are still audible, though they correspond to much more than 24,000 impacts per second. For the moment, however, it was impossible for me to push the research further, since the machines which would have been necessary are very expensive.

The role which the increase in the diameter plays in these phenomena can be shown in a very simple demonstration. One needs only to take a disk and provide it with a sufficiently large number of diametral grooves and then produce a sound by directing impacts against these grooves. If one brings the striking body close to the center of the disk, one produces a very weak and impure tone. As one gradually comes closer to the circumference the sound increases in intensity, and when the striking body reaches the outer end of the groove the sound reaches its maximum of purity and intensity. The result is in good agreement with all that is known about the intensity of sound.

Since the number of impacts that I got with the 82-centimeter diameter wheel was already very great and significantly surpassed the limits previously established for audible sound, it was therefore necessary on the one hand to have a precise means for the measurement of the revolutions of the wheel, and on the other hand to make sure that the thin bodies against which the teeth strike do not themselves fall into a vibratory motion in consequence of which they might for example meet only every other tooth or every third one.

In order to remove the first difficulty one can attach a counter to the axis of the wheel. However, this means of which I originally made use did not permit easy application as soon as the rotation speed became great. It appeared to me preferable to determine the number of revolutions by means of the tone of a second toothed wheel, which had a smaller diameter and had a 30 or 40 times smaller number of teeth on its periphery than the larger wheel. Since the pitch of the tone of this second wheel was much lower one could easily bring it into unison with the monochord and determine the number of oscillations from which the number of revolutions of the larger wheel could be found.

With respect to the second difficulty, that is, the presumption that the thin body will periodically miss being hit by a certain number of teeth, there are two different ways of getting around it. The first consists in mounting several toothed wheels on the same axis all of the same diameter and thickness, but the number of whose teeth stand in simple relation to each other, behave like the corresponding frequencies of several tones on the scale, and thus arranged so that one may see whether with the rotational speed, the accord obtained is really that which one intended to produce. Since four wheels, carrying 200, 250, 300, and 400 teeth, respectively, and which therefore should be in complete accord, actually produced this accord, it therefore followed that no one of the thin bodies against which the teeth were striking, was missing in any way, especially since the cards were held firmly with the fingers, a circumstance which in itself would make vibrations of the cards almost impossible.

The second scheme, which is much simpler, consists in blowing a thin stream of air against the teeth, directing it perpendicular to the plane of the wheel. It is clear that in this way one produces an affect analogous to that of the siren of Mr. Cagniard Latour, that is, a tone must be produced when the stream of air is interrupted by the teeth. The tone that is produced must consequently be the same as that produced by the striking of the teeth against the card, provided no teeth are missed as the wheel goes around. Now the experiment shows that by simultaneous application of both schemes one gets the same tone in both cases. So the siren here provides a means of confirming my investigation and at the same time conversely my research provides a complete test of the accuracy of the theory which Mr. Cagniard Latour has developed for his ingenious instrument.

The question of the limit beyond which the high pitch tones become imperceptible appears to be related naturally with the determination of the more or less long time the periodic impacts of the teeth must repeat themselves before one gets the sensation of definite and comparable tones. Indeed there must exist a kind of dependence between the degree of sensitivity which allows the audibility of very high pitch tones and that which we must have in order to perceive tones which persist only for an unusually short time, for the high pitch tones which are comparable must be considered as the result of a succession of sounds which last only a very short time but which nevertheless individually make an impression on the ear.

The new method for sound production set forth here is particularly well suited for the investigation of the question just mentioned. Let us assume, for example, that we have a wheel equipped with a thousand teeth that rotates once every second. One makes a note of the tone and then removes the teeth from a half of the periphery. It is clear that the tone will not thereby be altered, since in one of the half seconds there will be precisely the same number of tooth impacts as before, only now after the tone there will come a silent period of half a second, if indeed the effect on the ear does not last longer than the activity of the cause which produces it. Actually this is what happens, namely, that after the removal of a substantial number of teeth, one gets an interrupted tone which however possesses the same pitch as resulted when all teeth were present.

The question then arises for investigation: how many teeth can be removed without having the tone lose its essential character? To answer this I prepared a wheel in such a fashion that one could readily remove all teeth and also restore them at will. I also fabricated certain other apparatus, which deserves no special description, but which was all intended to serve the purpose of the experiment. With this arrangement I found that no matter how fast I drove the wheel and no matter how large the number of teeth, one could remove all teeth down to the number two without having the tone lose its pitch and that, with a little care it was always possible to establish unison between the tone and that on a musical instrument.

From this it follows: (1) that two successive impacts are sufficient to produce a comparable tone and that accordingly four single oscillations per second give the same result; (2) that the time which passes between the two impacts conditions the degree of frequency of the tone. For example, those which succeed each other in twice as long a time interval, yield the lower octave, whereas if they follow in three times shorter time they produce the upper fifth of the octave, etc.; (3) that the duration that a tone must have in order to be heard depends only on the time interval between two periodic impacts producing this tone. Consequently this time interval is the shorter, the higher the pitch.

Since we have found that 20,000 single vibrations or 10,000 impacts per second give a tone which the ear can identify, it follows that this organ is able to perceive all the characteristics of a phenomenon which lasts only 1/5000 second. However, this conclusion, which initially seemed to be strictly correct, is not really so, since it could happen that the vibration of the tooth which was last struck might continue for a very short time after the cause of the vibration had ceased to act. However, one may assume that this time is very small indeed as long as the striking body, as in the present case, has very small dimensions, especially lengthwise.

If we permit only one tooth to remain on the periphery of the wheel, the single impact corresponding to each rotation of the wheel, still produces a tone which, however, as far as high or low pitch is concerned, has no relation to that which is produced when there are two or more teeth on the wheel. It is always the same, no matter what the speed of rotation may be. [Editorial note by Poggendorf: One would nevertheless object that if the speed of rotation were great enough so that the time interval between two impacts of the one tooth is equal to that between impacts of two teeth, in both cases the tone must be the same.] One can understand that it must be so, because it always originates from the sound of two bodies against each other and because in all cases these bodies have the same dimension. We must remark only that if the wheel makes more than 32 revolutions per second the periodic repetition of the impacts on the tooth produces a characteristic tone which is the higher the more considerable the number of revolutions per second.

A single impact produces in and for itself a sound or a perceptible noise, and since on the other hand the ear, as we have seen above, can hear tones corresponding to about 24,000 impacts per second, it follows that sound which lasts only 1/24,000 of a second, is perceptible, though it is however no longer identifiable. Here, as we shall see at once, resides the source of error which corresponds to the continuation of the motion after the impact. However, even if from this standpoint these results leave something to be desired, we can pretty well assume it as demonstrated that a sound or noise lasting for a very small fraction of a second can be perceived and its frequency estimated.

It must be remarked here that the duration of the phenomenon which produces the sensation of the tone must be carefully distinguished from the duration of the sensation itself. For we know that the influence on a sense organ lasts for some time after the cause of this influence has ceased to function. We know, for example, that when a glowing piece of carbon is moved in a circle, if this takes place with sufficient velocity, we see a fiery line of circular form. One is inclined to believe that the persistence of the sensation which produces this phenomenon in the case of the eye, will also make itself evident in the case of the ear. I have therefore sought to determine how long the hearing sensation persists after the cause has ceased. For this purpose the tooth wheels seem to offer an efficient means.

Suppose that a wheel rotates with definitely known uniform speed and one removes one of the teeth. It is clear that there will thereby take place an interruption in the tone, assuming that the sensation does not persist after the terminated effect of the cause producing it. If it continues for a more or less long time, we can measure it by means of the number of teeth which one must remove in order to make the interruption perceptible. On various occasions I have investigated this and have thereby determined that without any doubt the sensation persists for some time after the cause has ceased to work. However, up to now it has been impossible for me to secure precise determinations in this matter because the sensation is only gradually extinguished and because when it has become very weak, one can hardly say whether it is still there or whether it has vanished completely. In addition it seemed to me that the sensitivity of my ear was not always the same. For on several occasions it happened that in order to perceive the interruption I had to remove a larger number of teeth than was necessary several hours or days previously. I have also noticed that several individuals who collaborated with me in my researches almost always made a judgment different from mine for the same set-up.

It cannot be doubted that if a tone is to continue, the sensation due to a given impact must persist with a definite intensity long enough until the sensation from the following impact has taken place. If this were not so, one would only hear the noise of the individual impacts separately. Therefore, if one has a wheel with a very small number of teeth and starts it rotating with a small velocity, but then accelerates it more and more, in the beginning the impacts will be heard separately and there will be no continuous tone. Thereafter one will indeed perceive a tone. This however will appear if I may use the expression, chopped, and this is due to the fact that the end of the sensation which the ear receives at every impact begins to join itself with the following sensation. Finally the impacts follow each other with greater speed; the tone becomes very pure and intense. However, the intensity ultimately decreases and the tone disappears entirely, as soon as the rotational speed becomes very great, without doubt because the impacts are no longer pure enough.

In a word, it appears indispensable if we are to receive the sensation of a full and persistent tone that the impressions made on the ear must stand in a certain relation to each other. This is probably the reason why we must increase the diameter of the wheel in order to achieve the higher tones, because it is only in this way that one can change the duration of the impression produced by each impact. Conversely it does not appear doubtful that one would perceive lower tones than those corresponding to 30 to 32 single oscillations per second, if one could find a means of producing impacts whose impression would last longer than a sixteenth of a second. I close with a remark that the tones which can be produced by toothed wheels can be applied to advantage to determine the number of revolutions performed by the axes of many machines, as well as to assure oneself of the uniformity of their rotation. The application of this scheme is so simple that I consider it superfluous to provide further details about it.

23

Acoustic Streaming Over Vibrating Plates

MICHAEL FARADAY

Michael Faraday (1791–1867), considered by many to have been the greatest of all natural philosophers, spent his entire professional life at the Royal Institution of Great Britain. He is most famous for his research in electromagnetism and his discovery of electromagnetic induction. But he was interested in all aspects of natural phenomena and made it his life work to uncover the relations among their different manifestations. It is not generally recognized that Faraday at about the time when he was engaged on his epoch-making investigations on electromagnetism was also much concerned with sound. The following article describes some acoustical experiments he made around 1830 on the vibrations of surfaces. He was led to connect the resulting acoustical radiation with the propagation of electromagnetic effects, thus foreshadowing in a certain sense Maxwell's work on electromagnetic waves.

Reprinted from *Philosophical Transactions of the Royal Society*, 299–318 (1831)

XVII. *On a peculiar class of Acoustical Figures; and on certain Forms assumed by groups of particles upon vibrating elastic Surfaces.* By M. FARADAY, *F.R.S. M.R.I., Corr. Mem. Royal Acad. Sciences of Paris, &c. &c.*

Read May 12, 1831.

1. THE beautiful series of forms assumed by sand, filings, or other grains, when lying upon vibrating plates, discovered and developed by CHLADNI, are so striking as to be recalled to the minds of those who have seen them by the slightest reference. They indicate the quiescent parts of the plates, and visibly figure out what are called the nodal lines.

2. Afterwards M. CHLADNI observed that shavings from the hairs of the exciting violin bow did not proceed to the nodal lines, but were gathered together on those parts of the plate the most violently agitated, i. e. at the centres of oscillation. Thus when a square plate of glass held horizontally was nipped above and below at the centre, and made to vibrate by the application of a violin bow to the middle of one edge, so as to produce the lowest possible sound, sand sprinkled on the plate assumed the form of a diagonal cross; but the light shavings were gathered together at those parts towards the middle of the four portions where the vibrations were most powerful and the excursions of the plate greatest.

3. Many other substances exhibited the same appearance. Lycopodium, which was used as a light powder by OERSTED, produced the effect very well. These motions of lycopodium are entirely distinct from those of the same substance upon plates or rods in which longitudinal vibrations are excited.

4. In August 1827, M. SAVART read a paper to the Royal Academy of Sciences *, in which he deduced certain important conclusions respecting the subdivision of vibrating sonorous bodies from the forms thus assumed by light powders. The arrangement of the sand into lines in CHLADNI's experiments

* Annales de Chimie, xxxvi. p. 187.

shows a division of the sounding plate into parts, all of which vibrate isochro-nously, and produce the same tone. This is the principal mode of division. The fine powder which can rest at the places where the sand rests, and also accumulate at other places, traces a more complicated figure than the sand alone, but which is so connected with the first, that, as M. SAVART states, " the first being given, the other may be anticipated with certainty ; from which it results that every time a body emits sounds, not only is it the seat of many modes of division which are superposed, but amongst all these modes there are always two which are more distinctly established than all the rest. My object in this memoir is to put this fact beyond a doubt, and to study the laws to which they appear subject."

5. M. SAVART then proceeds to establish a secondary mode of division in circular, rectangular, triangular and other plates ; and in rods, rings, and membranes. This secondary mode is pointed out by the figures delineated by the lycopodium or other light powder ; and as far as I can perceive, its existence is assumed, or rather proved, exclusively from these forms. Hence much of the importance which I attach to the present paper. A secondary mode of division, so subordinate to the principal as to be always superposed by it, might have great influence in reasonings upon other points in the philosophy of vibrating plates ; to prove its existence therefore is an important matter. But its exist-ence being assumed and supported by such high authority as the name of SAVART, to prove its non-existence, supposing it without foundation, is of equal consequence.

6. The essential appearances, as far as I have observed them, are as follows. Let the plate before mentioned (2), which may be three or four inches square, be nipped and held in a horizontal position by a pair of pincers of the proper form, and terminated, at the part touching the glass, by two pieces of cork ; let lycopodium powder be sprinkled over the plate, and a violin bow be drawn downwards against the middle of one edge so as to produce a clear full tone. Immediately the powder on those four parts of the plate towards the four edges will be agitated, whilst that towards the two diagonal cross lines will remain nearly or quite at rest. On repeating the application of the bow several times, a little of the loose powder, especially that in small masses, will collect upon the diagonal lines, and thus, showing one of the figures which CHLADNI dis-

covered, will also show the principal mode of division of the plate. Most of the powder which remains upon the plate will, however, be collected in four parcels ; one placed near to each edge of the plate, and evidently towards the place of greatest agitation. Whilst the plate is vibrating (and consequently sounding) strongly, these parcels will each form a rather diffuse cloud, moving rapidly within itself; but as the vibration diminishes, these clouds will first contract considerably in bulk, and then settle down into four groups, each consisting of one, two, or more hemispherical parcels (53), which are in an extraordinary condition ; for the powder of each parcel continues to rise up at the centre and flow down on every side to the bottom, where it enters the mass to ascend at the centre again, until the plate has nearly ceased to vibrate. If the plate be made to vibrate strongly, these parcels are immediately broken up, being thrown into the air, and form clouds, which settle down as before ; but if the plate be made to vibrate in a smaller degree, by a more moderate application of the bow, the little hemispherical parcels are thrown into commotion without being sensibly separated from the plate, and often slowly travel towards the quiescent lines. When one or more of them have thus receded from the place over which the clouds are always formed, and a powerful application of the bow is made, sufficient to raise the clouds, it will be seen that these heaps rapidly diminish, the particles of which they are composed being swept away from them, and passing back in a current over the glass to the cloud under formation, which ultimately settles as before into the same four groups of heaps. These effects may be repeated any number of times, and it is evident that the four parts into which the plate may be considered as divided by the diagonal lines are repetitions of one effect.

7. The form of the little heaps, and the involved motion they acquire, are no part of the phenomena under consideration at present. They depend upon the adhesion of the particles to each other and to the plate, combined with the action of the air or surrounding medium, and will be resumed hereafter (53). The point in question is the manner in which fine particles do not merely remain at the centres of oscillation, or places of greatest agitation, but are actually driven towards them, and that with so much the more force as the vibrations are more powerful.

8. That the agitated substance should be in very fine powder, or very light, appears to be the only condition necessary for success ; fine scrapings from a

common quill, even when the eighth of an inch in length or more, will show the effect. Chemically pure and finely divided silica rivals lycopodium in the beauty of its arrangement at the vibrating parts of the plate, although the same substance in sand or heavy particles proceeds to the lines of rest. Peroxide of tin, red lead, vermilion, sulphate of baryta, and other heavy powders when highly attenuated, collect also at the vibrating parts. Hence it is evident that the nature of the powder has nothing to do with its collection at the centres of agitation, provided it be dry and fine.

9. The cause of these effects appeared to me, from the first, to exist in the medium within which the vibrating plate and powder were placed, and every experiment which I have made, together with all those in M. SAVART's paper, either strongly confirm, or agree with this view. When a plate is made to vibrate (2), currents (24) are established in the air lying upon the surface of the plate, which pass from the quiescent lines towards the centres or lines of vibration, that is, towards those parts of the plates where the excursions are greatest, and then proceeding outwards from the plate to a greater or smaller distance, return towards the quiescent lines. The rapidity of these currents, the distance to which they rise from the plate at the centre of oscillation, or any other part, the blending of the progressing and returning air, their power of carrying light or heavy particles, and with more or less rapidity or force, are dependent upon the intensity or force of the vibrations, the medium in which the vibrating plate is placed, the vicinity of the centre of vibration to the limit or edge of the plate, and other circumstances, which a simple experiment or two will immediately show must exert much influence on the phenomena.

10. So strong and powerful are these currents, that when the vibrations were energetic, the plate might be inclined 5°, 6°, or 8° to the horizon and yet the gathering clouds retain their places. As the vibrations diminished in force, the little heaps formed from the cloud descended the hill; but on strengthening the vibrations they melted away, the particles ascending the inclined plane on those sides proceeding upwards, and passing again to the cloud. This took place when neither sand nor filings could rest on the quiescent or nodal lines. Nothing could remain upon the plate except those particles which were so fine as to be governed by the currents, which (if they exist at all) it is evident would exist in whatever situation the plate was placed.

11. M. Savart seems to consider that the reason why the powder gathers together at the centres of oscillation is, " that the amplitude of the oscillations being very great, the middle of each of those centres (of vibration) is the only place where the plate remains nearly plane and horizontal, and where, consequently, the powder may reunite, whilst the surface being inclined to the right or left of this point, the parcels of powder cannot stop there." But the inclination thus purposely given to the plate, was very many times that which any part acquires by vibration in a horizontal position, and consequently proves that the horizontality of any part of the plate is not the cause of the powder collecting there, although it may be favourable to its remaining there when collected.

12. Guided by the idea of what ought to happen, supposing the cause now assigned were the true one, the following amongst many other experiments were made. A piece of card about an inch long and a quarter of an inch wide was fixed by a little soft cement on the face of the plate near one edge, the plate held as before at the middle, lycopodium or fine silica strewed upon it, and the bow applied at the middle of another edge; the powder immediately advanced close to the card, and the place of the cloud was much nearer to the edge than before. Fig. 1 represents the arrangement; the diagonal lines being those which sand would have formed, the line at the top *a* representing the place of the card, and the × to the right the place where the bow was applied. On applying a second piece of card as at *b*, the powder seemed indifferent to it or nearly so, and ultimately collected as in the first figure: *c* represents the place of the cloud when no card is present.

Fig. 1.

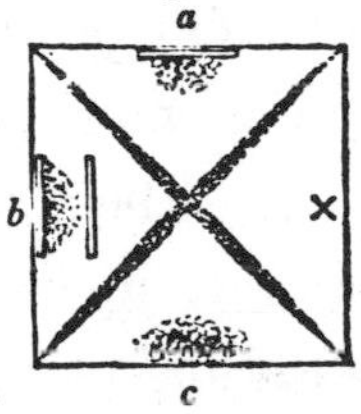

13. Pieces of card were then fixed on the glass in the three angular forms represented in fig. 2; upon vibrating the plate the fine powder always went into the angle, notwithstanding its difference of position in the three experiments, but perfectly in accordance with the idea of currents intercepted more or less by the card. When two pieces of card were fixed on the plate as in fig. 3. *a*, the powder proceeded into the angle but not to the edge of the glass, remaining about ⅛th of an inch from it; but on closing up that opening, as at *b*, the powder went quite up into the corner.

Fig. 2.

Fig. 3.

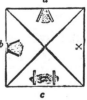

MDCCCXXXI.   2 R

14. Upon fixing two pieces of card on the plate as at *c* fig. 3, the powder between them collected in the middle very nearly as if no card had been present; but that on the outside of the cards gathered close up against them, being able to proceed so far in its way to the middle, but no further.

15. In all these experiments the sound was very little lowered, the form of the cross was not changed, and the light powders collected on the other three portions of the plate, exactly as if no card walls had been applied on the fourth; so that no reason appears for supposing that the mode in which the plate vibrated was altered, but the powders seem to have been carried forward by currents which could be opposed or directed at pleasure by the card stops.

16. A piece of gold-leaf being laid upon the plate, so that it did not overlap the edge, fig. 4, the current of air towards the centre of vibration was beautifully shown; for, by its force, the air crept in under the gold-leaf on all sides, and raised it up into the form of a blister; that part of the gold-leaf corresponding

Fig. 4.

to the centre of the locality of the cloud, when light powder was used, being frequently a sixteenth or twelfth of an inch from the glass. Lycopodium or other fine powder sprinkled round the edge of the gold-leaf, was carried in by the entering air, and accumulated underneath.

17. When silica was placed on the edge of another glass plate, or upon a book, or block of wood, and the edge of the vibrating plate brought as nearly as possible to the edge of the former, fig. 5, part of the silica was always driven on to the vibrating plate, and collected in the usual place; as if in the midst of all the agitation of the air in the neighbourhood of the two edges, there was still a current towards the centre of vibration, even from bodies not themselves vibrating.

Fig. 5.

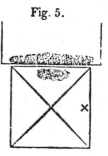

18. When a long glass plate is supported by bridges or strings at the two nodal lines represented in fig. 6, and made to vibrate, the lycopodium collects in three divisions; that between the nodal lines does not proceed at once into a line equidistant from the nodal lines and parallel to them, but advances from the

Fig. 6.

edges of the plate towards the middle by paths, which are a little curved and oblique to the edges where they occur near the nodal lines, but are almost perpendicular to it elsewhere, and the powder gradually forms a line along the middle of the plate; it is only by continuing the experiment for some time that it gathers up into a heap or cloud equidistant from the nodal lines. But upon fixing card walls upon this plate, as in fig. 7, the course of the powder within the cards was directly parallel to them and to the edge, instead of being perpendicular, and also directly towards the centre of oscillation.

Fig. 7.

To prove that it was not as a weight that the card acted, but as an obstacle to the currents of air formed, it was not moved from its place, but bent flat down outwards, and then the fine powder resumed the courses it took upon the plate when without the cards. Upon raising the cards the first effect was reproduced.

19. The lycopodium sprinkled over the extremities of such a plate proceeds towards places equidistant from the sides and near the ends, as at *a* fig. 8; but on cementing a piece of paper to the edge, so as to form a wall about one quarter or one third of an inch high, *b*, the powder immediately moved up to it, and

Fig. 8.

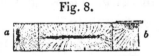

retained this new place. In a longer narrow plate, similarly arranged, the powder could be made to pass to either edge, or to the middle, according as paper interceptors to the currents of air were applied.

20. Plates of tin, four or five inches long, and from an inch to two inches wide, fixed firmly at one end in a horizontal position, and vibrated by applying the fingers, show the progress of the air and the light powders well. The vibrations are of comparatively enormous extent and the appearances are consequently more instructive.

21. If a tuning-fork be vibrated, then held horizontally with the broad surface of one leg uppermost, and a little lycopodium be sprinkled upon it, the collection of the powder in a cloud along the middle, and the formation of the involving heaps also in a line along the middle of the vibrating steel bar, may be beautifully observed. But if a piece of paper be attached by wax to the side of the limb, so as to form a fence projecting above it, as in the former experiments (19), then the powder will take up its place close to the paper; and if pieces of paper be attached on different parts of the same leg, the powder will go to the different sides, in the different parts, at the same time.

22. The effects under consideration are exceedingly well shown and illus-

2 R 2

trated by membranes.   A piece of parchment was stretched and tightly tied, whilst moist, over the aperture of a funnel five or six inches in diameter; a small hole was made in the middle, and a horse-hair passed through it, but with a knot at the extremity that it might thereby be retained.   Upon fixing the funnel in an upright position, and after applying a little powdered resin to the thumbs and fore-fingers, drawing them upward over the horse-hair, the membrane was thrown into vibration with more or less force at pleasure.   By supporting the funnel on a ring, passing the horse-hair in the opposite direction through the hole in the membrane, and drawing the fingers over it downwards, the direction in which the force was applied could be varied according to circumstances.

23. When lycopodium or light powders were sprinkled upon this surface, the rapidity with which they ran to the centre, the cloud formed there, the involving heaps, and many other circumstances, could be observed very advantageously.

24. The currents which I have considered as existing upon the surface of the plate, membranes, &c. from the quiescent parts towards the centres or lines of vibration (9), arise necessarily from the mechanical action of that surface upon the air.   As any particular part of the surface moves upwards in the course of its vibration, it propels the air and communicates a certain degree of force to it, perpendicular or nearly so to the vibrating surface; as it returns, in the course of its vibration, it recedes from the air so projected, and the latter consequently tends to return into the partial vacuum thus formed. But as of two neighbouring portions of air, that over the part of the plate nearest to the centre of oscillation has had more projectile force communicated to it than the other, because the part of the plate urging it was moving with greater velocity, and through a greater space, so it is in a more unfavourable condition for its immediate return, and the other, i. e. the portion next to it towards the quiescent line, presses into its place.   This effect is still further favoured, because the portion of air thus displaced is urged from similar causes at the same moment into the place left vacant by the air still nearer the centre of oscillation; so that each time the plate recedes from the air, an advance of the air immediately above it is made from the quiescent towards the vibrating parts of the plates.

25. It will be evident that this current is highly favourable for the transference of light powders towards the centre of vibration. Whilst the air is forced forward, the advance of the plate against the particles holds them tight ; but when the plate recedes, and the current exists, the particles are at that moment left unsupported except by the air, and are free to move with it.

26. The air which is thus thrown forward at and towards the centre of oscillation, must tend by the forces concerned to return towards the quiescent lines, forming a current in the opposite direction to the first, and blending more or less with it. I endeavoured, in various ways, to make the extent of this system of currents visible. In the experiment already referred to, where gold-leaf was placed over the centre of oscillation (16), the upward current at the most powerful part was able to raise the leaf about one tenth of an inch from the plate. The higher the sounds with the same plate or membrane, i. e. the greater the number of vibrations, the less extensive must be the series of currents ; the slower the vibrations, or the more extensive the excursion of the parts from increased force applied, the greater the extent of disturbance. With glass plates (2. 12) the cloud is higher and larger as the vibrations are stronger, but still not so extensive as they are upon the stretched membrane (22), where the cloud may frequently be seen rising up in the middle and flowing over towards the sides.

27. When the membrane stretched upon the funnel (22) was made to vibrate by the horse-hair proceeding downwards, and a large glass tube, as a cylindrical lamp-glass, was brought near to the centre of vibration, no evidence of a current entirely through the lamp-glass could be perceived ; but still the most striking proofs were obtained of the existence of carrying currents by the effects upon the light powder, for it flew more rapidly under the edge, and tended to collect towards the axis of the tube ; it could even be diverted somewhat from its course towards the centre of oscillation. A piece of upright paper, held with its edge equally near, did not produce the same effect; but immediately that it was rolled into a tube, it did. When the glass chimney was suspended very carefully, and at but a small distance from the membrane, the powder often collected at the edge, and revolved there ; a complicated action between the currents and the space under the thickness of the glass taking place, but still tending to show the influence of the air in arranging and disposing the powders.

28. A sheet of drawing-paper was stretched tightly over a frame so as to form a tense elastic surface nearly three feet by two feet in extent. Upon placing this in a horizontal position, throwing a spoonful of lycopodium upon it, and striking it smartly below with the fingers, the phenomena of collection at the centre of vibration, and of moving heaps, could be obtained upon a magnificent scale. When the lycopodium was uniformly spread over the surface, and any part of the paper slightly tapped by the hand, the lycopodium at any place chosen could be drawn together merely by holding the lamp-glass over it. It will be unnecessary to enter into the detail of the various actions combining to produce these effects; it is sufficiently evident, from the mode in which they may be varied, that they depend upon currents of air.

29. A very interesting set of effects occurred when the stretched parchment upon the funnel (22) was vibrated under plates; the horse-hair was directed downwards, and the membrane, after being sprinkled over with light powder, was covered by a plate of glass resting upon the edge of the funnel; upon throwing the membrane into a vibratory state, the powder collected with much greater rapidity than without the plate; and instead of forming the semi-globular moving heaps, it formed linear arrangements, all concentric to the centre of vibration. When the vibrations were strong, these assumed a revolving motion, rolling towards the centre at the part in contact with the membrane, and from it at the part nearest the glass; thus illustrating in the clearest manner the double currents caged up between the glass and the membrane. The effect was well shown by carbonate of magnesia.

30. Sometimes when the plate was held down very close and tight, and the vibrations were few and large, the powder was all blown out at the edge; for then the whole arrangement acted as a bellows; and as the entering air travelled with much less velocity than the expelled air, and as the forces of the currents are as the squares of the velocity, the issuing air carried the powder more forcibly than the air which passed in, and finally threw it out.

31. A thin plate of mica laid loosely upon the vibrating membrane showed the rotating concentric lines exceedingly well.

32. From these experiments on plates and surfaces vibrating in air, it appears that the forms assumed by the determination of light powders towards the places of most intense vibration, depend, not upon any secondary mode of

division, or upon any immediate and peculiar action of the plate, but upon the currents of air necessarily formed over its surface, in consequence of the extra-mechanical action of one part beyond another. In this point of view the nature of the medium in which those currents were formed ought to have great influence over the phenomena; for the only reason why silica as sand should pass towards the quiescent lines, whilst the same silica as fine powder went from them, is, that in its first form the particles are thrown up so high by the vibrations as to be above the currents, and that if they were not thus thrown out of their reach they would be too heavy to be governed by them; whilst in the second form they are not thrown out of the lower current, except near the principal place of oscillation, and are so light as to be carried by it in whatever direction it may proceed.

33. In the exhausted receiver of the air-pump therefore the phenomena ought not to occur as in air; for as the force of the currents would be there excessively weakened, the light powders ought to assume the part of heavier grains in the air. Again, in denser media than air, as in water for instance, there was every reason to expect that the heavier powder, as sand and filings, would perform the part of light powders in air, and be carried from the quiescent to the vibrating parts.

34. The experiments in the air-pump receiver were made in two ways. A round plate of glass was supported on four narrow cork legs upon a table, and then a thin glass rod with a rounded end held perpendicularly upon the middle of the glass. By passing the moistened fingers longitudinally along this rod the plate was thrown into a vibratory state; the cork legs were then adjusted in the circular nodal line occurring with this mode of vibration; and when their places were thus found they were permanently fixed. The plate was then trans-ferred into the receiver of an air-pump, and the glass rod by which it was to be thrown into vibration passed through collars in the upper part of the receiver, the entrance of air there being prevented by abundance of pomatum. When fine silica was sprinkled upon the plate, and the plate vibrated by the wet fingers applied to the rod, the receiver not being exhausted, the fine powder travelled from the nodal line, part collecting at the centre, and other part in a circle, between the nodal line and the edge. Both these situations were places of vibration, and exhibited themselves as such by the agitation of the powder. Upon again sprinkling fine silica uniformly over the plate, ex-

hausting the receiver to twenty-eight inches, and vibrating the plate, the silica went from the middle towards the nodal line or place of rest, performing exactly the part of sand in air. It did not move at the edges of the plate, and as the apparatus was inconvenient and broke during the experiment, the following arrangement was adopted in its place.

35. The mouth of a funnel was covered (22) with a well-stretched piece of fine parchment, and then fixed on a stand with the membrane horizontal; the horse-hair was passed loosely through a hole in a cork, fixed in a metallic tube on the top of the air-pump receiver; the tube above the cork was filled to the depth of half an inch with pomatum, and another perforated cork put over that; a cup was formed on the top of the second cork, which was filled with water. In this way the horse-hair passed first through pomatum and then water, and by giving a little pressure and rotatory motion to the upper cork during the time that the horse-hair was used to throw the membrane into vibration, it was easy to keep the pomatum below perfectly in contact with the hair, and even to make it exude upwards into the water above. Thus no possibility of the entrance of air by and along the horse-hair could exist, and the tightness of all the other and fixed parts of the apparatus was ascertained by the ordinary mode of examination. A little paper shelf was placed in the receiver under the cork to catch any portion of pomatum that might be forced through by the pressure, and prevent its falling on to the membrane.

36. This arrangement succeeded: when the receiver was full of air, the lycopodium gathered at the centre of the membrane with great facility and readiness, exhibiting the cloud, the currents, and the involving heaps. Upon exhausting the receiver until the barometrical gauge was at twenty-eight inches, the lycopodium, instead of collecting at the centre, passed across the membrane towards one side which was a little lower than the other. It passed by the middle just as it did over any other part; and when the force of the vibrations was much increased, although the powder was more agitated at the middle than elsewhere, it did not collect there, but went towards the edges or quiescent parts. Upon allowing air to enter until the barometer stood at twenty-six inches, and repeating the experiments, the effect was nearly the same. When the vibrations were very strong, there were faint appearances of a cloud, consisting of the very finest particles, collecting at the centre of vibration;

but no sensible accumulation of the powder took place. At twenty-four inches of the barometer the accumulation at the centre began to appear, and there was a sensible, though very slight effect visible of the return of the powder from the edges. At twenty-two inches these effects were stronger; and when the barometer was at twenty inches, the currents of air within the receiver had force enough to cause the collection of the principal part of the lycopodium at the centre of vibration. Upon again, however, restoring the exhaustion to twenty-eight inches, all the effects were reproduced as at first, and the lycopodium again proceeded to the lower or the quiescent parts of the membrane. These alternate effects were obtained several times in succession before the apparatus was dismounted.

37. In this form of experiment there were striking proofs of the existence of a current upwards from the middle of the membrane when vibrating in air, (24), and the extent of the system of currents (26) was partly indicated. The powder purposely collected at the middle by vibrations, when the receiver was full of air, was observed as to the height to which it was forced upwards by the vibrations; and then the receiver being exhausted, the height to which the powder was thrown by similar vibrations was again observed. In the latter cases it was nothing like so great as in the former, the height not being two-thirds, and barely one-half, the first height. Had the powder been thrown up by mere propulsion, it should have risen far higher in vacuo than in air: but the reverse took place; and the cause appears to be, that in air the current had force enough to carry the fine particles up to a height far beyond what the mere blow which they received from the vibrating membrane could effect.

38. For the experiments in a denser medium than air, water was chosen. A circular plate of glass was supported upon four feet in a horizontal position, surrounded by two or three inches of water, and thrown into vibration by applying a glass rod perpendicular to the middle, as in the first experiment in vacuo (34); the feet were shifted until the arrangement gave a clear sound, and the moistened brass filings sprinkled upon the plate formed regular lines or figures. These lines were not however lines of rest, as they would have been in the air, but were the places of greatest vibration; as was abundantly evident from their being distant from that nodal line determined and indicated by the contact of the feet, and also from the violent agitation of the filings.

MDCCCXXXI.   2 s

In fact, the filings proceeded from the quiescent to the moving parts, and there were gathered together; not only forming the cloud of particles over the places of intense vibration, but also settling down, when the vibrations were weaker, into the same involving groups, and in every respect imitating the action of light powders in air. Sand was affected exactly in the same manner; and even grains of platina could be in this way collected by the currents formed in so dense a medium as water.

39. The experiments were then made under water with the membranes stretched over funnels (22) and thrown into vibration by horse-hairs drawn between the fingers. The space beneath the membrane could be retained, filled with air, whilst the upper surface was covered two or three inches deep with water; or the space below could also be filled with water, or the force applied to the membrane by the horse-hair could be upwards or downwards at pleasure. In all these experiments the sand or filings could be made to pass with the utmost facility to the most powerfully vibrating part, that being either at the centre only, or in addition, in circular lines, according to the mode in which the membrane vibrated. The edge of the funnel was always a line of rest; but circular nodal lines were also formed, which were indicated, not by the accumulation of filings upon them, but by the tranquil state of those filings which happened to be there, and also by being between those parts where the filings, by their accumulation and violent agitation, indicated the parts in the most powerful vibratory state.

40. Even when by the relaxation of the parchment from moisture, and the force upwards applied by the horse-hair, the central part of the membrane was raised the eighth of an inch or more above the edges, the circle not being four inches in diameter, still the filings would collect there.

41. When in place of parchment common linen was used, as becoming tighter rather than looser when wetted, the same effects were obtained.

42. Both the reasoning adopted and the effects described were such as to lead to the expectation that if the plate vibrating in air was covered with a layer of liquid instead of sand or lycopodium, that liquid ought to be determined from the quiescent to the vibrating parts and be accumulated there. A square plate was therefore covered with water, and vibrated as in the former experiments (2. 6.); but all endeavours to ascertain whether accumulation

occurred at the centres of oscillation, either by direct observation, or the reflection from its surface of right-lined figures, or by looking through the parts, as through a lens, at small print and other objects, failed.

43. As however when the plate was strongly vibrated, the well-known and peculiar crispations which form on water at the centres of vibration, occurred and prevented any possible decision as to accumulation, it was only when these were absent and the vibration weak, and the accumulation therefore small, that any satisfactory result could be expected; but as even then no appearance was perceived, it was concluded that the force of gravity combined with the mobility of the fluid was sufficient to restore the uniform condition of the layer of water after the bow was withdrawn, and before the eye had time to observe the convexity expected.

44. To remove in part the effect of gravity, or rather to make it coincide with, instead of oppose the convexity, the under surface of the plate was moist-ened instead of the upper, and by inclining the plate a little,

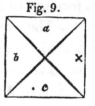

Fig. 9.

the water made to hang in drops at *a* or *b* or *c*, fig. 9, at plea-sure. On applying the bow at ×, and causing the plate to vibrate, the drops instantly disappeared, the water being gathered up and expanded laterally over the parts of the plate from which it had flowed.   On stopping the vibration, it again accumulated in hanging drops, which instantly disappeared as before on causing the plate to vibrate, the force of gravity being entirely overpowered by the superior forces excited by the vibrating plate.   Still, no visible evidence of convexity at the centres of vibra-tion were obtained, and the water appeared rather to be urged from the vibrating parts than to them.

45. The tenacity of oil led to the expectation that better results would be obtained with it than with water.   A round plate, held horizon-tally by the middle (6. 42), was covered with oil over the upper surface, so as to be flooded, except at ×, fig. 10, and the bow ap-plied at × as before, to produce strong vibration. No crispation occurred in the oil, but it immediately accumulated at *a*, *b*, and *c*,

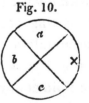

Fig. 10.

forming fluid lenses there, rendered evident by their magnifying power when print was looked at through them.   The accumulations were also visible on putting a sheet of white paper beneath, in consequence of the colour of the oil

2 s 2

being deeper at the accumulations than elsewhere; and they were also rend ered beautifully evident by making the experiment in sunshine, or by putting a candle beneath the plate, and placing a screen on the opposite side to receive the images formed at the focal distance.

46. When the vibration of the plate ceased, the oil gradually flowed back until of uniform depth. On renewing the vibration, the accumulations were re-formed, the phenomena of accumulation occurring with as much certainty and beauty as if lycopodium powder had been used.

47. To remove every doubt of the fluid passing from the quiescent to the agitated parts, centres of vibration were used, nearly surrounded by nodal lines. A square plate, fig. 11, being held at c, and the bow applied at ✕, gave with sand, nodal lines, resembling those in the figure. Then clearing off the sand, putting oil in its place, and producing the same mode of vibration as before, the oil accumulated at a and b, forming two heaps or lenses as in the former experiment (45).

Fig. 11.

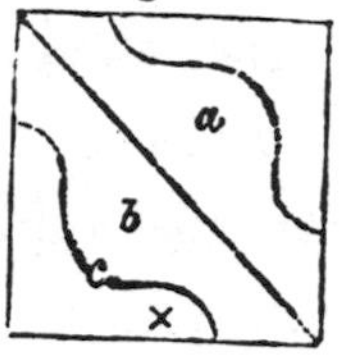

48. The experiment made with water on the under surface (44) was now repeated with oil, the round plate being used (45). The hanging drop of oil rose up as the water did before, but the lateral diffusion was soon limited; for lenses were formed at the centres of vibration just as when the oil was upon the upper surface, and, as far as could be ascertained by general examination, of the same form and power. On stopping the vibration, the oil gathered again into hanging drops; and on renewing it, it was again disposed in the lens-like accumulations.

49. With white of egg the same observable accumulation at the centres of vibration could be produced.

50. Hence it is evident that when a surface vibrating normally, is covered with a layer of liquid, that liquid is determined from the quiescent to the vibrating parts, producing accumulation at the latter places; and that this accumulation is limited, so that if purposely rendered too great by gravity or other means, it will quickly be diminished by the vibrations until the depth of fluid at any one part has a certain and constant relation to the velocity there and to the depth elsewhere.

51. From the accumulated evidence which these experiments afford, I think there can remain no doubt of the cause of the collection of fine powders at the

centres or lines of vibration of plates, membranes, &c. under common circumstances; and that no secondary mode of division need be assumed to account
for them.   I have been the more desirous of accumulating experimental evidence, because I have thought on the one hand that the authority of SAVART
should not be doubted on slight grounds, and on the other, that if by accident
it be placed in the wrong scale, the weight of evidence against it should be
such as fully to establish the truth and prevent a repetition of the error by
others.

52. It must be evident that the phenomena of collection at the centres or
lines of greatest vibration are exhibited in their purest form at those places
which are surrounded by nodal lines; and that where the centre or place of
vibration is at or near to an edge, the effects must be very much modified by
the manner in which the air is there agitated.   It is this influence, which, in
the square plates (6. 12) and other arrangements, prevents the clouds being at
the very edge of the glass.   They may be well illustrated by vibrating tin
plates under water over a white bottom, and sprinkling dark-coloured sand or
filings upon various parts of the plates.

On the peculiar Arrangement and Motions of the heaps formed by particles lying on vibrating surfaces.

53. The peculiar manner in which the fine powder upon a vibrating surface
is accumulated into little heaps, either hemispherical or merely rounded, and
larger or smaller in size, has already been described (6. 28), as well also as the
singular motion which they possess, as long as the plate continues in vibration.   These heaps form on any part of the surface which is in a vibratory
state, and not merely under the clouds produced at the centres of vibration,
although the particles of the clouds always settle into similar heaps.   They
have a tendency, as heaps, to proceed to the nodal or quiescent lines, but are
often swept away in powder by the currents already described (6).   When on
a place of rest, they do not acquire the involving motion.   When two or more
are near together or touch, they will frequently coalesce and form but one
heap, which quickly acquires a rounded outline.   When in their most perfect
and final form, they are always round.

54. The moving heaps formed by lycopodium on large stretched drawing-

paper (28), are on so large a scale as to be very proper for critical examination. The phenomena can be exhibited also even by dry sand on such a membrane, the sand being in large quantity and the vibrations slow. When the surface is thickly covered by sand from a sieve, and the paper tapped with the finger, the manner in which the sand draws up into moving heaps is very beautiful.

55. When a single heap is examined, which is conveniently done by holding a vibrating tuning-fork in a horizontal position, and dropping some lycopodium upon it, it will be seen that the particles of the heap rise up at the centre, overflow, fall down upon all sides, and disappear at the bottom, apparently proceeding inwards; and this evolving and involving motion continues until the vibrations have become very weak.

56. That the medium in which the experiment is made has an important influence, is shown by the circumstance of heavy particles, such as filings, exhibiting all these peculiarities when they are placed upon surfaces vibrating in water (39); the heaps being even higher at the centre than a heap of equal diameter formed of light powder in the air. In water, too, they are formed indifferently upon any part of the plate or membrane which is in a vibratory state. They do not tend to the quiescent lines; but that is merely from the great force of the currents formed in water as already described (38), and the power with which they urge obstacles to the place of greatest vibration.

57. If a glass plate be supported and vibrated (6), its surface having been covered with sand enough to hide the plate, and water enough to moisten and flow over the sand, the sand will draw together in heaps, and these will exhibit the peculiar and characteristic motion of the particles in a very striking manner.

58. The aggregation and motion of these heaps, either in air or other fluids, is a very simple consequence of the mechanical impulse communicated to them by the joint action of the vibrating surface and the surrounding medium. Thus in air, when, in the course of a vibration, the part of a plate under a heap rises, it communicates a propelling force upwards to that heap, mingled as it is with air, greater than that communicated to the surrounding atmosphere, because of the superior specific gravity of the former; upon receding from the heap, therefore, in performing the other half of its vibration, it forms a partial

vacuum, into which the air, round the heap, enters with more readiness than the heap itself; and as it enters, carries in the powder at the bottom edge of the heap with it.   This action is repeated at every vibration, and as they occur in such rapid succession that the eye cannot distinguish them, the centre part of the heap is continually progressing upwards; and as the powder thus accumulates above, whilst the base is continually lessened by what is swept in underneath, the particles necessarily fall over and roll down on every side.

59. Although this statement is made upon the relation of the heap, as a mass, to the air surrounding it, yet it will be seen at once that the same relation exists between any two parts of the heap at different distances from the centre; for the one nearest the centre will be propelled upward with the greatest force, and the other will be in the most favourable state for occupying the partial vacuum left by the receding plate.

60. This view of the effect will immediately account for all the appearances; the circular form, the fusion together of two or more heaps, their involving motion, and their existence upon any vibrating part of the plate.   The manner in which the neighbouring particles would be absorbed by the heaps is also evident; and as to their first formation, the slightest irregularities in the powder or surface would determine a commencement, which would then instantly favour the increase.

61. It is quite true, that if the powder were coherent, that force alone would tend to produce the same effect, but only in a very feeble degree.   This is sufficiently shown by the experiments made in the exhausted receiver (36).   When the barometer of the air-pump was at twenty-eight inches, that in the air being about 29.2 inches, the heaps, or rather parcels, formed very beautifully over the whole surface of the membrane; but they were very flat and extensive compared with the heaps in air, and the involving motion was very weak. As the air was admitted, the vibration being continued, the heaps rose in height, contracted in diameter, and moved more rapidly.   Again, in the experiments with filings and sand in water, no cohesive action could assist in producing the effect; it must have been entirely due to the manner in which the particles were mechanically urged in a medium of less density than themselves.

62. The conversion of these round heaps into linear concentric involving parcels, in the experiment already described (29. 31), when the membrane was

covered by a plate of glass, is a necessary consequence of the arrangements there made, and tends to show how influential the action of the air or other including medium is in all the phenomena considered in this paper.   No incompatible principles are assumed in the explication given of the arrangement of the forces producing the two classes of effects in question, and though by variation of the force of vibration and other circumstances, the one effect can be made, within certain limits, to pass into the other, no anomaly or contradiction is thus involved, nor any result produced, which, as it appears to me, cannot be immediately accounted for by reference to the principles laid down.

Royal Institution,
March 21, 1831.

24

On the Reflexion and Refraction of Sound

GEORGE GREEN

George Green (1793–1841) was one of the great mathematicians of the 19th century. The self-taught "Miller of Nottingham," he made an important contribution to the theory of electricity and magnetism by introducing the concept of potential and solving the differential equations involving it. He was also interested in wave propagation and made the first theoretical study of the reflection of plane sound waves incident obliquely on a plane interface. His paper on this subject published in the *Transactions of the Cambridge Philosophical Society* is presented here in full.

Reprinted from *Transactions of the Cambridge Philosophical Society*, **6**, 403–412 (1838)

24

ON THE REFLEXION AND REFRACTION OF SOUND.

THE object of the communication which I have now the honour of laying before the Society, is to present, in as simple a form as possible, the laws of the reflexion and refraction of sound, and of similar phenomena which take place at the surface of separation of any two fluid media when a disturbance is propagated from one medium to the other. The subject has already been considered by Poisson (*Mém. de l'Acad.*, &c. Tome x. p. 317, &c.). The method employed by this celebrated analyst is one that he has used on many occasions with great success, and which he has explained very fully in several of his works, and recently in a digression on the Integrals of Partial Differential Equations (*Théorie de la Chaleur*, p. 129, &c.). In this way, the question is made to depend on sextuple definite integrals. Afterwards, by supposing the initial disturbance to be confined to a small sphere in one of the fluids, and to be everywhere the same at the same distance from its centre, the formulæ are made to depend on double definite integrals; from which are ultimately deduced the laws of the propagation of the motion at great distances from the centre of the sphere originally disturbed.

The chance of error in every very long analytical process, more particularly when it becomes necessary to use Definite Integrals affected with several signs of integration, induced me to think, that by employing a more simple method we should possibly be led to some useful result, which might easily be overlooked in a more complicated investigation. With this impression I endeavoured to ascertain how a plane wave of infinite extent, accompanied by its reflected and refracted waves, would be propagated in any two indefinitely extended media of which the surface of separation in a state of equilibrium should also be in a plane of infinite extent.

The suppositions just made simplify the question extremely. They may also be considered as rigorously satisfied when light is reflected. In which case the unit of space properly belonging to the problem is a quantity of the same order as $\lambda = \dfrac{1}{50,000}$ inch, and the unit of time that which would be employed by light itself in passing over this small space. Very often too, when sound is reflected, these suppositions will lead to sensibly correct results. On this last account, the problem has here been considered generally for all fluids whether *elastic* or *non-elastic* in the usual acceptation of these terms; more especially, as thus its solution is not rendered more complicated. One result of our analysis is so simple that I may perhaps be allowed to mention it here. It is this: If A be the ratio of the density of the reflecting medium to the density of the other, and B the ratio of the cotangent of the angle of refraction to the cotangent of the angle of incidence, then for all fluids

$$\frac{\text{the intensity of the reflected vibration}}{\text{the intensity of the incident vibration}} = \frac{A - B}{A + B}.$$

If now we apply this to the reflexion of sound at the surface of still water, we have $A > 800$, and the maximum value of $B < \frac{1}{4}$. Hence the intensity of the reflected wave will in every case be sensibly equal to that of the incident one. This is what we should naturally have anticipated. It is however noticed here because M. Poisson has inadvertently been led to a result entirely different.

When the velocity of transmission of a wave in the second medium, is greater than that in the first, we may, by sufficiently increasing the angle of incidence in the first medium, cause the refracted wave in the second to disappear. In this case the change in the intensity of the reflected wave is here shown to be such that, at the moment the refracted wave disappears, the intensity of the reflected becomes exactly equal to that of the incident one. If we moreover suppose the vibrations of the incident wave to follow a law similar to that of the cycloidal pendulum, as is usual in the Theory of Light, it is proved that on farther increasing the angle of incidence, the intensity of the reflected wave remains unaltered whilst the phase of the vibration gradually changes. The laws of the change of intensity, and of the subsequent alteration of phase, are given here for all media, *elastic* or *non-elastic.* When, however, both the media are *elastic,* it is remarkable that these laws are precisely the same as those for light polarized in a plane perpendicular to the plane of incidence. Moreover, the disturbance excited in the second medium, when, in the case of total reflexion, it ceases to transmit a wave in the regular way, is represented by a quantity of which one factor is a negative exponential. This factor, for light, decreases with very great rapidity, and thus the disturbance is not propagated to a sensible depth in the second medium.

Let the plane surface of separation of the two media be taken as that of (yz), and let the axis of z be parallel to the line of intersection of the plane *front* of the wave with (yz), the axis of x being supposed vertical for instance, and directed downwards; then, if Δ and Δ_1 are the densities of the two media under the constant pressure P and s, s_1 the condensations, we must have

$$\begin{cases} \Delta\,(1+s) = \text{density in the upper medium,} \\ \Delta_1(1+s_1) = \text{density in the lower medium.} \end{cases}$$

$$\begin{cases} P\,(1+As) \;= \text{pressure in the upper medium,} \\ P\,(1+A_1 s_1) = \text{pressure in the lower medium.} \end{cases}$$

Also, as usual, let ϕ be such a function of x, y, z, that the resolved parts of the velocity of any fluid particle parallel to the axes, may be represented by

$$\frac{d\phi}{dx}, \quad \frac{d\phi}{dy}, \quad \frac{d\phi}{dz}.$$

In the particular case, here considered, ϕ will be independent of z, and the general equations of motion in the upper fluid will be

$$0 = \frac{ds}{dt} + \frac{d^2\phi}{dx^2} + \frac{d^2\phi}{dy^2},$$

$$0 = \frac{d\phi}{dt} + \gamma^2 s;$$

where we have

$$\gamma^2 = \frac{PA}{\Delta},$$

or by eliminating s

$$\frac{d^2\phi}{dt^2} = \gamma^2 \left(\frac{d^2\phi}{dx^2} + \frac{d^2\phi}{dy^2}\right) \dots\dots\dots\dots (1).$$

Similarly, in the lower medium

$$\frac{d^2\phi_{\prime}}{dt^2} = \gamma_{\prime}^2 \left(\frac{d^2\phi_{\prime}}{dx^2} + \frac{d^2\phi_{\prime}}{dy^2}\right) \dots\dots\dots\dots (2),$$

where

$$s_{\prime} = \frac{-d\phi_{\prime}}{\gamma_{\prime}^2 dt}, \text{ and } \gamma_{\prime}^2 = \frac{PA_{\prime}}{\Delta_{\prime}}.$$

The above are the known general equations of fluid motion, which must be satisfied for all the internal points of both fluids; but at the surface of separation, the velocities of the particles perpendicular to this surface and the pressure there must be the same for both fluids. Hence we have the particular conditions

$$\left.\begin{aligned} \frac{d\phi}{dx} &= \frac{d\phi_{\prime}}{dx} \\ As &= A_{\prime} s_{\prime} \end{aligned}\right\} \text{ (where } x = 0),$$

neglecting such quantities as are very small compared with those retained, or by eliminating s and $s_{\prime}$, we get

$$\left.\begin{aligned} \frac{d\phi}{dx} &= \frac{d\phi_{\prime}}{dx} \\ \Delta \frac{d\phi}{dt} &= \Delta_{\prime} \frac{d\phi_{\prime}}{dt} \end{aligned}\right\} \text{ (when } x = 0) \dots\dots\dots (A).$$

The general equations (1) and (2), joined to the particular conditions (A) which belong to the surface of separation (yz), only, are sufficient for completely determining the motion of our two fluids, when the velocities and condensations are independent of the co-ordinate z, whatever the initial disturbance may be. We shall not here attempt to give their complete solution, which would be complicated, but merely consider the propagation of a plane wave of indefinite extent, which is accompanied by its reflected and refracted wave.

Since the disturbance of all the particles, in any *front* of the incident plane wave, is the same at the same instant, we shall have for the incident wave

$$\phi = f\,(ax + by + ct),$$

retaining b and c unaltered, we may give to the *fronts* of the reflected and refracted waves, any position by making for them

$$\phi = F\,(a'x + by + ct),$$
$$\phi_{,} = f_{,}\,(a_{,}x + by + ct).$$

Hence, we have in the upper medium,

$$\phi = f(ax + by + ct) + F(a'x + by + ct) \ldots\ldots\ldots (4),$$

and in the lower one

$$\phi_{,} = f_{,}\,(a_{,}x + by + ct) \ldots\ldots\ldots\ldots\ldots\ldots\ldots (5).$$

These, substituted in the general equations (1) and (2), give

$$\left. \begin{array}{l} c^2 = \gamma^2\,(a^2 + b^2) \\ c^2 = \gamma^2\,(a'^2 + b^2) \\ c^2 = \gamma_{,}^{2}\,(a_{,}^{2} + b^2) \end{array} \right\} \ldots\ldots\ldots\ldots\ldots\ldots (6).$$

Hence, $a' = \pm\,a$, where the lower signs must evidently be taken to represent the reflected wave. This value proves, that the angle of incidence is equal to that of reflexion. In like manner, the value of $a_{,}$, will give the known relation of sines for the incident and refracted wave, as will be seen afterwards.

Having satisfied the general equations (1) and (2), it only remains to satisfy the conditions (A), due to the surface of separation of the two media. But these by substitution give

$$af'\,(by + ct) - aF'\,(by + ct) = a_{,}f_{,}'\,(by + ct),$$

$$\Delta\,\{\,f'\,(by + ct) + F'\,(by + ct)\} = \Delta_{,}f_{,}'\,(by + ct),$$

because $a' = -\,a$, and $x = 0$.

Hence by writing, to abridge, the characteristics only of the functions

$$\left. \begin{array}{l} f' = \dfrac{1}{2}\left(\dfrac{\Delta_{,}}{\Delta} + \dfrac{a_{,}}{a}\right)f_{,}' \\[2ex] F' = \dfrac{1}{2}\left(\dfrac{\Delta_{,}}{\Delta} - \dfrac{a_{,}}{a}\right)f_{,}' \end{array} \right\} \ldots\ldots\ldots\ldots\ldots (7),$$

or if we introduce θ, $\theta_{,}$, the angle of incidence and refraction, since

$$\cot \theta = \frac{a}{b}\,,$$

$$\cot \theta_{,} = \frac{a_{,}}{b}\,,$$

$$f' = \frac{1}{2}\left(\frac{\Delta_{,}}{\Delta} + \frac{\cot \theta_{,}}{\cot \theta}\right)f_{,}',$$

$$F' = \frac{1}{2}\left(\frac{\Delta_{,}}{\Delta} - \frac{\cot \theta_{,}}{\cot \theta}\right)f_{,}',$$

and therefore $\dfrac{F''}{f'} = \dfrac{\dfrac{\Delta_{,}}{\Delta} - \dfrac{\cot\theta_{,}}{\cot \theta}}{\dfrac{\Delta_{,}}{\Delta} + \dfrac{\cot\theta_{,}}{\cot \theta}}\,,$

which exhibits under a very simple form, the ratio between the intensities of the disturbances, in the incident and reflected wave.

But the equations (6) give

$$\gamma^{2}\left(\frac{a^{2}}{b^{2}} + 1\right) = \gamma_{,}^{2}\left(\frac{a_{,}^{2}}{b^{2}} + 1\right);$$

and hence

$$\frac{\gamma}{\sin \theta} = \frac{\gamma_{,}}{\sin \theta_{,}}\,,$$

the ordinary law of sines.

The reflected wave will vanish when

$$0 = \frac{\Delta_{,}}{\Delta} - \frac{\cot \theta'}{\cot \theta}\,;$$

which with the above gives

$$\cot \theta = \Delta \sqrt{\frac{\gamma^{2} - \gamma_{,}^{2}}{(\gamma_{,}\Delta_{,})^{2} - (\Delta\gamma)^{2}}}\,.$$

Hence the reflected wave may be made to vanish if $\gamma^{2} - \gamma_{,}^{2}$ and $(\gamma_{,}\Delta)^{2} - (\gamma_{,}\Delta_{,})$ have different signs.

For the ordinary elastic fluids, at least if we neglect the change of temperature due to the condensation, A is independent of the nature of the gas, and therefore

$$A = A_{,}\ \text{or}\ \gamma^{2}\Delta = \gamma_{,}^{2}\Delta_{,}.$$

Hence

$$\tan \theta = \frac{\gamma}{\gamma_i},$$

which is the precise angle at which light polarized perpendicular to the plane of reflexion is wholly transmitted.

But it is not only at this particular angle that the reflexion of sound agrees in intensity with light polarized perpendicular to the plane of reflexion. For the same holds true for every angle of incidence. In fact, since

$$\gamma^2 \Delta = \gamma_i^2 \Delta_i; \quad \therefore \quad \frac{\Delta'}{\Delta} = \frac{\gamma^2}{\gamma_i^2} = \frac{\sin^2 \theta}{\sin^2 \theta_i},$$

and the formulæ (7) give

$$\frac{F'}{f'} = \frac{\dfrac{\sin^2 \theta}{\sin^2 \theta_i} - \dfrac{\tan \theta}{\tan \theta_i}}{\dfrac{\sin^2 \theta}{\sin^2 \theta_i} + \dfrac{\tan \theta}{\tan \theta_i}} = \frac{\tan (\theta - \theta_i)}{\tan (\theta + \theta_i)};$$

which is the same ratio as that given for light polarized perpendicular to the plane of incidence. (Vide Airy's *Tracts*, p. 356)*.

What precedes is applicable to all waves of which the *front* is plane. In what follows we shall consider more particularly the case in which the vibrations follow the law of the cycloidal pendulum, and therefore in the upper medium we shall have,

$$\phi = \alpha \sin (ax + by + ct) + \beta \sin (- ax + by + ct) \ldots\ldots\ldots(8).$$

Also, in the lower one,

$$\phi_i = \alpha_i \sin (a_i x + by + ct):$$

and as this is only a particular case of the more general one, before considered, the equation (7) will give

$$\alpha = \frac{1}{2} \left(\frac{\Delta'}{\Delta} + \frac{a_i}{a} \right) \alpha_i,$$

$$\beta = \frac{1}{2} \left(\frac{\Delta'}{\Delta} - \frac{a_i}{a} \right) \alpha_i.$$

If $\gamma_i > \gamma$, or the velocity of transmission of a wave, be greater in the lower than in the upper medium, we may by decreasing a render a_i imaginary. This last result merely indicates that the form of our integral must be changed, and that as far as

* [Airy on *The Undulatory Theory of Optics*, p. 111, Art. 129.]

regards the co-ordinate x an exponential must take the place of the circular function. In fact the equation,

$$\frac{d^2\phi_{,}}{dt^2} = \gamma_{,}^2 \left\{ \frac{d^2\phi_{,}}{dx^2} + \frac{d^2\phi_{,}}{dy^2} \right\},$$

may be satisfied by

$$\phi_{,} = \epsilon^{-a_{,}'x} . B \sin \psi,$$

(where, to abridge, ψ is put for $by + ct$) provided

$$c^2 = \gamma_{,}^2 (- a_{,}'^2 + b^2);$$

when this is done it will not be possible to satisfy the conditions (A) due to the surface of separation, without adding constants to the quantities under the circular functions in ϕ. We must therefore take, instead of (8), the formula,

$$\phi = \alpha \sin (ax + by + ct + e) + \beta \sin (- ax + by + ct + e_{,}) \ldots (9).$$

Hence when $x = 0$, we get

$$\frac{d\phi}{dx} = a\alpha \cos (\psi + e) - a\beta \cos (\psi + e_{,}),$$

$$\frac{d\phi}{dt} = c\alpha \cos (\psi + e) - c\beta \cos (\psi + e_{,}),$$

$$\frac{d\phi_{,}}{dx} = - a_{,}' B \sin \psi,$$

$$\frac{d\phi}{dt} = cB \cos \psi;$$

these substituted in the conditions (A), give

$$\alpha \cos (\psi + e) - \beta \cos (\psi + e_{,}) = - \frac{a_{,}'}{a} B \sin \psi,$$

$$\alpha \cos (\psi + e) + \beta \cos (\psi + e_{,}) = \frac{\Delta_{,}}{\Delta} B \cos \psi;$$

these expanded, give

$$\alpha \cos e - \beta \cos e_{,} = 0,$$

$$- \alpha \sin e + \beta \sin e_{,} = - \frac{a_{,}'}{a} B,$$

$$\alpha \cos e + \beta \cos e_{,} = \frac{\Delta_{,}}{\Delta} B,$$

$$\alpha \sin e + \beta \sin e_{,} = 0.$$

Hence, we get

$$2\alpha \sin e = \frac{a_{,}'}{a} B \quad \dots\dots\dots\dots\dots(10),$$

$$2\alpha \cos e = \frac{\Delta_{,}}{\Delta} B,$$

$$2\beta \sin e_{,} = -\frac{a_{,}'}{a} B,$$

$$2\beta \cos e_{,} = \frac{\Delta_{,}}{\Delta} B,$$

and, consequently,

$$e = -e_{,}, \quad \beta = \alpha,$$

and

$$\tan e = + \frac{a_{,}'\Delta}{a\Delta_{,}}.$$

This result is general for all fluids, but if we would apply it to those only which are usually called *elastic*, we have, because in this case $\gamma^2\Delta = \gamma_{,}^2\Delta_{,}$,

$$\tan e = \frac{a_{,}'\Delta}{a\Delta_{,}} = \frac{a_{,}'\gamma_{,}^2}{a\gamma^2}.$$

But generally

$$c^2 = \gamma_{,}^2(-a_{,}'^2 + b^2) = \gamma^2(a^2 + b^2) \quad \dots\dots\dots\dots(11);$$

and therefore, by substitution,

$$\tan e = \frac{a_{,}'\gamma_{,}^2}{a\gamma^2} = \frac{\gamma_{,}\sqrt{\gamma_{,}^2 b^2 - (a^2 + b^2)\gamma^2}}{a\gamma^2} = \mu\sqrt{\mu^2 \tan^2\theta - \sec^2\theta},$$

because $\mu - \frac{\gamma_{,}}{\gamma}$, and $\frac{b}{a} = \tan\theta$.

As $e = -e_{,}$, we see from equation (9), that $2e$ is the change of phase which takes place in the reflected wave; and this is precisely the same value as that which belongs to light polarized perpendicularly to the plane of incidence; (Vide Airy's *Tracts*, p. 362*.) We thus see, that not only the intensity of the reflected wave, but the change of phase also, when reflexion takes place at the surface of separation of two elastic media, is precisely the same as for light thus polarized.

* Airy, *ubi sup.* p. 114, Art. 133.

16

As $\alpha = \beta$, we see that when there is no transmitted wave the intensity of the reflected wave is precisely equal to that of the incident one. This is what might be expected: it is, however, noticed here because a most illustrious analyst has obtained a different result. (Poisson, *Mémoires de l'Academie des Sciences*, Tome x.) The result which this celebrated mathematician arrives at is, That at the moment the transmitted wave ceases to exist, the intensity of the reflected becomes precisely equal to that of the incident wave. On increasing the angle of incidence this intensity again diminishes, until it vanish at a certain angle. On still farther increasing this angle the intensity continues to increase, and again becomes equal to that of the incident wave, when the angle of incidence becomes a right angle.

It may not be altogether uninteresting to examine the nature of the disturbance excited in that medium which has ceased to transmit a wave in the regular way. For this purpose, we will resume the expression,

$$\phi_{,} = B\epsilon^{-a'_{,}x} \sin \psi = B\epsilon^{-a'_{,}x} \sin (by + ct);$$

or if we substitute for B, its value given by the last of the equations (10); and for $a'_{,}$, its value from (11); this expression, in the case of ordinary elastic fluids where $\gamma^2\Delta = \gamma'^2, \Delta_{,}$, will reduce to

$$\phi_{,} = 2a\mu^2 \cos e . \epsilon^{-\frac{2\pi x}{\lambda}\sqrt{\frac{\mu^2\sin^2\theta - 1}{\mu}}} \sin (by + ct),$$

λ being the length of the incident wave measured perpendicular to its own front, and θ the angle of incidence. We thus see with what rapidity in the case of light, the disturbance diminishes as the depth x below the surface of separation of the two media increases; and also that the rate of diminution becomes less as θ approaches the *critical* angle, and entirely ceases when θ is exactly equal to this angle, and the transmission of a wave in the ordinary way becomes possible.

25

On the Definition of a Tone with the Associated Theory of the Siren and Similar Sound Producing Devices

GEORG SIMON OHM

Georg Simon Ohm (1787–1854), the German physicist whose name is forever associated with the law connecting voltage and current in an electric circuit and with the unit of electrical resistance, was professor of physics at the University of Munich from 1849 until his death. In 1843 he put forward a theory of audition, leading to another Ohm's law, this time in acoustics. His paper on this subject in Poggendorf's *Annalen* for 1843 played a significant role in the foundation of psychological acoustics. The introduction and certain other sections are presented here in English translation with editorial notes.

Translated by R. Bruce Lindsay from
Poggendorf's *Annalen der Physik und Chemie*,
59, 497ff (1843)

1. When a short time ago I undertook to bring my earlier investigations on combination tones and impulses into clearer definition, I was suddenly assailed by a not insignificant doubt. For I had always previously assumed as a self-evident fact that the components of a tone, whose frequency is said to be m, must retain the form $a \cdot \sin 2\pi mt$ or $a \cdot \cos 2\pi mt$, where t denotes the time and a the amplitude for the successive tone elements. And conversely I had taken it for granted that a succession of stimuli on our ear which consists uninterruptedly of the form here specified must produce the sensation of a tone. But the introduction by Savart and Cagniard de Latour of special methods of producing tones appears likely to drive this assumption originating in earlier days out of its secure position. At any rate the words of several well-known acousticians appear to point in this direction. In the *Reportorium der Physik* (vol. 3, p. 30) Röher expresses his opinion as follows:

> For this purpose it may be permitted to touch on the well-known investigations of Savart and Cagniard de Latour on the production of tones. If we seek out the common elements in these investigations as well as of those of Savart on the tones of lowest and highest pitch that are audible, as well as those of Cagniard de Latour concerning the use of the siren, we find that the production of tones is brought about only by the regular repetition of any type of impulse affecting the hearing mechanism. In all cases the same dependence of the pitch of the tone on the number of impulses per second is found to prevail. In the investigations mentioned here this impulse appears either as a combination of a condensation and a rarefaction or merely as a condensation. In particular Savart's investigations on the lowest audible tones show that the separation of the two maxima in a succession of condensations and rarefactions in no way depends on the duration of the individual impulses. In other words it does not depend

on the separation of the maxima of two successive condensation and rarefaction couples. From this it accordingly follows that the conventional production of musical tones for which the different maxima of condensation and rarefaction follow each other in equal time intervals must be considered only a special case of the general repetition of an impulse made up of a condensation and rarefaction.

And further on (p. 53) he says:

It was only after I had developed the earlier description that I became acquainted with Seebeck's publication in which on the same grounds as in the previous discussion he explains the nature of a tone as the regular repetition of any impulse and demonstrates that the appropriate designation of the pitch of a tone consists in the indication of the number of simple and compound impulses per unit time, but not in the usual indication of condensations and rarefactions.

2. Since then experiments with the siren performed by Seebeck have been reported in Poggendorf's *Annalen* (vol. 53, p. 417ff, 1841). Through these even the equality of the time intervals in which the individual intervals follow each other appears not to be essential for the formation of a tone. Seebeck himself has drawn from his experiments (see p. 423) the following conclusion: "From this investigation we see that the ear on the one hand possesses the ability to analyze a system of impulses into two or three systems of isochronous impulses, but that on the other hand it receives from something only approximately isochronous the impression of a definite pitch, as if it were completely isochronous." Since these investigations of Seebeck contain all that can be brought against the old definition of a tone so far as facts are concerned, and my present article is fundamentally nothing more than a continuing commentary on Seebeck's work, for the convenience of the reader and since not too much space is required, I shall set down Seebeck's results here and refer to this outline in the sequel.

a) We mount two tubes, one on each side of a row of holes in a siren and make them perpendicular to the siren disk in such a way that if one of the tubes faces a hole the other is also opposite a hole. If one blows air through *one* of the tubes as the siren disk rotates, one gets a definite tone. However, if we blow through both tubes at the same time, the tone disappears and one hears only the rushing sound of the air streaming through the openings. However, if one arranges the tubes so that the air impacts from the two tubes do not take place at the same time but follow each other in alternation, one hears the original tone and it is indeed more intense.

b) If one introduces on a disk two concentric rows of holes, one of which has double the number of holes of the other, the latter gives [Editor's note: with the same speed of rotation] the octave above the tone from the smaller number of holes. If both sets of holes are blown at once, in general both tones are heard together.

It is only when blowing takes place from opposite sides of the disk and takes place in such a way that every air stream impact of the lower tone coincides precisely with one of the higher that the higher tone disappears and one hears only the lower one.

c) If a series of impulses in quick succession, but not isochronistically, reach the ear, in such a way that the interval between the two alternates from t to t', a tone is produced corresponding to the interval $t + t'$ and if t and t' are not too different, one hears at the same time a tone of period $(t + t')/2$, that is an octave above the previous one. Seebeck remarks that in this case, the higher or the lower tone tends to predominate accordingly as the values t and t' become more nearly equal or unequal to each other.

d) If a series of impacts follow in succession in such a way that the intervals are successively t, t', and t'', one hears a tone with the period $t + t' + t''$, and at the same time, if the three times are not too far apart, a tone with period $(t + t' + t'')/3$. Here again the remark is made that the higher or the lower tone predominates accordingly as the three times are closer together or further apart.

e) If one lets the hole separations vary in an irregular fashion, but in such a way that they do not depart too widely from an average value, one hears only one more or less incomplete tone, whose frequency corresponds to that average value.

f) If in the investigations described under c) t' is an integral multiple of t, the tone with the period $t + t'$ is still heard, but instead of the tone $(t + t')/2$, that with period t makes its appearance.

g) If t' was very much greater than t, but not an exact multiple of t, it seemed to Seebeck that in addition to the tone $t + t'$, the tone corresponding to period t was also present; but that occasionally there was also present a still higher tone, which sounded with the tone $(t + t')/2$. However, Seebeck was in some doubt about this.

h) If the time intervals between any three impulses were alternated between t, t', and t'', on every occasion the tone corresponding to t was heard, no matter whether t was greater or smaller than t'.

3. From the statements just presented from qualified scientists and based on experience concerning the true nature of tones, it seems that everything previously accepted as true in this matter has been overthrown. At the same time nothing else of any degree of reliability has been put in its place. It occurred to me that a new definition of tone is demanded. I had to keep in mind the old rule that in the explanation of a natural phenomenon no other causes should be assumed than are both necessary and sufficient. Moreover my own personal feelings did not incline me to dismiss at once previously established results in favor of the attraction of a new viewpoint. So I set myself the task of determining whether the definition of a tone as it has come down to us from our predecessors does not after all contain what is necessary and sufficient for an explanation of the newly discovered facts. However, as a result of this investigation the previously accepted definition of a tone has been placed in its proper place in a fashion which has seemed to me to justify publication. I now propose to show how by penetrating into the obscurity of Seebeck's researches with their help I believe I have arrived at a real understanding of them. The old definition of a tone, however, I now formulate as follows:

a) For the formation of a tone of frequency m, the necessary impulses must follow each other in time intervals of the length $1/m$, and in every one of these intervals the impulses must carry in themselves the pure form $a \cdot \sin 2\pi(mt + p)$ or at any rate this form must be able to be separated out of these impulses.

b) These forms whether they are actually contained in the individual impulses or can be separated out from them must be connected in such a way that the phases of the successive forms maintain the same relation to each other, or in other words the quantity p must be the same for all.

c) The quantity a must either always be positive or always negative, for a change in sign of a is equivalent to a change in the value of p, because $-a \sin 2\pi(mt + p) = +a \sin 2\pi(mt + p \pm \frac{1}{2})$.

It will of course be understood, though it has not been expressly included in the foregoing definition, that the perceptibility of the pitch of a tone demands the invariability of the quantity p in successive impulses only in so far as the ear requires it for the precise estimation of the pitch. Indeed, if after so many waves have struck the ear that it is in a position to estimate precisely the pitch, by some means the value of p were changed and then persisted that way for a long enough time that the ear, other things being equal, was able once again to hear the same pitch, no occasion would be given thereby for any change in its judgment as to the pitch of the tone being received. The tone, therefore, so far as pitch is concerned can be considered as unchanged.

4. As a means of judging whether in a given impulse the form $a \cdot \sin 2\pi(mt + p)$ is contained as a real component or not, I use the theorem of Fourier, which has become famous for its many important applications. If F^1 denotes any arbitrary continuous or discontinuous function of t, which has arbitrary real values in the interval from $t = -l$ to $t = +l$, then between these limits, the theorem states that

$$F^{(1)} = A_0 + A_1 \cos \pi \frac{t}{l} + A_2 \cos \pi \frac{2t}{l}$$
$$+ A_3 \cos \pi \frac{3t}{l} + \cdots$$
$$+ B_1 \sin \pi \frac{t}{l} + B_2 \sin \pi \frac{2t}{l}$$
$$+ B_3 \sin \pi \frac{3t}{l} + \cdots$$

[Editor's note: Ohm then quotes the values of the coefficients A_n and B_n, which we need not reproduce here.]

If we now represent by F^1 any sound impulse striking the ear at time t, then Fourier's theorem says that this impulse is analyzable into the components

$$A_0, A_1 \cos \pi \frac{t}{l} + B_1 \sin \pi \frac{t}{l},$$
$$A_2 \cos \pi \frac{2t}{l} + B_2 \sin \pi \frac{2t}{l},$$
$$A_3 \cos \pi \frac{3t}{l} + B_3 \sin \pi \frac{3t}{l}, \text{ etc.}$$

A_0 corresponds to no oscillation but represents merely a displacement of the oscillating parts of the ear. The other components, however, all correspond to oscillations which take place around the displaced position. The expression $A_1 \cos \pi(t/l) + B_1 \sin \pi(t/l)$ produces an oscillation belonging to the tone whose frequency is $1/2l$, and so on for the other components with frequencies $2/2l$, $3/2l$, etc.

[Editor's note: Ohm now takes $F^{(1)} = a \sin 2\pi(mt + p)$ and substitutes into the expressions for the Fourier component amplitudes. He thus finds:

$$\text{for } m = \frac{1}{2l}, A_1 = a \sin 2\pi p, \ B_1 = a \cos 2\pi p$$

$$\text{for } m = \frac{2}{2l}, A_2 = a \sin 2\pi p, \ B_2 = a \cos 2\pi p, \text{etc.}$$

with $a^2 = A_1{}^2 + B_1{}^2$ etc.

and $\dfrac{A_1}{B_1} = \tan 2\pi p$

Then he goes on.]

If now a succession of impulses is given, each one of which lasts during the same time interval $2l$, and if we want to know whether the tone of frequency $1/2l$ is contained in this succession of impulses, we merely have to compute the values of A_1, B_1, etc. corresponding to the given impulse and from these we can get at once the corresponding value of p from

$$\tan 2\pi p = A_1/B_1$$

One can also find a in terms of A_1 and B_1.

[Editor's note: The author continues his long paper with an application of the foregoing theory to the various uses of sound produced by the siren as reported by Seebeck. He begins by assuming that in every successive time interval $2l$ a sinusoidal impulse strikes the ear. He then shows how this is analyzed into component tones. He then derives for this case the following theorems.]

a) Impulses which are repeated in time intervals of length $2l$ produce a tone of frequency $1/2l$ if the impulses in the successive intervals maintain one and the same pattern at least as long as the ear needs in order to identify a tone.

b) Impulses which repeat each other in intervals of length $2l$ produce *no* tone of frequency $1/2l$ if the pattern of the interval changes from interval to interval.

[Editor's note: The remainder of the article is devoted to a detailed examination of the cases in which there are two or three impulses per given time interval. The mathematical details become very complicated, but the essential nature of the results remains fundamentally the same as for the simpler case. In essence what Ohm sought to do was to establish the law that the human ear is able to analyze any complex sound into a set of simple harmonic tones in terms of which it may be expanded by means of the theorem of Fourier. This law has engaged the attention of acoustical scientists from the time of Helmholtz to the present. Helmholtz in his famous volume, *Sensations of Tone*, showed how the Ohm's law could be justified and used it as a basis for his theory of hearing (see the reproduction of his article later in this volume).]

26

On the Effects of Magnetism Upon the Dimensions of Iron and Steel Bars

JAMES PRESCOTT JOULE

James Prescott Joule (1818–1889) was a celebrated British amateur physicist whose remarkably careful and precise measurement of the mehcanical equivalent of heat established this quantity as one of the important physical constants and the key quantity in the theory of thermodynamics. Long interested also in magnetism, in 1841 he began a series of experiments on the effect of magnetization on the dimensions of iron and steel. His results were summarized in a paper published in 1847, an extract from which appears here. Though Joule's work was followed up and extended by continental researchers, he is usually given the credit for establishing the effect now known as magnetostriction. It was soon found that a magnetic field could produce an oscillation in a bar of magnetizable material. Hence out of Joule's discovery came the development of the magnetostrictive transducer, which has proved of great value in underwater sound technology, such as sonar.

Reprinted from *Philosophical Magazine*, series 3, **30**, 76–87 (1847)

XVII. *On the Effects of Magnetism upon the Dimensions of Iron and Steel Bars. By* J. P. Joule, *Esq.**

ABOUT the close of the year 1841, Mr. F. D. Arstall, an ingenious machinist of this town, suggested to me a new form of electro-magnetic engine. He was of opinion that a bar of iron experienced an increase of bulk by receiving the magnetic condition; and that, by reversing its polarity rapidly by means of alternating currents of electricity, an available and useful motive power might be realized. At Mr. Arstall's request I undertook some experiments in order to decide how far his opinions were well-founded.

The results of my inquiries were brought before the public on the occasion of a *conversazione* held at the Royal Victoria Gallery of Manchester on the 16th of February 1842, and are printed in the eighth volume of Sturgeon's Annals of Electricity, p. 219. As many of my readers may not have access to that work, I will, with the permission of the Editors of the Philosophical Magazine, quote a small portion of the paper, which is necessary to complete the history of this interesting branch of investigation.

"A length of thirty feet of copper wire, one-twentieth of an inch thick, and covered with cotton thread, was formed into a

* Communicated by the Author.

coil twenty-two inches long and one-third of an inch in interior diameter. This coil was secured in a perpendicular position; and the rod of iron, of which I wished to ascertain the increment, was suspended in its axis so as to receive the magnetic influence whenever a current of electricity was passed through the coil. Lastly, the upper extremity of the rod was fixed, and the lower extremity was attached to a system of levers which multiplied its motion three thousand times.

" A bar of rectangular iron wire, two feet long, one quarter of an inch broad, and one-eighth of an inch thick, caused the index of the multiplying apparatus to spring from its position and vibrate about a point one-tenth of an inch in advance, when the coil was made to complete the circuit of a battery capable of magnetizing the iron to saturation, or nearly so. After a short interval of time, the index ceased to vibrate, and began to advance very gradually in consequence of the expansion of the bar from the heat which was radiated from the coil. On breaking the circuit, the index immediately began to vibrate about a point, exactly one-tenth of an inch lower than that to which it had attained.

" By multiplying the advance of the index by the power of the levers, we have $\frac{1}{30000}$th of an inch, the increment of the bar, which may be otherwise stated as $\frac{1}{720000}$th of its whole length.

" Similar results were obtained by the use of an iron wire, two feet long and one-twelfth of an inch in diameter. Five pairs of the nitric acid battery produced an increment of the thirty-thousandth part of an inch ; and when only one pair of the battery was employed, I had an increment very slightly less, viz., the thirty-three thousandth part of an inch.

" This increment does not appear to depend upon the thickness of the bar; for an electro-magnet made of iron, three feet long and one inch square, was found to expand under the magnetic influence to nearly the same extent, compared with its length, as the wires did in the previous experiments.

" I made some experiments in order to ascertain the *law* of the increment. Their results proved it *to be very nearly proportional to the intensity of the magnetism and the length of the bar*.

" Trial was made whether any effect could be produced by using a *copper* wire, two feet long and about one-tenth of an inch in diameter ; but I need scarcely observe that the attempt was unattended with the slightest success.

" A very good way of observing the above phænomena is to examine one end of an electro-magnet with a powerful microscope while the other end is fixed. The increment is then observed to take place with extreme suddenness, as if it had been occasioned by a blow at the other extremity. The expansion, though very minute, is indeed so very rapid that it may be felt by the touch; and if the electro-magnet be placed perpendicularly on a hard elastic body, such as glass, the ear can readily detect the fact that it makes a slight jump every time that contact is made with the battery.

" When one end of the electro-magnet is applied to the ear, a distinct musical sound is heard every time that contact is made with, or broken from, the battery; another proof of the suddenness with which the particles of iron are disturbed."

In another part of the lecture I stated my reasons for supposing that whilst the bar of iron was increased in length by the magnetic influence, it experienced a *contraction* at right angles to the magnetic axis, so as to prevent any change taking place in the bulk of the bar. I intended as soon as possible to bring this conjecture to the test of experiment, and I prepared some apparatus for the purpose; but owing to other occupations I was obliged to relinquish the experiments until the beginning of last summer. In the meantime the inquiry has been taken up by De la Rive, Matteucci, Wertheim, Wartmann, Marrian, Beatson and others, whose ingenious experiments have invested the subject with additional interest. The researches of Beatson have taken a similar direction to mine; and he appears also to have employed a somewhat similar apparatus to that which I shall presently describe. I have confirmed several of the results at which this gentleman has arrived, and have added new facts, which I hope will throw further light upon this rather obscure department of physics.

In order to ascertain how far my opinion as to the invariability of the *bulk* of a bar of iron under magnetic influence was well-founded, I devised the following apparatus. Ten copper wires, each 110 yards long and one-twentieth of an inch in diameter, were bound together by tape so as to form a good, and at the same time very flexible conductor. The bundle of wires thus formed was coiled upon a glass tube 40 inches long and $1\frac{1}{2}$ inch in diameter. One end of the tube was hermetically sealed, and the other end was furnished with a glass stopper, which was itself perforated so as to admit of the insertion of a graduated capillary tube. In making the experiments, a bar of annealed iron, one yard long and half an inch square, was placed in the tube, which was then filled up with water. The stopper was then adjusted, and the capillary tube inserted so as to force the water to a convenient height within it.

The bulk of the iron bar was about 4,500,000 times the

capacity of each division of the graduated tube; consequently a very minute expansion of the former would have produced a perceptible motion of the water in the capillary tube : but, on connecting the coil with a Daniell's battery of five or six cells (a voltaic apparatus quite adequate to saturate the iron), no perceptible effect whatever was produced either in making or breaking contact with the battery, whether the water was stationary in the stem, or gradually rising or falling from a change of temperature.   Now had the usual increase of length been unaccompanied by a corresponding diminution of the diameter of the bar, the water would have been forced through twenty divisions of the capillary tube every time that contact was made with the battery.

Having thus ascertained that the bulk of the bar was invariable, I proceeded to repeat my first experiments with a more delicate apparatus, in order, by a more careful investigation of the laws of the increment of length, to ascend to the probable cause of the phænomenon.

A coiled glass tube, similar to that already described, was fixed vertically in a wooden frame.   Its length was such that when a bar one yard long was introduced so as to rest on the sealed end, each extremity of the bar was a full inch within the corresponding extremity of the coil.   The apparatus for observing the increment of length consisted of two levers of the first order, and a powerful microscope situated at the extremity of the second lever.   These levers were furnished with brass knife edges resting upon glass.   The connexion between the free extremity of the bar of iron and the first lever, and that between the two levers, were established by means of exceedingly fine platinum wires.

The first lever multiplied the motion of the extremity of the bar 7·8 times, the second multiplied the motion of the first 8 times, and the microscope was furnished with a micrometer divided into parts each corresponding to $\frac{1}{2220}$th of an inch.   Consequently each division of the micrometer passed over by the index indicated an increment of the length of the bar amounting to $\frac{1}{138528}$th of an inch.

The quantities of electricity passing through the coil were measured by an accurate galvanometer of tangents, consisting of a circle of thick copper wire one foot in diameter, and a needle half an inch long furnished with a suitable index.

The quantities of magnetic polarity communicated to the iron bar were measured by a finely suspended magnet 18 inches long, placed at the distance of one foot from the centre of the coil.   This magnetic bar was furnished with scales precisely in the manner of an ordinary balance, and the weight required to bring it to a horizontal position indicated the intensity of the magnetism of the iron bar under examination.

After a few preliminary trials, a great advantage was found to result from filling the tube with water. The effect of the water was, as De la Rive had already remarked, to prevent the sound. It also checked the oscillations of the index, and had the important effect of preventing any considerable irregularities in the temperature of the bar.

The first experiment which I shall record was made with a bar consisting of two pieces of very well-annealed rectangular iron wire, each one yard long, a quarter of an inch broad, and about one-eighth of an inch thick. The pieces were fastened together so as to form a bar of nearly a quarter of an inch square. The coil was placed in connexion with a single constant cell, the resistance being further increased by the addition of a few feet of fine wire. The instant that the circuit was closed, the index passed over one division of the micrometer. The needle of the galvanometer was then observed to stand at 7° 20′, while the magnetic balance required 0·52 of a grain to bring it to an equilibrium. It had been found by proper experiments that a current of 7° 20′ passing through the coil was itself capable of exerting a force of 0·03 of a grain upon the balance; consequently the magnetic intensity of the bar was represented by 0·49 of a grain. On breaking the circuit, the index was observed to retire 0·3 of a division, leaving a permanent elongation of 0·7, and a permanent polarity of 0·42 of a grain. More powerful currents were now passed through the coil, and the observations repeated as before, with the results tabulated below.

Experiment 1.

Deflection of galvanometer.	Tangent of deflection.	Elongation or shortening of bar.	Total elongation.	Magnetic intensity of bar.	Square of magnetic intensity divided by total elongation.
− 7 20	128	1·0 E.	1·0	−0·49	240
0	0	0·3 S.	0·7	−0·42	252
− 9 30	167	2·9 E.	3·6	−0·93	240
0	0	1·2 S.	2·4	−0·74	228
−14 48	264	5·9 E.	8·3	−1·42	243
0	0	3·8 S.	4·5	−1·00	222
−23 10	428	10·3 E.	14·8	−1·87	236
0	0	7·6 S.	7·2	−1·26	220
−47 25	1088	16·1 E.	23·3	−2·22	211
0	0	13·9 S.	9·4	−1·35	194
−58 50	1653	14·8 E.	24·2	−2·21	202
0	0	13·3 S.	10·9	−1·35	168

Everything being left in the same position, the electricity was now transmitted in the + or contrary direction, so as first to remove the polarity acquired by the — current, and then to induce the opposite polarity. The total elongation of 10·9, acquired in the last experiment, is carried forward in the 4th column of the following second table of results.

Experiment 1 (continued).

Deflection of galvanometer.	Tangent of deflection.	Elongation or shortening of bar.	Total elongation.	Magnetic intensity of bar.	Square of magnetic intensity divided by total elongation.
+ 6 15	109	3·4 S.	7·5	−0·12	2
0	0	0	7·5	−0·17	4
+ 9 55	175	0·1 E.	7·6	+0·57	43
0	0	1·0 S.	6·6	+0·25	9
+15 40	280	3·7 E.	10·3	+1·30	164
0	0	4·0 S.	6·3	+0·78	97
+38 45	802	16·8 E.	23·1	+2·30	229
0	0	14·5 S.	8·6	+1·12	148
+51 30	1257	16·7 E.	25·3	+2·35	218
0	0	16·3 S.	9·0	+1·05	122

The next series of results was obtained with a fresh bar of exactly the same size and temper as the preceding. To avoid an unnecessary occupation of space, I shall omit the fresh heading of the table for a change in the direction of the current, and simply designate the commencement of a new condition by ruling a line.

Experiment 2.

Deflection of galvanometer.	Tangent of deflection.	Elongation or shortening of bar.	Total elongation.	Magnetic intensity of bar.	Square of magnetic intensity divided by total elongation.
+ 5 0	87	0·1 E.	0·1	+0·08	64
0	0	0	0·1	+0·03	9
+ 8 27	148	1·9 E.	2·0	+0·50	125
0	0	1·0 S.	1·0	+0·30	90
+13 27	239	5·8 E.	6·8	+1·16	198
0	0	3·1 S.	3·7	+0·69	129
+33 50	670	18·8 E.	22·5	+2·20	215
0	0	14·3 S.	8·2	+1·01	124
+53 50	1368	19·0 E.	27·2	+2·32	198
0	0	17·1 S.	10·1	+1·03	105
− 7 5	124	2·0 S.	8·1	−0·15	3
0	0	0·1 S.	8·0	−0·07	0
−55 10	1437	20·0 E.	28·0	−2·20	173
0	0	14·6 S.	13·4	−1·39	144

The next experiment was with a bar of well-annealed iron, one yard long and about half an inch square. Its weight was $45\frac{1}{4}$ oz. I have introduced an additional column into the table, in order to reduce the action on the magnetic balance to the section of the former bars, the weight of each of which was 8 oz.

Experiment 3.

Deflection of galvanometer.	Tangent of deflection.	Elongation or shortening of bar.	Total elongation.	Magnetic intensity of bar.	Magnetic intensity per unity of section.	Square of magnetic intensity per unity, divided by elongation.
+ 5 10	90	0·4 E.	0·4	+ 1·18	0·21	110
0	0	0·1 S.	0·3	+ 0·45	0·08	21
+ 8 2	141	0·7 E.	1·0	+ 1·82	0·32	102
0	0	0·2 S.	0·8	+ 0·67	0·12	18
+14 43	262	2·0 E.	2·8	+ 4·10	0·72	185
0	0	1·0 S.	1·8	+ 0·90	0·16	14
+40 3	840	12·0 E.	13·8	+11·08	1·95	275
0	0	8·4 S.	5·4	+ 1·20	0·21	8
+54 0	1376	13·8 E.	19·2	+13·53	2·38	295
0	0	12·0 S.	7·2	+ 1·20	0·21	6
+62 5	1887	14·4 E.	21·6	+14·13	2·48	285
0	0	13·5 S.	8·1	+ 1·20	0·21	5
− 6 30	114	1·2 S.	6·9	− 0·70	0·12	2
0	0	0	6·9	− 0·30	0·05	0
−14 25	257	0·7 E.	7·6	− 3·80	0·67	59
0	0	1·3 S.	6·3	− 1·15	0·20	7
−41 15	877	11·0 E.	17·3	−11·33	1·99	229
0	0	8·8 S.	8·5	− 1·50	0·26	8
−62 45	1941	16·0 E.	24·5	−13·71	2·41	237
0	0	13·0 S.	11·5	− 1·55	0·27	6
+ 5 35	98	0·8 S.	10·7	+ 0·16	0·03	0
0	0	0	10·7	− 0·40	0·07	0
+ 9 0	158	0·2 S.	10·5	+ 1·17	0·21	4
0	0	0·2 S.	10·3	+ 0·15	0·03	0
+14 20	255	0·3 E.	10·6	+ 3·30	0·58	32
0	0	1·2 S.	9·4	+ 0·50	0·09	1
+24 45	461	3·3 E.	12·7	+ 7·16	1·26	125
0	0	3·4 S.	9·3	+ 0·82	0·14	2
+39 50	834	9·6 E.	18·9	+11·43	2·01	214
0	0	8·0 S.	10·9	+ 0·95	0·17	2
+54 15	1389	12·6 E.	23·5	+13·47	2·37	239
0	0	11·6 S.	11·9	+ 1·00	0·18	3
+60 45	1785	13·2 E.	25·1	+13·84	2·43	235
0	0	12·4 S.	12·7	+ 1·00	0·18	3

Experiment 3 (continued).

Deflection of galvano-meter.	Tangent of deflection.	Elongation or short-ening of bar.	Total elongation.	Magnetic intensity of bar.	Magnetic intensity per unity of section.	Square of magnetic intensity per unity, divided by elongation.
$-\overset{\circ}{7}\ \overset{\prime}{13}$	127	1·0 S.	11·7	$-$ 1·13	0·20	3
0	0	0·1 S.	11·6	$-$ 0·50	0·09	1
$-10\ \ 25$	184	0	11·6	$-$ 2·16	0·38	12
0	0	0·2 S.	11·4	$-$ 1·00	0·18	3
$-15\ \ 57$	286	0·5 E.	11·9	$-$ 4·14	0·73	45
0	0	1·1 S.	10·8	$-$ 1·25	0·22	4
$-26\ \ \ \ 0$	488	3·5 E.	14·3	$-$ 7·45	1·31	120
0	0	3·2 S.	11·1	$-$ 1·50	0·26	6
$-40\ \ 55$	867	9·6 E.	20·7	$-11·48$	2·02	197
0	0	8·0 S.	12·7	$-$ 1·70	0·30	7
$-62\ \ 48$	1946	14·6 E.	27·3	$-13·76$	2·42	214
0	0	13·0 S.	14·3	$-$ 1·73	0·30	6

From the last column of each of the preceding tables we may, I think, safely infer that *the elongation is in the duplicate ratio of the magnetic intensity of the bar*, both when the magnetism is maintained by the influence of the coil, and in the case of the permanent magnetism after the current has been cut off. The discrepancies observable will, I think, be satisfactorily accounted for when we consider the nature of the magnetic actions taking place. When a bar experiences the inductive influence of a coil traversed by an electrical current, the particles near its axis do not receive as much polarity as those near its surface, because the former have to withstand the opposing inductive influence of a greater number of magnetic particles than the latter. This phænomenon will be diminished in the extent of its manifestation with an increase of the electrical force, and will finally disappear when the current is sufficiently powerful to saturate the iron. Again, when the iron, after having been magnetized by the coil, is abandoned to its own retentive powers by cutting off the electrical current, the magnetism of the interior particles will suffer a greater amount of deterioration than that of the exterior particles. The polarity of the former may indeed be sometimes actually reversed, as Dr. Scoresby found it to be in some extensive combinations of steel bars. Now whenever such influences as the above occur, so as to make the different parts of the bar magnetic to a various extent, the elongation will necessarily bear a greater proportion to the square of the magnetic intensity measured by the balance than would otherwise be the case.

For similar causes the interior of the bar will in general receive the neutralization and reversion of its polarity before the exterior, and hence we see in the tables that there is a considerable elongation of the bar after the reversion of the current; even when the effect upon the balance has become imperceptible, owing to the opposite effects of the interior and exterior magnetic particles.

The bars employed in the preceding experiments were annealed as perfectly as possible. The next series was made with a bar of exactly the same dimensions and quality as the bars employed in Experiments 1 and 2, excepting that it was not annealed.

Experiment 4.

Deflection of galvanometer.	Tangent of deflection.	Elongation or shortening of bar.	Total elongation.	Magnetic intensity of bar.	Square of magnetic intensity divided by total elongation.
+ 9 20	164	0·2 E	0·2	+0·15	112
0	0	0	0·2	+0·08	32
+15 20	274	0·9 E.	1·1	+0·50	227
0	0	0·7 S.	0·4	+0·33	272
+38 32	796	7·1 E.	7·5	+1·36	247
0	0	5·2 S.	2·3	+0·80	278
+50 30	1213	10·2 E.	12·5	+1·76	247
0	0	9·6 S.	2·9	+0·97	324
+57 40	1580	13·0 E.	15·9	+1·94	236
0	0	11·8 S.	4·1	+1·00	244
+62 20	1907	14·0 E.	18·1	+2·10	243
0	0	14·0 S.	4·1	+1·01	249
− 6 50	120	1·2 S.	2·9	+0·58	116
0	0	0	2·9	+0·65	145
−10 35	168	0·4 S.	2·5	+0·21	17
0	0	0	2·5	+0·35	49
−14 57	267	0	2·5	−0·30	36
0	0	0·2 S.	2·3	−0·13	7
−40 10	844	5·7 E.	8·0	−1·36	231
0	0	4·6 S.	3·4	−0·88	228
−53 30	1351	10·0 E.	13·4	−1·70	215
0	0	9·5 S.	3·9	−0·95	231
+ 9 27	166	1·3 S.	2·6	−0·36	50
0	0	0·1 E.	2·7	−0·40	59
+22 30	414	0·1 S.	2·6	+0·38	55
0	0	0	2·6	+0·22	18
+38 27	794	4·9 E.	7·5	+1·50	300
0	0	4·6 S.	2·9	+0·97	324

In the foregoing series the discrepancies before adverted to do not make their appearance to any considerable extent, except in the case of the reversion of the magnetic polarity. Taken altogether the series is strikingly confirmatory of the law of elongation already announced.

The next series of observations was obtained with a piece of soft steel wire one yard long and a quarter of an inch in diameter. Its weight being exactly 8 oz., no reduction of magnetic intensity had to be made for unity of section.

Experiment 5.

Deflection of galvanometer.	Tangent of deflection.	Elongation or shortening of bar.	Total elongation.	Magnetic intensity of bar.	Square of magnetic intensity divided by total elongation.
+38° 10	786	1·4 E.	1·4	+0·94	631
0	0	0·6 S.	0·8	+0·65	528
+50 45	1224	2·8 E.	3·6	+1·43	568
0	0	1·8 S.	1·8	+0·98	533
+60 25	1761	3·8 E.	5·6	+1·71	521
0	0	3·1 S.	2·5	+1·12	502
+67 50	2454	5·0 E.	7·5	+1·88	471
0	0	4·5 S.	3·0	+1·23	504
+69 20	2651	5·5 E.	8·5	+1·97	456
0	0	4·5 S.	4·0	+1·28	409
−41 40	890	1·3 S.	2·7	−0·76	214
0	0	1·0 S.	1·7	−0·35	72

Experiments 6 and 7 were made with fresh steel bars of precisely the same sort as that employed in the last series.

Experiment 6.

Deflection of galvanometer.	Tangent of deflection.	Elongation or shortening of bar.	Total elongation.	Magnetic intensity of bar.	Square of magnetic intensity divided by total elongation.
+38° 25	793	0·8 E.	0·8	+0·78	760
0	0	0·5 S.	0·3	+0·46	705
+60 50	1792	5·2 E.	5·5	+1·60	466
0	0	3·4 S.	2·1	+0·99	467
+70 30	2824	7·0 E.	9·1	+1·88	388
0	0	5·8 S.	3·3	+1·16	408
−16 28	295	1·8 S.	1·5	+0·82	448
0	0	0·2 S.	1·3	+0·94	680
−38 50	805	0	1·3	−0·64	315
0	0	0·3 S.	1·0	−0·33	108

Experiment 7.

Deflection of galvanometer.	Tangent of deflection.	Elongation or shortening of bar.	Total elongation.	Magnetic intensity of bar.	Square of magnetic intensity divided by total elongation.
+38 20	790	1·4 E.	1·4	+0·74	391
0	0	0·7 S.	0·7	+0·46	302
+61 5	1810	5·3 E.	6·0	+1·64	448
0	0	3·2 S.	2·8	+1·07	409
+69 55	2735	4·7 E.	7·5	+1·90	481
0	0	4·5 S.	3·0	+1·20	480
−26 40	502	3·0 S.	0	+0·20	
0	0	0·2 S.	−0·2	+0·32	

The uniformity of the numbers in the last columns of the preceding tables shows that where the metal possesses a considerable degree of retentive power the anomalies occasioned by the reaction of the magnetic particles upon one another, which have been already adverted to, do not exist to any considerable extent, and presents an additional confirmation of the law I have stated, viz. *the elongation is proportional, in a given bar, to the square of the magnetic intensity.*

I now made trial of a bar of steel one yard long, half an inch broad, and a quarter of an inch thick, weighing 23 oz. It was hardened to a certain extent throughout its whole length, but not to such a degree as entirely to resist the action of the file.

Experiment 8.

Deflection of galvanometer.	Tangent of deflection.	Elongation or shortening of bar.	Total elongation.	Magnetic intensity of bar.	Magnetic intensity per unity of section.	Square of magnetic intensity per unity, divided by elongation.
+39 6	810	0	0	+1·11	0·38	
0	0	0·2 E.	0·2	+1·36	0·47	1104
+52 35	1307	0·8 E.	1·0	+4·09	1·42	2016
0	0	0·3 E.	1·3	+2·85	0·99	754
+60 15	1750	0·5 E.	1·8	+5·10	1·77	1740
0	0	0·1 E.	1·9	+3·52	1·22	783
+69 45	2710	0·6 E.	2·5	+5·91	2·06	1697
0	0	0·2 E.	2·7	+4·20	1·46	790
−41 15	877	1·6 S.	1·1	−0·43	0·15	20
0	0	0·1 E.	1·2	+0·35	0·12	12
−56 5	1487	1·4 E.	2·6	−3·90	1·36	711
0	0	0·1 E.	2·7	−2·63	0·91	307

In the above table it will be observed that the hard steel bar was slightly increased in length every time that contact with the battery was broken, although a considerable diminution of the magnetism of the bar took place at the same time. I am disposed to attribute this effect to the state of tension in the hardened steel, for I find that soft iron wire presents a similar anomaly when stretched tightly.

On inspecting the tables, it will be remarked that the elongation is, for the same intensity of magnetism, greater in proportion to the softness of the metal. It is greatest of all in the well-annealed iron bars, and least in the hardened steel. This circumstance appears to me to favour the hypothesis that the phænomena are produced by the attractions taking place between the magnetic particles of the bar, an hypothesis in perfect accordance with the law of elongation which I have pointed out.

[To be continued.]

27

Sound Attenuation Due to Viscosity

GEORGE GABRIEL STOKES

George Gabriel Stokes (1819–1903), British mathematician and physicist, is well known for his theorem in vector analysis and for his work on viscous fluids and on fluorescence. It was in connection with his research on fluid motion that he discovered the effect of fluid viscosity on sound propagation. An extract from his 1845 article on this subject follows. It was the failure of the experimentally measured attenuation of sound in a fluid to agree with Stokes' theoretical prediction that was the starting point of the modern attempts to account for attenuation in terms of molecular relaxation.

[From the *Transactions of the Cambridge Philosophical Society,*
Vol. VIII. p. 287.]

ON THE THEORIES OF THE INTERNAL FRICTION OF FLUIDS IN MOTION, AND OF THE EQUILIBRIUM AND MOTION OF ELASTIC SOLIDS.

[Read April 14, 1845.]

THE equations of Fluid Motion commonly employed depend upon the fundamental hypothesis that the mutual action of two adjacent elements of the fluid is normal to the surface which separates them. From this assumption the equality of pressure in all directions is easily deduced, and then the equations of motion are formed according to D'Alembert's principle. This appears to me the most natural light in which to view the subject; for the two principles of the absence of tangential action, and of the equality of pressure in all directions ought not to be assumed as independent hypotheses, as is sometimes done, inasmuch as the latter is a necessary consequence of the former*. The equations of motion so formed are very complicated, but yet they admit of solution in some instances, especially in the case of small oscillations. The results of the theory agree on the whole with observation, so far as the time of oscillation is concerned. But there is a whole class of motions of which the common theory takes no cognizance whatever, namely, those which depend on the tangential action called into play by the sliding of one portion of a fluid along another, or of a fluid along the surface of a solid, or of a different fluid, that action in fact which performs the same part with fluids that friction does with solids.

* This may be easily shewn by the consideration of a tetrahedron of the fluid, as in Art. 4.

Thus, when a ball pendulum oscillates in an indefinitely extended fluid, the common theory gives the arc of oscillation constant. Observation however shews that it diminishes very rapidly in the case of a liquid, and diminishes, but less rapidly, in the case of an elastic fluid. It has indeed been attempted to explain this diminution by supposing a friction to act on the ball, and this hypothesis may be approximately true, but the imperfection of the theory is shewn from the circumstance that no account is taken of the equal and opposite friction of the ball on the fluid.

Again, suppose that water is flowing down a straight aqueduct of uniform slope, what will be the discharge corresponding to a given slope, and a given form of the bed ? Of what magnitude must an aqueduct be, in order to supply a given place with a given quantity of water? Of what form must it be, in order to ensure a given supply of water with the least expense of materials in the construction ? These, and similar questions are wholly out of the reach of the common theory of Fluid Motion, since they entirely depend on the laws of the transmission of that tangential action which in it is wholly neglected. In fact, according to the common theory the water ought to flow on with uniformly accelerated velocity; for even the supposition of a certain friction against the bed would be of no avail, for such friction could not be transmitted through the mass. The practical importance of such questions as those above mentioned has made them the object of numerous experiments, from which empirical formulæ have been constructed. But such formulæ, although fulfilling well enough the purposes for which they were constructed, can hardly be considered as affording us any material insight into the laws of nature; nor will they enable us to pass from the consideration of the phenomena from which they were derived to that of others of a different class, although depending on the same causes.

In reflecting on the principles according to which the motion of a fluid ought to be calculated when account is taken of the tangential force, and consequently the pressure not supposed the same in all directions, I was led to construct the theory explained in the first section of this paper, or at least the main part of it, which consists of equations (13), and of the principles on which they are formed. I afterwards found that Poisson had written a memoir on the same subject, and on referring to it I found that he had arrived at the same equations. The method which he employed was however so different from mine that I feel justified in laying the latter before this Society*. The leading principles of my theory will be found in the hypotheses of Art. 1, and in Art. 3.

The second section forms a digression from the main object of this paper, and at first sight may appear to have little connexion with it. In this section I have, I think, succeeded in shewing that Lagrange's proof of an important theorem in the ordinary theory of Hydrodynamics is untenable. The theorem to which I refer is the one of which the object is to shew that $udx + vdy + wdz$, (using the common notation,) is always an exact differential when it is so at one instant. I have mentioned the principles of M. Cauchy's proof, a proof, I think, liable to no sort of objection. I have also given a new proof of the theorem, which would have served to establish it had M. Cauchy not been so fortunate as to obtain three first integrals of the general equations of motion. As it is, this proof may possibly be not altogether useless.

Poisson, in the memoir to which I have referred, begins with establishing, according to his theory, the equations of equilibrium and motion of elastic solids, and makes the equations of motion of fluids depend on this theory. On reading his memoir, I was led to apply to the theory of elastic solids principles precisely analogous to those which I have employed in the case of fluids. The formation of the equations, according to these principles, forms the subject of Sect. III.

The equations at which I have thus arrived contain two arbitrary constants, whereas Poisson's equations contain but one. In Sect. IV. I have explained the principles of Poisson's theories of elastic solids, and of the motion of fluids, and pointed out what appear to me serious objections against the truth of one of the hypotheses which he employs in the former. This theory seems to be very generally received, and in consequence it is usual to deduce the measure of the cubical compressibility of elastic solids from that of their extensibility, when formed into rods or wires,

---

* The same equations have also been obtained by Navier in the case of an incompressible fluid (*Mém. de l'Académie*, t. vi. p. 389), but his principles differ from mine still more than do Poisson's.

or from some quantity of the same nature. If the views which I have explained in this section be correct, the cubical compressibility deduced in this manner is too great, much too great in the case of the softer substances, and even the softer metals. The equations of Sect. III. have, I find, been already obtained by M. Cauchy in his *Exercises Mathématiques*, except that he has not considered the effect of the heat developed by sudden compression. The method which I have employed is different from his, although in some respects it much resembles it.

The equations of motion of elastic solids given in Sect. III. are the same as those to which different authors have been led, as being the equations of motion of the luminiferous ether in vacuum. It may seem strange that the same equations should have been arrived at for cases so different; and I believe this has appeared to some a serious objection to the employment of those equations in the case of light. I think the reflections which I have made at the end of Sect. IV., where I have examined the consequences of the law of continuity, a law which seems to pervade nature, may tend to remove the difficulty.

SECTION I.

Explanation of the Theory of Fluid Motion proposed. Formation of the Differential Equations. Application of these Equations to a few simple cases.

1.   Before entering on the explanation of this theory, it will be necessary to define, or fix the precise meaning of a few terms which I shall have occasion to employ.

In the first place, the expression " the velocity of a fluid at any particular point" will require some notice. If we suppose a fluid to be made up of ultimate molecules, it is easy to see that these molecules must, in general, move among one another in an irregular manner, through spaces comparable with the distances between them, when the fluid is in motion. But since there is no doubt that the distance between two adjacent molecules is quite insensible, we may neglect the irregular part of the velocity, compared with the common velocity with which all the molecules in the neighbourhood of the one considered are moving. Or, we may consider the mean velocity of the molecules in the neighbourhood of the one considered, apart from the velocity due to the irregular motion. It is this regular velocity which I shall understand by the *velocity of a fluid at any point*, and I shall accordingly regard it as varying continuously with the co-ordinates of the point.

Let P be any material point in the fluid, and consider the instantaneous motion of a very small element E of the fluid about P. This motion is compounded of a motion of translation, the same as that of P, and of the motion of the several points of E relatively to P. If we conceive a velocity equal and opposite to that of P impressed on the whole element, the remaining velocities form what I shall call the *relative velocities* of the points of the fluid about P; and the motion expressed by these velocities is what I shall call the *relative motion* in the neighbourhood of P.

It is an undoubted result of observation that the molecular forces, whether in solids, liquids, or gases, are forces of enormous intensity, but which are sensible at only insensible distances. Let E' be a very small element of the fluid circumscribing E, and of a thickness greater than the distance to which the molecular forces are sensible. The forces acting on the element E are the external forces, and the pressures arising from the molecular action of E'. If the molecules of E were in positions in which they could remain at rest if E were acted on by no external force and the molecules of E' were held in their actual positions, they would be in what I shall call a state of *relative equilibrium*. Of course they may be far from being in a state of actual equilibrium. Thus, an element of fluid at the top of a wave may be sensibly in a state of relative equilibrium, although far removed from its position of equilibrium. Now, in consequence of the intensity of the molecular forces, the pressures arising from the molecular action on E will be very great compared with the external moving forces acting on E. Consequently the state of relative equilibrium, or of relative motion, of the molecules of E will not be sensibly affected by the external forces acting on E. But the pressures in different directions about the point P depend on that state of relative equilibrium or motion, and consequently will not be sensibly affected by the external moving forces acting on E. For the same reason they will not be sensibly affected by any motion of rotation common to all the points of E; and it is a direct consequence of the second law of motion, that they will

80   ON THE FRICTION OF FLUIDS IN MOTION,

not be affected by any motion of translation common to the whole element. If the molecules of E were in a state of relative equilibrium, the pressure would be equal in all directions about P, as in the case of fluids at rest. Hence I shall assume the following principle :—

That the difference between the pressure on a plane in a given direction passing through any point P of a fluid in motion and the pressure which would exist in all directions about P if the fluid in its neighbourhood were in a state of relative equilibrium depends only on the relative motion of the fluid immediately about P; and that the relative motion due to any motion of rotation may be eliminated without affecting the differences of the pressures above mentioned.

Let us see how far this principle will lead us when it is carried out.

2. It will be necessary now to examine the nature of the most general instantaneous motion of an element of a fluid. The proposition in this article is however purely geometrical, and may be thus enunciated :—" Supposing space, or any portion of space, to be filled with an infinite number of points which move in any continuous manner, retaining their identity, to examine the nature of the instantaneous motion of any elementary portion of these points."

Let u, v, w be the resolved parts, parallel to the rectangular axes, Ox, Oy, Oz, of the velocity of the point P, whose co-ordinates at the instant considered are x, y, z. Then the relative velocities at the point P', whose co-ordinates are $x + x', y + y', z + z'$, will be

$$\frac{du}{dx} x' + \frac{du}{dy} y' + \frac{du}{dz} z' \text{ parallel to } x,$$

$$\frac{dv}{dx} x' + \frac{dv}{dy} y' + \frac{dv}{dz} z' \ \ldots\ldots\ldots\ldots y,$$

$$\frac{dw}{dx} x' + \frac{dw}{dy} y' + \frac{dw}{dz} z' \ \ldots\ldots\ldots\ldots z,$$

neglecting squares and products of x', y', z'. Let these velocities be compounded of those due to the angular velocities $\omega', \omega'', \omega'''$ about the axes of x, y, z, and of the velocities U, V, W parallel

267

to x, y, z. The linear velocities due to the angular velocities being $\omega''z' - \omega'''y'$, $\omega'''x' - \omega'z'$, $\omega'y' - \omega''x'$ parallel to the axes of x, y, z, we shall therefore have

$$U = \frac{du}{dx}x' + \left(\frac{du}{dy} + \omega'''\right)y' + \left(\frac{du}{dz} - \omega''\right)z',$$

$$V = \left(\frac{dv}{dx} - \omega'''\right)x' + \frac{dv}{dy}y' + \left(\frac{dv}{dz} + \omega'\right)z',$$

$$W = \left(\frac{dw}{dx} + \omega''\right)x' + \left(\frac{dw}{dy} - \omega'\right)y' + \frac{dw}{dz}z'.$$

Since ω', ω'', ω''' are arbitrary, let them be so assumed that

$$\frac{dU}{dy'} = \frac{dV}{dx'}, \quad \frac{dV}{dz'} = \frac{dW}{dy'}, \quad \frac{dW}{dx'} = \frac{dU}{dz'},$$

which gives

$$\omega' = \tfrac{1}{2}\left(\frac{dw}{dy} - \frac{dv}{dz}\right), \quad \omega'' = \tfrac{1}{2}\left(\frac{du}{dz} - \frac{dw}{dx}\right), \quad \omega''' = \tfrac{1}{2}\left(\frac{dv}{dx} - \frac{du}{dy}\right), \quad \dots(1),$$

$$\left.\begin{aligned}
U &= \frac{du}{dx}x' + \tfrac{1}{2}\left(\frac{du}{dy} + \frac{dv}{dx}\right)y' + \tfrac{1}{2}\left(\frac{du}{dz} + \frac{dw}{dx}\right)z', \\
V &= \tfrac{1}{2}\left(\frac{dv}{dx} + \frac{du}{dy}\right)x' + \frac{dv}{dy}y' + \tfrac{1}{2}\left(\frac{dv}{dz} + \frac{dw}{dy}\right)z', \\
W &= \tfrac{1}{2}\left(\frac{dw}{dx} + \frac{du}{dz}\right)x' + \tfrac{1}{2}\left(\frac{dw}{dy} + \frac{dv}{dz}\right)y' + \frac{dw}{dz}z',
\end{aligned}\right\} \dots(2).$$

The quantities ω', ω'', ω''' are what I shall call the *angular velocities of the fluid* at the point considered. This is evidently an allowable definition, since, in the particular case in which the element considered moves as a solid might do, these quantities coincide with the angular velocities considered in rigid dynamics. A further reason for this definition will appear in Sect. III.

Let us now investigate whether it is possible to determine x', y', z' so that, considering only the relative velocities U, V, W, the line joining the points P, P' shall have no angular motion. The conditions to be satisfied, in order that this may be the case, are evidently that the increments of the relative co-ordinates x, y, z' of the second point shall be ultimately proportional to those co-ordinates. If e be the rate of extension of the line joining the two points considered, we shall therefore have

$$\left.\begin{aligned}
Fx' + hy' + gz' &= ex', \\
hx' + Gy' + fz' &= ey', \\
gx' + fy' + Hz' &= ez',
\end{aligned}\right\} \dots\dots(3);$$

S.

6

where

$$F = \frac{du}{dx}, \quad G = \frac{dv}{dy}, \quad H = \frac{dw}{dz}, \quad 2f = \frac{dv}{dz} + \frac{dw}{dy},$$

$$2g = \frac{dw}{dx} + \frac{du}{dz}, \quad 2h = \frac{du}{dy} + \frac{dv}{dx}.$$

If we eliminate from equations (3) the two ratios which exist between the three quantities x', y', z', we get the well known cubic equation

$$(e-F)(e-G)(e-H) - f^2(e-F) - g^2(e-G) - h^2(e-H) - 2fgh = 0 \ldots (4),$$

which occurs in the investigation of the principal axes of a rigid body, and in various others. As in these investigations, it may be shewn that there are in general three directions, at right angles to each other, in which the point P' may be situated so as to satisfy the required conditions. If two of the roots of (4) are equal, there is one such direction corresponding to the third root, and an infinite number of others situated in a plane perpendicular to the former; and if the three roots of (4) are equal, a line drawn in any direction will satisfy the required conditions.

The three directions which have just been determined I shall call *axes of extension*. They will in general vary from one point to another, and from one instant of time to another. If we denote the three roots of (4) by e', e'', e''', and if we take new rectangular axes $Ox_{,}$, $Oy_{,}$, $Oz_{,}$, parallel to the axes of extension, and denote by $u_{,}$, $U_{,}$, &c. the quantities referred to these axes corresponding to u, U, &c., equations (3) must be satisfied by $y_{,}' = 0$, $z_{,}' = 0$, $e = e'$, by $x_{,}' = 0$, $z_{,}' = 0$, $e = e''$, and by $x_{,}' = 0$, $y_{,}' = 0$, $e = e'''$, which requires that $f_{,} = 0$, $g_{,} = 0$, $h_{,} = 0$, and we have

$$e' = F_{,} = \frac{du_{,}}{dx_{,}}, \quad e'' = G_{,} = \frac{dv_{,}}{dy_{,}}, \quad e'' = H_{,} = \frac{dw_{,}}{dz_{,}}.$$

The values of $U_{,}$, $V_{,}$, $W_{,}$, which correspond to the residual motion after the elimination of the motion of rotation corresponding to ω', ω'' and ω''', are

$$U_{,} = e'x_{,}', \quad V_{,} = e''y_{,}', \quad W_{,} = e''z_{,}'.$$

The angular velocity of which ω', ω'', ω''' are the components is independent of the arbitrary directions of the co-ordinate axes: the same is true of the directions of the axes of extension, and of the values of the roots of equation (4). This might be proved in various ways; perhaps the following is the simplest. The conditions by which ω', ω'', ω''' are determined are those which express that the relative velocities U, V, W, which remain after eliminating a certain angular velocity, are such that $Udx' + Vdy' + Wdz'$ is ultimately an exact differential, that is to say when squares and products of x', y' and z' are neglected. It appears moreover from the solution that there is only one way in which these conditions can be satisfied for a given system of co-ordinate axes. Let us take new rectangular axes, Ox, Oy, Oz, and let U, V, W be the resolved parts along these axes of the velocities U, V, W, and x', y', z', the relative co-ordinates of P'; then

$$U = \text{U} \cos x\mathbf{x} + \text{V} \cos x\mathbf{y} + \text{W} \cos x\mathbf{z},$$

$$dx' = \cos x\mathbf{x}\,dx' + \cos x\mathbf{y}\,dy' + \cos x\mathbf{z}\,dz', \&c.;$$

whence, taking account of the well known relations between the cosines involved in these equations, we easily find

$$Udx' + Vdy' + Wdz' = \text{U}dx' + \text{V}dy' + \text{W}dz'.$$

It appears therefore that the relative velocities U, V, W, which remain after eliminating a certain angular velocity, are such that $\text{U}dx' + \text{V}dy' + \text{W}dz'$ is ultimately an exact differential. Hence the values of U, V, W are the same as would have been obtained from equations (2) applied directly to the new axes, whence the truth of the proposition enunciated at the head of this paragraph is manifest.

The motion corresponding to the velocities $U_{,}$, $V_{,}$, $W_{,}$ may be further decomposed into a motion of dilatation, positive or negative, which is alike in all directions, and two motions which I shall call *motions of shifting*, each of the latter being in two dimensions, and not affecting the density. For let δ be the rate of linear extension corresponding to a uniform dilatation; let $\sigma x_{,}' - \sigma y_{,}'$ be the velocities parallel to $x_{,}$, $y_{,}$, corresponding to a motion of shifting parallel to the plane $x_{,}y_{,}$, and let $\sigma'x_{,}'$, $-\sigma'z_{,}'$ be the velocities parallel to $x_{,}$, $z_{,}$, corresponding to a similar motion of shifting parallel to the plane $x_{,}z_{,}$. The velocities parallel to $x_{,}$, $y_{,}$, $z_{,}$ respectively corresponding to the quantities δ, σ and σ' will be $(\delta + \sigma + \sigma')x_{,}'$, $(\delta - \sigma)y_{,}'$, $(\delta - \sigma')z_{,}'$, and equating these to $U_{,}$, $V_{,}$, $W_{,}$ we shall get

$$\delta = \tfrac{1}{3}(e' + e'' + e'''), \quad \sigma = \tfrac{1}{3}(e' + e''' - 2e''), \quad \sigma' = \tfrac{1}{3}(e' + e'' - 2e''').$$

Hence the most general instantaneous motion of an elementary portion of a fluid is compounded of a motion of translation, a

6—2

motion of rotation, a motion of uniform dilatation, and two motions of shifting of the kind just mentioned.

3. Having determined the nature of the most general instantaneous motion of an élement of a fluid, we are now prepared to consider the normal pressures and tangential forces called into play by the relative displacements of the particles. Let p be the pressure which would exist about the point P if the neighbouring molecules were in a state of relative equilibrium: let $p + p_{,}$ be the normal pressure, and $t_{,}$ the tangential action, both referred to a unit of surface, on a plane passing through P and having a given direction. By the hypotheses of Art. 1. the quantities $p_{,}, t_{,}$ will be independent of the angular velocities ω', ω'', ω''', depending only on the residual relative velocities U, V, W, or, which comes to the same, on e', e'' and e''', or on σ, σ' and δ. Since this residual motion is symmetrical with respect to the axes of extension, it follows that if the plane considered is perpendicular to any one of these axes the tangential action on it is zero, since there is no reason why it should act in one direction rather than in the opposite; for by the hypotheses of Art. 1 the change of density and temperature about the point P is to be neglected, the constitution of the fluid being ultimately uniform about that point. Denoting then by $p + p'$, $p + p''$, $p + p'''$ the pressures on planes perpendicular to the axes of $x_{,}, y_{,}, z_{,}$ we must have

$$p' = \phi(e', e'', e'''), \quad p'' = \phi(e'', e''', e'), \quad p''' = \phi(e''', e', e''),$$

$\phi(e', e'', e''')$ denoting a function of e', e'' and e''' which is symmetrical with respect to the two latter quantities. The question is now to determine, on whatever may seem the most probable hypothesis, the form of the function ϕ.

Let us first take the simpler case in which there is no dilatation, and only one motion of shifting, or in which $e'' = -e'$, $e''' = 0$, and let us consider what would take place if the fluid consisted of smooth molecules acting on each other by actual contact. On this supposition, it is clear, considering the magnitude of the pressures acting on the molecules compared with their masses, that they would be sensibly in a position of relative equilibrium, except when the equilibrium of any one of them became impossible from the displacement of the adjoining ones, in which case the molecule in question would start into a new position of equilibrium. This start would cause a corresponding displacement in the molecules

271

immediately about the one which started, and this disturbance would be propagated immediately in all directions, the nature of the displacement however being different in different directions, and would soon become insensible. During the continuance of this disturbance, the pressure on a small plane drawn through the element considered would not be the same in all directions, nor normal to the plane: or, which comes to the same, we may suppose a uniform normal pressure p to act, together with a normal pressure $p_{,,}$, and a tangential force $t_{,,}$, $p_{,,}$ and $t_{,,}$ being forces of great intensity and short duration, that is being of the nature of impulsive forces. As the number of molecules comprised in the element considered has been supposed extremely great, we may take a time τ so short that all summations with respect to such intervals of time may be replaced without sensible error by integrations, and yet so long that a very great number of starts shall take place in it. Consequently we have only to consider the average effect of such starts, and moreover we may without sensible error replace the impulsive forces such as $p_{,,}$ and $t_{,,}$, which succeed one another with great rapidity, by continuous forces. For planes perpendicular to the axes of extension these continuous forces will be the normal pressures p', p'', p'''.

Let us now consider a motion of shifting differing from the former only in having e' increased in the ratio of m to 1. Then, if we suppose each start completed before the starts which would be sensibly affected by it are begun, it is clear that the same series of starts will take place in the second case as in the first, but at intervals of time which are less in the ratio of 1 to m. Consequently the continuous pressures by which the impulsive actions due to these starts may be replaced must be increased in the ratio of m to 1. Hence the pressures p', p'', p''' must be proportional to e', or we must have

$$p' = Ce', \quad p'' = C'e', \quad p''' = C''e'.$$

It is natural to suppose that these formulæ hold good for negative as well as positive values of e'. Assuming this to be true, let the sign of e' be changed. This comes to interchanging x and y, and consequently p''' must remain the same, and p' and p'' must be interchanged. We must therefore have $C'' = 0$, $C' = -C$. Putting then $C = -2\mu$ we have

$$p' = -2\mu e', \quad p'' = 2\mu e', \quad p''' = 0.$$

272

It has hitherto been supposed that the molecules of a fluid are in actual contact. We have every reason to suppose that this is not the case. But precisely the same reasoning will apply if they are separated by intervals as great as we please compared with their magnitudes, provided only we suppose the force of restitution called into play by a small displacement of *any one* molecule to be very great.

Let us now take the case of two motions of shifting which co-exist, and let us suppose $e' = \sigma + \sigma'$, $e'' = -\sigma$, $e''' = -\sigma'$. Let the small time τ be divided into $2n$ equal portions, and let us suppose that in the first interval a shifting motion corresponding to $e' = 2\sigma$, $e'' = -2\sigma$ takes place parallel to the plane $x_{,}y_{,}$, and that in the second interval a shifting motion corresponding to $e' = 2\sigma'$, $e''' = -2\sigma'$ takes place parallel to the plane $x_{,}z_{,}$, and so on alternately. On this supposition it is clear that if we suppose the time $\tau/2n$ to be extremely small, the continuous forces by which the effect of the starts may be replaced will be $p' = -2\mu(\sigma + \sigma')$, $p'' = 2\mu\sigma$, $p''' = 2\mu\sigma'$. By supposing n indefinitely increased, we might make the motion considered approach as near as we please to that in which the two motions of shifting coexist; but we are not at liberty to do so, for in order to apply the above reasoning we must suppose the time $\tau/2n$ to be so large that the average effect of the starts which occur in it may be taken. Consequently it must be taken as an additional assumption, and not a matter of absolute demonstration, that the effects of the two motions of shifting are superimposed.

Hence if $\delta = 0$, *i.e.* if $e' + e'' + e''' = 0$, we shall have in general
$$p' = -2\mu e', \quad p'' = -2\mu e'', \quad p''' = -2\mu e''' \dots\dots\dots(5).$$
It was by this hypothesis of starts that I first arrived at these equations, and the differential equations of motion which result from them. On reading Poisson's memoir however, to which I shall have occasion to refer in Section IV., I was led to reflect that however intense we may suppose the molecular forces to be, and however near we may suppose the molecules to be to their posi-tions of relative equilibrium, we are not therefore at liberty to suppose them *in* those positions, and consequently not at liberty to suppose the pressure equal in all directions in the intervals of time between the starts. In fact, by supposing the molecular forces indefinitely increased, retaining the same ratios to each other, we may suppose the displacements of the molecules from

their positions of relative equilibrium indefinitely diminished, but on the other hand the force of restitution called into action by a given displacement is indefinitely increased in the same proportion. But be these displacements what they may, we know that the forces of restitution make equilibrium with forces equal and opposite to the effective forces ; and in calculating the effective forces we may neglect the above displacements, or suppose the molecules to move in the paths in which they would move if the shifting motion took place with indefinite slowness. Let us first consider a single motion of shifting, or one for which $e'' = -e'$, $e''' = 0$, and let $p_{,}$ and $t_{,}$ denote the same quantities as before. If we now suppose e' increased in the ratio of m to 1, all the effective forces will be increased in that ratio, and consequently $p_{,}$ and $t_{,}$ will be increased in the same ratio. We may deduce the values of $p'\ p''$, and p''' just as before, and then pass by the same reasoning to the case of two motions of shifting which coexist, only that in this case the reasoning will be demonstrative, since we *may* suppose the time $\tau/2n$ indefinitely diminished. If we suppose the state of things considered in this paragraph to exist along with the motions of starting already considered, it is easy to see that the expressions for p', p'' and p''' will still retain the same form.

There remains yet to be considered the effect of the dilatation. Let us first suppose the dilatation to exist without any shifting : then it is easily seen that the relative motion of the fluid at the point considered is the same in all directions. Consequently the only effect which such a dilatation could have would be to introduce a normal pressure $p_{,}$, alike in all directions, in addition to that due to the action of the molecules supposed to be in a state of relative equilibrium. Now the pressure $p_{,}$ could only arise from the aggregate of the molecular actions called into play by the displacements of the molecules from their positions of relative equilibrium; but since these displacements take place, on an average, indifferently in all directions, it follows that the actions of which $p_{,}$ is composed neutralize each other, so that $p_{,} = 0$. The same conclusion might be drawn from the hypothesis of starts, supposing, as it is natural to do, that each start calls into action as much increase of pressure in some directions as diminution of pressure in others.

If the motion of uniform dilatation coexists with two motions of shifting, I shall suppose, for the same reason as before, that the effects of these different motions are superimposed. Hence subtracting δ from each of the three quantities e', e'' and e''', and putting the remainders in the place of e', e'' and e''' in equations (5), we have

$$p' = \tfrac{2}{3}\mu(e'' + e''' - 2e'), \quad p'' = \tfrac{2}{3}\mu(e''' + e' - 2e''),$$
$$p''' = \tfrac{2}{3}\mu(e' + e'' - 2e''') \dots\dots\dots(6).$$

If we had started with assuming $\phi(e', e'', e''')$ to be a linear function of e', e'' and e''', avoiding all speculation as to the molecular constitution of a fluid, we should have had at once $p' = Ce' + C'(e'' + e''')$, since p' is symmetrical with respect to e'' and e'''; or, changing the constants, $p' = \tfrac{2}{3}\mu(e'' + e''' - 2e') + \kappa(e' + e'' + e''')$. The expressions for p'' and p''' would be obtained by interchanging the requisite quantities. Of course we may at once put $\kappa = 0$ if we assume that in the case of a uniform motion of dilatation the pressure at any instant depends only on the actual density and temperature at that instant, and not on the rate at which the former changes with the time. In most cases to which it would be interesting to apply the theory of the friction of fluids the density of the fluid is either constant, or may without sensible error be regarded as constant, or else changes slowly with the time. In the first two cases the results would be the same, and in the third case nearly the same, whether κ were equal to zero or not. Consequently, if theory and experiment should in such cases agree, the experiments must not be regarded as confirming that part of the theory which relates to supposing κ to be equal to zero.

4. It will be easy now to determine the oblique pressure, or resultant of the normal pressure and tangential action, on any plane. Let us first consider a plane drawn through the point P parallel to the plane yz. Let Ox, make with the axes of x, y, z angles whose cosines are l', m', n'; let l'', m'', n'' be the same for Oy, and l''', m''', n''' the same for Oz. Let P_1 be the pressure, and (xty), (xtz) the resolved parts, parallel to y, z respectively, of the tangential force on the plane considered, all referred to a unit of surface, (xty) being reckoned positive when the part of the fluid towards $-x$ urges that towards $+x$ in the positive direction of y, and similarly for (xtz). Consider the portion of the fluid comprised within a tetrahedron having its vertex in the point P, its base parallel to the plane yz, and its three sides parallel to the

planes $x_{,}y_{,}$, $y_{,}z_{,}$, $z_{,}x_{,}$ respectively.   Let A be the area of the base, and therefore $l'A$, $l''A$, $l'''A$ the areas of the faces perpendicular to the axes of $x_{,}$, $y_{,}$, $z_{,}$.   By D'Alembert's principle, the pressures and tangential actions on the faces of this tetrahedron, the moving forces arising from the external attractions, not including the molecular forces, and forces equal and opposite to the effective moving forces will be in equilibrium, and therefore the sums of the resolved parts of these forces in the directions of x, y and z will each be zero.   Suppose now the dimensions of the tetrahedron indefinitely diminished, then the resolved parts of the external, and of the effective moving forces will vary ultimately as the cubes, and those of the pressures and tangential forces on the sides as the squares of homologous lines.   Dividing therefore the three equations arising from equating to zero the resolved parts of the above forces by A, and taking the limit, we have

$$P_{,} = \Sigma l'^2 (p + p'), \quad (xty) = \Sigma l'm' (p + p'), \quad (xtz) = \Sigma l'n' (p + p'),$$

the sign Σ denoting the sum obtained by taking the quantities corresponding to the three axes of extension in succession.   Putting for p', p'' and p''' their values given by (6), putting $e' + e'' + e''' = 3\delta$, and observing that $\Sigma l'^2 = 1$, $\Sigma l'm' = 0$, $\Sigma l'n' = 0$, the above equations become

$$P_{,} = p - 2\mu \Sigma l'^2 e' + 2\mu\delta, \quad (xty) = - 2\mu \Sigma l'm'e', \quad (xtz) = - 2\mu \Sigma l'n'e'.$$

The method of determining the pressure on any plane from the pressures on three planes at right angles to each other, which has just been given, has already been employed by MM. Cauchy and Poisson.

The most direct way of obtaining the values of $\Sigma l'^2 e'$ &c. would be to express l, m' and n' in terms of e' by any two of equations (3), in which x', y', z' are proportional to l', m', n', together with the equation $l'^2 + m'^2 + n'^2 = 1$, and then to express the resulting symmetrical function of the roots of the cubic equation (4) in terms of the coefficients.   But this method would be excessively laborious, and need not be resorted to.   For after eliminating the angular motion of the element of fluid considered the remaining velocities are $e'x_{,}'$, $e''y_{,}'$, $e'''z_{,}'$, parallel to the axes of $x_{,}$, $y_{,}$, $z_{,}$.   The sum of the resolved parts of these parallel to the axis of x is $l'e'x_{,}' + l''e''y_{,}' + l'''e'''z_{,}'$.   Putting for $x_{,}'$, $y_{,}'$, $z_{,}'$ their values $lx' + m'y' + n'z'$ &c., the above sum becomes

$$x'\Sigma l'^2 e' + y'\Sigma l'm'e' + z'\Sigma l'n'e';$$

but this sum is the same thing as the velocity U in equation (2), and therefore we have

$$\Sigma l^2 e' = \frac{du}{dx}, \quad \Sigma l'm'e' = \tfrac{1}{2}\left(\frac{du}{dy} + \frac{dv}{dx}\right), \quad \Sigma l'n'e' = \tfrac{1}{2}\left(\frac{du}{dz} + \frac{dw}{dx}\right).$$

It may also be very easily proved directly that the value of 3δ, the rate of cubical dilatation, satisfies the equation

$$3\delta = \frac{du}{dx} + \frac{dv}{dy} + \frac{dw}{dz} \quad\dots\dots\dots\dots\dots\dots (7).$$

Let P_2, (ytz), (ytx) be the quantities referring to the axis of y, and P_3, (ztx), (zty) those referring to the axis of z, which correspond to P_1 &c. referring to the axis of x. Then we see that $(ytz) = (zty)$, $(ztx) = (xtz)$, $(xty) = (ytx)$. Denoting these three quantities by T_1, T_2, T_3, and making the requisite substitutions and interchanges, we have

$$\left.\begin{aligned}
P_1 &= p - 2\mu\left(\frac{du}{dx} - \delta\right), \\[2mm]
P_2 &= p - 2\mu\left(\frac{dv}{dy} - \delta\right), \\[2mm]
P_3 &= p - 2\mu\left(\frac{dw}{dz} - \delta\right), \\[2mm]
T_1 &= -\mu\left(\frac{dv}{dz} + \frac{dw}{dy}\right), \\[2mm]
T_2 &= -\mu\left(\frac{dw}{dx} + \frac{du}{dz}\right), \\[2mm]
T_3 &= -\mu\left(\frac{du}{dy} + \frac{dv}{dx}\right),
\end{aligned}\right\} \dots\dots\dots\dots (8).$$

It may also be useful to know the components, parallel to x, y, z, of the oblique pressure on a plane passing through the point P, and having a given direction. Let l, m, n be the cosines of the angles which a normal to the given plane makes with the axes of x, y, z; let P, Q, R be the components, referred to a unit of surface, of the oblique pressure on this plane, P, Q, R being reckoned positive when the part of the fluid in which is situated the normal to which l, m and n refer is urged by the other part in the positive directions of x, y, z, when l, m and n are positive. Then considering as before a tetrahedron of which the base is parallel to the given plane, the vertex in the point P, and the sides parallel to the co-ordinate planes, we shall have

$$\left.\begin{aligned} P &= lP_1 + mT_3 + nT_2, \\ Q &= lT_3 + mP_2 + nT_1, \\ R &= lT_2 + mT_1 + nP_3, \end{aligned}\right\} \quad \dots\dots\dots\dots (9).$$

In the simple case of a sliding motion for which $u = 0$, $v = f(x)$, $w = 0$, the only forces, besides the pressure p, which act on planes parallel to the co-ordinate planes are the two tangential forces T_3, the value of which in this case is $-\mu \, dv/dx$. In this case it is easy to shew that the axes of extension are, one of them parallel to Oz, and the two others in a plane parallel to xy, and inclined at angles of $45°$ to Ox. We see also that it is necessary to suppose μ to be positive, since otherwise the tendency of the forces would be to increase the relative motion of the parts of the fluid, and the equilibrium of the fluid would be unstable.

5.   Having found the pressures about the point P on planes parallel to the co-ordinate planes, it will be easy to form the equations of motion.   Let X, Y, Z be the resolved parts, parallel to the axes, of the external force, not including the molecular force; let ρ be the density, t the time.   Consider an elementary parallelepiped of the fluid, formed by planes parallel to the co-ordinate planes, and drawn through the point (x, y, z) and the point $(x + \Delta x, y + \Delta y, z + \Delta z)$.   The mass of this element will be ultimately $\rho \Delta x \Delta y \Delta z$, and the moving force parallel to x arising from the external forces will be ultimately $\rho X \Delta x \Delta y \Delta z$; the effective moving force parallel to x will be ultimately $\rho \, Du/Dt \, . \, \Delta x \Delta y \Delta z$, where D is used, as it will be in the rest of this paper, to denote differentiation in which the independent variables are t and three parameters of the particle considered, (such for instance as its initial cordinates,) and not t, x, y, z.   It is easy also to shew that the moving force acting on the element considered arising from the oblique pressures on the faces is ultimately

$$\left(\frac{dP_1}{dx} + \frac{dT_3}{dy} + \frac{dT_2}{dz}\right) \Delta x \, \Delta y \, \Delta z,$$

acting in the negative direction.   Hence we have by D'Alembert's principle

$$\rho \left(\frac{Du}{Dt} - X\right) + \frac{dP_1}{dx} + \frac{dT_3}{dy} + \frac{dT_2}{dz} = 0, \ \&\text{c.} \ \dots\dots (10),$$

278

in which equations we must put for Du/Dt its value

$$\frac{du}{dt} + u\,\frac{du}{dx} + v\,\frac{du}{dy} + w\,\frac{du}{dz},$$

and similarly for Dv/dt and Dw/dt. In considering the general equations of motion it will be needless to write down more than one, since the other two may be at once derived from it by interchanging the requisite quantities. The equations (10), the ordinary equation of continuity, as it is called,

$$\frac{d\rho}{dt} + \frac{d\rho u}{dx} + \frac{d\rho v}{dy} + \frac{d\rho w}{dz} = 0 \;\ldots\ldots\ldots\ldots\ldots(11),$$

which expresses the condition that there is no generation or destruction of mass in the interior of a fluid, the equation connecting p and ρ, or in the case of an incompressible fluid the equivalent equation $D\rho/Dt = 0$, and the equation for the propagation of heat, if we choose to take account of that propagation, are the only equations to be satisfied at every point of the interior of the fluid mass.

As it is quite useless to consider cases of the utmost degree of generality, I shall suppose the fluid to be homogeneous, and of a uniform temperature throughout, except in so far as the temperature may be raised by sudden compression in the case of small vibrations. Hence in equations (10) μ may be supposed to be constant as far as regards the temperature; for, in the case of small vibrations, the terms introduced by supposing it to vary with the temperature would involve the square of the velocity, which is supposed to be neglected. If we suppose μ to be independent of the pressure also, and substitute in (10) the values of P_1 &c. given by (8), the former equations become

$$\rho\left(\frac{Du}{Dt} - X\right) + \frac{dp}{dx} - \mu\left(\frac{d^2u}{dx^2} + \frac{d^2u}{dy^2} + \frac{d^2u}{dz^2}\right)$$

$$-\frac{\mu}{3}\,\frac{d}{dx}\left(\frac{du}{dx} + \frac{dv}{dy} + \frac{dw}{dz}\right) = 0,\; \&c. \;\ldots\ldots (12).$$

Let us now consider in what cases it is allowable to suppose μ to be independent of the pressure. It has been concluded by Dubuat, from his experiments on the motion of water in pipes and canals, that the total retardation of the velocity due to friction is not increased by increasing the pressure. The total retardation depends, partly on the friction of the water against the sides of the pipe or canal, and partly on the mutual friction, or tangential action, of the different portions of the water. Now if these two parts of the whole retardation were separately variable with p, it is very unlikely that they should when combined give a result independent of p. The amount of the internal friction of the water depends on the value of μ. I shall therefore suppose that for water, and by analogy for other incompressible fluids, μ is independent of the pressure. On this supposition, we have from equations (11) and (12)

$$\rho \left(\frac{Du}{Dt} - X\right) + \frac{dp}{dx} - \mu \left(\frac{d^2u}{dx^2} + \frac{d^2u}{dy^2} + \frac{d^2u}{dz^2}\right) = 0, \&c....(13),$$

$$\frac{du}{dx} + \frac{dv}{dy} + \frac{dw}{dz} = 0.$$

These equations are applicable to the determination of the motion of water in pipes and canals, to the calculation of the effect of friction on the motions of tides and waves, and such questions.

If the motion is very small, so that we may neglect the square of the velocity, we may put $Du/Dt = du/dt$, &c. in equations (13). The equations thus simplified are applicable to the determination of the motion of a pendulum oscillating in water, or of that of a vessel filled with water and made to oscillate. They are also applicable to the determination of the motion of a pendulum oscillating in air, for in this case we may, with hardly any error, neglect the compressibility of the air.

The case of the small vibrations by which sound is propagated in a fluid, whether a liquid or a gas, is another in which $d\mu/dp$ may be neglected. For in the case of a liquid reasons have been shewn for supposing μ to be independent of p, and in the case of a gas we may neglect $d\mu/dp$, if we neglect the small change in the value of μ, arising from the small variation of pressure due to the forces X, Y, Z.

6. Besides the equations which must hold good at any point in the interior of the mass, it will be necessary to form also the equations which must be satisfied at its boundaries. Let M be a point in the boundary of the fluid. Let a normal to the surface at M, drawn on the outside of the fluid, make with the axes angles whose cosines are l, m, n. Let P', Q', R' be the components

of the pressure of the fluid about M on the solid or fluid with which it is in contact, these quantities being reckoned positive when the fluid considered presses the solid or fluid outside it in the positive directions of x, y, z, supposing l, m and n positive. Let S be a very small element of the surface about M, which will be ultimately plane, S' a plane parallel and equal to S, and directly opposite to it, taken within the fluid. Let the distance between S and S' be supposed to vanish in the limit compared with the breadth of S, a supposition which may be made if we neglect the effect of the curvature of the surface at M; and let us consider the forces acting on the element of fluid comprised between S and S', and the motion of this element. If we suppose equations (8) to hold good to within an insensible distance from the surface of the fluid, we shall evidently have forces ultimately equal to PS, QS, RS, (P, Q and R being given by equations (9),) acting on the inner side of the element in the positive directions of the axes, and forces ultimately equal to $P'S$, $Q'S$, $R'S$ acting on the outer side in the negative directions. The moving forces arising from the external forces acting on the element, and the effective moving forces will vanish in the limit compared with the forces PS, &c.: the same will be true of the pressures acting about the edge of the element, if we neglect capillary attraction, and all forces of the same nature. Hence, taking the limit, we shall have

$$P' = P, \quad Q' = Q, \quad R' = R.$$

The method of proceeding will be different according as the bounding surface considered is a free surface, the surface of a solid, or the surface of separation of two fluids, and it will be necessary to consider these cases separately. Of course the surface of a liquid exposed to the air is really the surface of separation of two fluids, but it may in many cases be regarded as a free surface if we neglect the inertia of the air: it may always be so regarded if we neglect the friction of the air as well as its inertia.

Let us first take the case of a free surface exposed to a pressure Π, which is supposed to be the same at all points, but may vary with the time; and let $L = 0$ be the equation to the surface. In this case we shall have $P' = l\Pi$, $Q' = m\Pi$, $R' = n\Pi$; and putting for P, Q, R their values given by (9), and for P_1 &c. their values given by (8), and observing that in this case $\delta = 0$, we shall have

$$l\left(\Pi - p\right) + \mu\left\{2l\frac{du}{dx} + m\left(\frac{du}{dy} + \frac{dv}{dx}\right) + n\left(\frac{du}{dz} + \frac{dw}{dx}\right)\right\} = 0, \&c.\dots(14),$$

in which equations l, m, n will have to be replaced by dL/dx, dL/dy, dL/dz, to which they are proportional.

If we choose to take account of capillary attraction, we have only to diminish the pressure Π by the quantity $H\left(\dfrac{1}{r_1} + \dfrac{1}{r_2}\right)$, where H is a positive constant depending on the nature of the fluid, and r_1, r_2, are the principal radii of curvature at the point considered, reckoned positive when the fluid is concave outwards. Equations (14) with the ordinary equation

$$\frac{dL}{dt} + u\frac{dL}{dx} + v\frac{dL}{dy} + w\frac{dL}{dz} = 0\dots(15),$$

are the conditions to be satisfied for points at the free surface. Equations (14) are for such points what the three equations of motion are for internal points, and (15) is for the former what (11) is for the latter, expressing in fact that there is no generation or destruction of fluid at the free surface.

The equations (14) admit of being differently expressed, in a way which may sometimes be useful. If we suppose the origin to be in the point considered, and the axis of z to be the external normal to the surface, we have $l = m = 0$, $n = 1$, and the equations become

$$\frac{dw}{dx} + \frac{du}{dz} = 0, \quad \frac{dw}{dy} + \frac{dv}{dz} = 0, \quad \Pi - p + 2\mu\frac{dw}{dz} = 0\dots(16).$$

The relative velocity parallel to z of a point $(x', y', 0)$ in the free surface, indefinitely near the origin, is $dw/dx \cdot x' + dw/dy \cdot y'$: hence we see that dw/dx, dw/dy are the angular velocities, reckoned from x to z and from y to z respectively, of an element of the free surface. Subtracting the linear velocities due to these angular velocities from the relative velocities of the point (x', y', z'), and calling the remaining relative velocities U, V, W, we shall have

$$U = \frac{du}{dx}\,x' + \frac{du}{dy}\,y' + \left(\frac{du}{dz} + \frac{dw}{dx}\right) z',$$

$$V = \frac{dv}{dx}\,x' + \frac{dv}{dy}\,y' + \left(\frac{dv}{dz} + \frac{dw}{dy}\right) z',$$

$$W = \frac{dw}{dz}\,z'.$$

Hence we see that the first two of equations (16) express the conditions that $dU/dz' = 0$ and $dV/dz' = 0$, which are evidently the conditions to be satisfied in order that there may be no sliding motion in a direction parallel to the free surface. It would be easy to prove that these are the conditions to be satisfied in order that the axis of z may be an axis of extension.

The next case to consider is that of a fluid in contact with a solid. The condition which first occurred to me to assume for this case was, that the film of fluid immediately in contact with the solid did not move relatively to the surface of the solid. I was led to try this condition from the following considerations. According to the hypotheses adopted, if there was a very large relative motion of the fluid particles immediately about any imaginary surface dividing the fluid, the tangential forces called into action would be very large, so that the amount of relative motion would be rapidly diminished. Passing to the limit, we might suppose that if at any instant the velocities altered discontinuously in passing across any imaginary surface, the tangential force called into action would immediately destroy the finite relative motion of particles indefinitely close to each other, so as to render the motion continuous; and from analogy the same might be supposed to be true for the surface of junction of a fluid and solid. But having calculated, according to the conditions which I have mentioned, the discharge of long straight circular pipes and rectangular canals, and compared the resulting formulæ with some of the experiments of Bossut and Dubuat, I found that the formulæ did not at all agree with experiment. I then tried Poisson's conditions in the case of a circular pipe, but with no better success. In fact, it appears from experiment that the tangential force varies nearly as the square of the velocity with which the fluid flows past the surface of a solid, at least when the velocity is not very small. It appears however from experiments on pendulums that the total

283

friction varies as the first power of the velocity, and consequently we may suppose that Poisson's conditions, which include as a particular case those which I first tried, hold good for very small velocities. I proceed therefore to deduce these conditions in a manner conformable with the views explained in this paper.

First, suppose the solid at rest, and let $L = 0$ be the equation to its surface. Let M' be a point within the fluid, at an insensible distance h from M. Let ϖ be the pressure which would exist about M if there were no motion of the particles in its neighbourhood, and let p, be the additional normal pressure, and t, the tangential force, due to the relative velocities of the particles, both with respect to one another and with respect to the surface of the solid. If the motion is so slow that the starts take place independently of each other, on the hypothesis of starts, or that the molecules are very nearly in their positions of relative equilibrium, and if we suppose as before that the effects of different relative velocities are superimposed, it is easy to shew that p, and t, are linear functions of u, v, w and their differential coefficients with respect to x, y and z; u, v, &c. denoting here the velocities of the fluid about the point M', in the expressions for which however the co-ordinates of M may be used for those of M', since h is neglected. Now the relative velocities about the points M and M' depending on du/dx, &c. are comparable with $du/dx \cdot h$, while those depending on u, v and w are comparable with these quantities, and therefore in considering the action of the fluid on the solid it is only necessary to consider the quantities u, v and w. Now since, neglecting h, the velocity at M' is tangential to the surface at M, u, v, and w are the components of a certain velocity V tangential to the surface. The pressure p, must be zero; for changing the signs u, v, and w the circumstances concerned in its production remain the same, whereas its analytical expression changes sign. The tangential force at M will be in the direction of V, and proportional to it, and consequently its components along the axes of x, y, z will be proportional to u, v, w. Reckoning the tangential force positive when, l, m, and n being positive, the solid is urged in the positive directions of x, y, z, the resolved parts of the tangential force will therefore be νu, νv, νw, where ν must evidently be positive, since the effect of the forces must be to check the relative motion of the fluid and solid. The normal pressure of the fluid on the solid being equal to ϖ, its components will be evidently $l\varpi$, $m\varpi$, $n\varpi$.

s.   7

Suppose now the solid to be in motion, and let u', v', w' be the resolved parts of the velocity of the point M of the solid, and ω', ω'', ω''' the angular velocities of the solid. By hypothesis, the forces by which the pressure at any point differs from the normal pressure due to the action of the molecules supposed to be in a state of relative equilibrium about that point are independent of any velocity of translation or rotation. Supposing then linear and angular velocities equal and opposite to those of the solid impressed both on the solid and on the fluid, the former will be for an instant at rest, and we have only to treat the resulting velocities of the fluid as in the first case. Hence $P' = l\varpi + \nu(u - u')$, &c.; and in the equations (8) we may employ the actual velocities u, v, w, since the pressures P, Q, R are independent of any motion of translation and rotation common to the whole fluid. Hence the equations $P' = P$, &c. gives us

$$l(\varpi - p) + \nu(u - u')$$

$$+ \mu \left\{ 2l\left(\frac{du}{dx} - \delta\right) + m\left(\frac{du}{dy} + \frac{dv}{dx}\right) + n\left(\frac{du}{dz} + \frac{dw}{dx}\right) \right\} = 0, \&c., \quad \ldots\ldots(17),$$

which three equations with (15) are those which must be satisfied at the surface of a solid, together with the equation $L = 0$. It will be observed that in the case of a free surface the pressures P', Q', R' are given, whereas in the case of the surface of a solid they are known only by the solution of the problem. But on the other hand the form of the surface of the solid is given, whereas the form of the free surface is known only by the solution of the problem.

Dubuat found by experiment that when the mean velocity of water flowing through a pipe is less than about one inch in a second, the water near the inner surface of the pipe is at rest. If these experiments may be trusted, the conditions to be satisfied in the case of small velocities are those which first occurred to me, and which are included in those just given by supposing $\nu = \infty$.

I have said that when the velocity is not very small the tangential force called into action by the sliding of water over the inner surface of a pipe varies nearly as the square of the velocity. This fact appears to admit of a natural explanation. When a current of water flows past an obstacle, it produces a resistance varying nearly as the square of the velocity. Now even if the inner surface of a pipe is polished we may suppose that little irregularities exist, forming so many obstacles to the current. Each little protuberance will experience a resistance varying nearly as the square of the velocity, from whence there will result a tangential action of the fluid on the surface of the pipe, which will vary nearly as the square of the velocity; and the same will be true of the equal and opposite reaction of the pipe on the fluid. The tangential force due to this cause will be combined with that by which the fluid close to the pipe is kept at rest when the velocity is sufficiently small*.

[* Except in the case of capillary tubes, or, in case the tube be somewhat wider, of excessively slow motions, the main part of the resistance depends upon the formation of eddies. This much appears clear; but the precise way in which the eddies act is less evident. The explanation in the text gives probably the correct account of what takes place in the case of a river flowing over a rough stony bed; but in the case of a pipe of fairly smooth interior surface the minute protuberances would be too small to produce much resistance of the same kind as that contemplated in the paragraph beginning near the foot of p. 53.

What actually happens appears to be this. The rolling motion of the fluid belonging to the eddies is continually bringing the more swiftly moving fluid which is found nearer to the centre of the pipe close to the surface. And in consequence the gliding or shifting motion of the fluid in the immediate neighbourhood of the surface in such places is very greatly increased, and with it the tangential pressure.

Thus while in some respects these two classes of resistances are similar, in others they are materially different. As typical examples of the two classes we may take, for the first, that of a polished sphere of glass of some size descending by its weight in deep water; for the second, that of a very long circular glass pipe down which water is flowing. In both cases alike eddies are produced, and the eddies once produced ultimately die away. In both cases alike the internal friction of the fluid, and the friction between the fluid and the solid, are intimately connected with the formation of eddies, and it is by friction that the eddies die away, and the kinetic energy of the mass is converted into molecular kinetic energy, that is, heat. But in the first case the resistance depends mainly on the difference of the pressure p in front and rear, the resultant of the other forces of which the expressions are given in equations (8) being comparatively insignificant, while in the second case it is these latter pressures that we are concerned with, the resultant of the pressure p in the direction of the axis of the tube being practically nil, even though the polish of the surface be not mathematically perfect.

Hence if, the motion being what it actually is, the fluidity of the fluid were suddenly to become perfect, the immediate effect on the resistance in the first case would be insignificant, while in the second case the resistance would practically vanish. Of course if the fluidity were to *remain* perfect, the motion after some time would be very different from what it had been before; but that is not a point under consideration.

Some questions connected with the effect of friction in altering the motion of a nearly perfect fluid will be considered further on in discussing the case of motion given in Art. 55 of a paper *On the Critical Values of the Sums of Periodic Series.*]

7—2

There remains to be considered the case of two fluids having a common surface. Let u', v', w', μ', δ' denote the quantities belonging to the second fluid corresponding to u, &c. belonging to the first. Together with the two equations $L = 0$ and (15) we shall have in this case the equation derived from (15) by putting u', v', w' for u, v, w; or, which comes to the same, we shall have the two former equations with

$$l(u - u') + m(v - v') + n(w - w') = 0 \ldots\ldots\ldots\ldots(18).$$

If we consider the principles on which equations (17) were formed to be applicable to the present case, we shall have six more equations to be satisfied, namely (17), and the three equations derived from (17) by interchanging the quantities referring to the two fluids, and changing the signs l, m, n. These equations give the value of ϖ, and leave five equations of condition. If we must suppose $\nu = \infty$, as appears most probable, the six equations above mentioned must be replaced by the six $u' = u$, $v' = v$, $w' = w$, and

$$lp - \mu f(u, v, w) = lp' - \mu' f(u', v', w'), \text{ &c.,}$$

$f(u, v, w)$ denoting the coefficient of μ in the first of equations (17). We have here six equations of condition instead of five, but then the equation (18) becomes identical.

7.   The most interesting questions connected with this subject require for their solution a knowledge of the conditions which must be satisfied at the surface of a solid in contact with the fluid, which, except perhaps in case of very small motions, are unknown. It may be well however to give some applications of the preceding equations which are independent of these conditions. Let us then in the first place consider in what manner the transmission of sound in a fluid is affected by the tangential action. To take the simplest case, suppose that no forces act on the fluid, so that the pressure and density are constant in the state of equilibrium, and conceive a series of plane waves to be propagated in the direction of the axis of x, so that $u = f(x, t)$, $v = 0$, $w = 0$. Let p_i be the pressure, and ρ_i the density of the fluid when it is in equilibrium, and put $p = p_i + p'$. Then we have from equations (11) and (12), omitting the square of the disturbance.

$$\frac{1}{\rho_i}\frac{d\rho}{dt} + \frac{du}{dx} = 0, \quad \rho_i\frac{du}{dt} + \frac{dp'}{dx} - \frac{4}{3}\mu\frac{d^2u}{dx^2} = 0\ldots\ldots\ldots\ldots(19),$$

Let $A\Delta\rho$ be the increment of pressure due to a very small increment $\Delta\rho$ of density, the temperature being unaltered, and let m be the ratio of the specific heat of the fluid when the pressure is constant to its specific heat when the volume is constant; then the relation between p' and ρ will be

$$p' = mA(\rho - \rho_,)\dots\dots\dots\dots\dots\dots\dots\dots\dots(20).$$

Eliminating p' and ρ from (19) and (20) we get

$$\frac{d^2u}{dt^2} - mA\frac{d^2u}{dx^2} - \frac{4\mu}{3\rho_,}\frac{d^3u}{dt\,dx^2} = 0.$$

To obtain a particular solution of this equation, let

$$u = \phi(t)\cos\frac{2\pi x}{\lambda} + \psi(t)\sin\frac{2\pi x}{\lambda}.$$

Substituting in the above equation, we see that $\phi(t)$ and $\psi(t)$ must satisfy the same equation, namely,

$$\phi''(t) + \frac{4\pi^2}{\lambda^2}mA\phi(t) + \frac{16\pi^2\mu}{3\lambda^2\rho_,}\phi'(t) = 0,$$

the integral of which is

$$\phi(t) = \epsilon^{-ct}\left(C\cos\frac{2\pi bt}{\lambda} + C'\sin\frac{2\pi bt}{\lambda}\right),$$

where

$$c = \frac{8\pi^2\mu}{3\lambda^2\rho_,}, \quad b^2 = mA - \frac{16\pi^2\mu^2}{9\lambda^2\rho_,^2},$$

C and C' being arbitrary constants. Taking the same expression with different arbitrary constants for $\psi(t)$, replacing products of sines and cosines by sums and differences, and combining the resulting sines and cosines two and two, we see that the resulting value of u represents two series of waves propagated in opposite directions. Considering only those waves which are propagated in the positive direction of x, we have

$$u = C_1\epsilon^{-ct}\cos\left\{\frac{2\pi}{\lambda}(bt - x) + C_2\right\}\dots\dots\dots\dots(21).$$

We see then that the effect of the tangential force is to make the intensity of the sound diminish as the time increases, and to render the velocity of propagation less than what it would otherwise be. Both effects are greater for high, than for low notes; but the former depends on the first power of μ, while the latter depends only on μ^2. It appears from the experiments of M. Biot, made on empty water pipes in Paris, that the velocity of propagation of sound is sensibly the same whatever be its pitch. Hence it is necessary to suppose that for air $\mu^2/\lambda^2\rho_{,}^2$ is insensible compared with A or $p_{,}/\rho_{,}$. I am not aware of any similar experiments made on water, but the ratio of $(\mu/\lambda\rho_{,})^2$ to A would probably be insensible for water also. The diminution of intensity as the time increases is, in the case of plane waves, due *entirely* to friction; but as we do not possess any means of measuring the intensity of sound the theory cannot be tested, nor the numerical value of μ determined, in this way.

The velocity of sound in air, deduced from the note given by a known tube, is sensibly less than that determined by direct observation. Poisson thought that this might be due to the retardation of the air by friction against the sides of the tube. But from the above investigation it seems unlikely that the effect produced by that cause would be sensible.

The equation (21) may be considered as expressing in all cases the effect of friction; for we may represent an arbitrary disturbance of the medium as the aggregate of series of plane waves propagated in all directions.

28

On the Mathematical Theory of Sound

SAMUEL EARNSHAW

Samuel Earnshaw (1805–1888), British mathematician and clergyman, was a graduate of Cambridge University in 1831, where he was senior wrangler. He became an Anglican clergyman and served the church in Sheffield for the larger part of his life. As an amateur mathematician of great competence he applied his ability in this field to statics and dynamics and also to problems in acoustics. He became a pioneer in what is now termed nonlinear acoustics, that is, the study of sound waves of large amplitude, the branch of acoustics also called macrosonics. We present in full his most celebrated paper in this field.

Reprinted from *Philosophical Transactions of the Royal Society, London*, **150**, 133–148 (1858)

[133]

VIII. *On the Mathematical Theory of Sound.* By the Rev. S. EARNSHAW, *M.A.*, *Sheffield.* *Communicated by Professor* W. H. MILLER, *F.R.S.*

Received November 20, 1858,—Read January 6, 1859*.

IN making certain investigations on the properties of the sound-wave, transmitted through a small horizontal tube of uniform bore, I found reason for thinking that the equation

$$\frac{dy}{dt} = F\left(\frac{dy}{dx}\right) \qquad \cdots \cdots \cdots \cdots (1.)$$

must always be satisfied; F being a function of a form to be determined. Differentiating this equation with regard to t, we find

$$\frac{d^2y}{dt^2} = \left\{ F'\left(\frac{dy}{dx}\right) \right\}^2 \cdot \frac{d^2y}{dx^2}, \qquad \cdots \cdots \cdots (2.)$$

which by means of the arbitrary function F can be made to coincide, not only with the ordinary dynamical equation of sound, but with any dynamical equation in which the ratio of $\frac{d^2y}{dt^2}$ and $\frac{d^2y}{dx^2}$ can be expressed in terms of $\frac{dy}{dx}$.

Equation (1.) is a partial first integral of (2.), and by means of it we shall be able to obtain a final integral of (2.), which will be shown to be the general integral of (2.) for wave-motion, propagated in one direction only in such a tube as we have supposed, by its satisfying all the conditions of such wave-motion.

It will be convenient to begin with the simplest case of sound,—that in which the development of heat and cold is neglected.

I. WAVE-MOTION WHEN CHANGE OF TEMPERATURE IS NEGLECTED.

1. The equations for this case of motion are, the dynamical equation

$$\left(\frac{dy}{dx}\right)^2 \cdot \frac{d^2y}{dt^2} = \mu \frac{d^2y}{dx^2}, \qquad \cdots \cdots \cdots (3.)$$

and the equation of continuity,

$$\frac{dy}{dx} = \frac{\varrho_0}{\varrho}; \qquad \cdots \cdots \cdots \cdots (4.)$$

ϱ_0, p_0 are the equilibrium density and pressure at any point of the fluid; ϱ, p the same for a particle in motion; x the equilibrium distance of the same particle from a fixed plane cutting the tube at right angles; and t is the time when the same particle, being in motion, is at the distance y from the same plane; μ is the constant which connects ϱ and p by BOYLE's law $p = \mu \varrho$.

---

* Subsequently recast and abridged by the author, but without introducing new matter.

MDCCCLX.　　　　　　　　　　　　　T

On comparing (3.) with (2.), we find $\frac{dy}{dx}\cdot F'\left(\frac{dy}{dx}\right)=\pm\sqrt{\mu}$; or, for brevity, writing α for $\frac{dy}{dx}$, we have

$$F'(\alpha)=\pm\frac{\sqrt{\mu}}{\alpha},$$

and $\qquad\qquad\qquad\qquad \therefore\quad F(\alpha)=C\pm\sqrt{\mu}\log_{\iota}(\alpha).$

But as $\frac{dy}{dt}=F(\alpha)$ from (1.), it follows that

$$dy=\frac{dy}{dx}\cdot dx+\frac{dy}{dt}\cdot dt$$
$$=\alpha dx+F(\alpha)\cdot dt,$$

which being integrated in the usual manner, substituting at the same time for $F(\alpha)$ its value, gives

$$\left.\begin{array}{l} y=\alpha x+(C\pm\sqrt{\mu}\log_{\iota}\alpha)t+\varphi(\alpha) \\ 0=\alpha x\pm\sqrt{\mu}t+\alpha\varphi'(\alpha) \end{array}\right\}\quad\cdots\cdots\cdots \quad(5.)$$

Between these equations, if we eliminate α, we have then the integral of equation (3.).

2. From equation (4.) we see that $\alpha=\frac{\varrho_0}{\varrho}$; and if we represent by u the velocity of the particle whose place is y, we find

$$u=\frac{dy}{dt}=C\pm\sqrt{\mu}\log_{\iota}\alpha,$$
$$=C\pm\sqrt{\mu}\log_{\iota}\left(\frac{\varrho_0}{\varrho}\right);$$

and $\qquad\qquad\qquad\qquad \therefore\quad \varrho=\varrho_0\varepsilon^{\mp\frac{u-C}{\sqrt{\mu}}}.$

3. To determine the arbitrary constant C, we observe that $\varrho=\varrho_0$ and $u=C$ are always simultaneous equations. But the former belongs to the confines of the wave, where in fact $u=0$; and therefore $C=0$. Hence for a wave transmitted through a medium which is itself at rest beyond the limits of the wave, we have these equations*:—

$$\varrho=\varrho_0\varepsilon^{\mp\frac{u}{\sqrt{\mu}}}.\quad\cdots\cdots\cdots\cdots \quad(6.)$$

$$\left.\begin{array}{l} y=\alpha x\pm\sqrt{\mu}\log_{\iota}\alpha\cdot t+\varphi(\alpha) \\ 0=\alpha x\pm\sqrt{\mu}t+\alpha\varphi'(\alpha) \end{array}\right\}\quad\cdots\cdots\cdots \quad(7.)$$

---

* If x and α be eliminated between the equations (7.) and $u=\pm\sqrt{\mu}\log_{\iota}\alpha$, we shall obtain the equation

$$u=f\{y-(u\mp\sqrt{\mu})t\},$$

which was first obtained, though in a very different manner from that employed in this paper, by M. Poisson, and printed in the Journal of the Polytechnique School, tome vii. It seems not to have occurred to him, however, that by means of this equation he might effect another integration of the equations of fluid motion, and thus discover the relation between ϱ and u, whereby his solution would have been completed.

Several of the properties of wave-motion, depending on the gradual change of type, which are included in this equation of M. Poisson's, were first brought forward and discussed by Professor Stokes in the Philosophical Magazine for November 1848, and by the Astronomer Royal in June 1849. In the latter

4. We have now to express these results in terms of the original genesis of the motion. Let us suppose the motion generated by a piston pushed forwards in the tube in a given manner. Let the piston at the time T (having the same origin as t) be at the distance Y from the plane of reference, and moving forwards with the velocity U; and by R denote the density of the air in contact with the piston at that moment. For all particles in contact with the piston $x=0$ (we suppose the piston to commence its motion at the origin of x). Then since at the time T the particles in contact with the piston are within the limits of the wave, equations (6.) and (7.) must be satisfied;

$$\therefore \quad R = \rho_0 \varepsilon^{\mp \frac{U}{\sqrt{\mu}}}$$
$$Y = \pm \sqrt{\mu} \log_\varepsilon \alpha' . T + \varphi(\alpha') \left.\right\}$$
$$0 = \pm \sqrt{\mu} T + \alpha' \varphi'(\alpha') \quad \quad \quad \quad \quad \quad \quad (8.)$$

In these equations $\alpha' = \dfrac{\rho_0}{R}$, and at present we have not sufficiently connected the two systems of equations (7.) and (8.). We shall further connect them by assuming $R=\rho$, which gives $\alpha'=\alpha$; the effect of which assumption is to limit the meaning of T, Y, U as follows:—

> T is the time of genesis of the density ρ which at the time t has been transmitted to the place denoted by y;
>
> Y is the place where the density ρ was generated;
>
> U is the velocity of the piston when ρ was generated by it.

We may now write α for α', and then eliminate α, $\varphi(\alpha)$, and $\varphi'(\alpha)$ between the four equations (7.), (8.). By this means we obtain

$$y = Y + (U \mp \sqrt{\mu})(t-T). \quad \quad \quad \quad \quad (9.)$$

$$x = \mp \sqrt{\mu} \varepsilon^{\mp \frac{U}{\sqrt{\mu}}}(t-T). \quad \quad \quad \quad \quad (10.)$$

$$\rho = \rho_0 \varepsilon^{\mp \frac{U}{\sqrt{\mu}}} = \rho_0 \varepsilon^{\mp \frac{u}{\sqrt{\mu}}}. \quad \quad \quad \quad \quad (11.)$$

5. By these equations the state of a wave at any moment is connected with its genesis; and they contain in fact the complete solution of the problem of every kind of motion, in a tube, which can be generated by a piston.

6. From (11.) it appears that $u=U$; that is, that the particle-velocity generated by the piston is transmitted through the medium without suffering any alteration. The same equation (11.) shows that between the density and the velocity there is an invariable relation, which is independent of the law of original genesis of the motion; so that in the same wave, or in different waves, wherever there is the same density, there will also be the same velocity.

7. One of the most obvious facts on looking at the equations just found is, that for the same genesis there are *two* values of x, two of y, and two of ρ. The signification of

Number of the Magazine it also appears that Professor DE MORGAN had discovered and communicated to the Astronomer Royal two particular forms of the function F; without perceiving, however, that a slight generalization of his results would put him in the way to the integral expressed by the equations (5.).

T 2

this is, that a single disturbance generates *two* waves; and (11.) shows that for one of them ϱ is *greater*, and for the other *less* than ϱ_0. Equation (10.) shows that they are propagated in contrary directions on opposite sides of the piston, and are therefore not parts of the *same* wave.

8. In the genesis of the wave we have supposed the piston pushed forwards, that is, in the direction of $+x$. Hence for the wave generated on that side of the piston we must, as appears from (10.), take the lower sign, which in (11.) gives ϱ greater than ϱ_0. This wave we call the *positive* wave, and the wave of *condensation*. For the wave generated on the other side of the piston we must take the upper sign, which gives ϱ less than ϱ_0; and this wave we call the *negative* wave, and the wave of *rarefaction*.

9. As it will be useful to have a definition of these two waves, which shall be independent of their position with regard to the generating piston, we may state that in general,—

a *positive* wave is one in which the motions of the particles are in the direction of wave-transmission: and

a *negative* wave is one in which the motions of the particles are in a direction opposite to that of wave transmission.

10. If ϱ_1 and ϱ_2 be the densities of the air in contact with the piston before and behind at any moment, and if p_1 and p_2 be the corresponding pressures; then from (11.) we have

$$\varrho_1 = \varrho_0 \varepsilon^{\frac{U}{\sqrt{\mu}}}, \text{ and } \varrho_2 = \varrho_0 \varepsilon^{-\frac{U}{\sqrt{\mu}}},$$

and $\therefore \quad \varrho_1 \varrho_2 = \varrho_0^2;$

which may be thus expressed in words:—if a piston move in a tube, filled with air, in any manner whatever, the densities of the air in contact with it at its front and back are such that the equilibrium density is a mean proportional between them. And since $p = \mu \varrho$, we have $p_1 p_2 = p_0^2$, which furnishes us with a similar property for the *pressures* on the piston.

11. Since $p_1 = p_0 \varepsilon^{\frac{U}{\sqrt{\mu}}}$ and $p_2 = p_0 \varepsilon^{-\frac{U}{\sqrt{\mu}}}$, it follows that the resistance to the motion of the piston (calling S its area) is

$$(p_1 - p_2)S = \left(\varepsilon^{\frac{U}{\sqrt{\mu}}} - \varepsilon^{-\frac{U}{\sqrt{\mu}}}\right) p_0 S.$$

Hence in different gases, if p_0 be the same in all, those will offer the greatest resistance to the piston for which μ is the least.

It will be convenient from this point to consider the two kinds of waves separately.

1. The Wave of Condensation.

12. The equations for this wave are

$$y = Y + (\sqrt{\mu} + U)(t - T)$$

$$x = \sqrt{\mu} \varepsilon^{\frac{U}{\sqrt{\mu}}}(t - T)$$

$$\varrho = \varrho_0 \varepsilon^{\frac{U}{\sqrt{\mu}}}.$$

13. Now with respect to the genesis of this wave, we have seen that U must satisfy the same conditions as u, and Y as y. But $u = \frac{dy}{dt}$, therefore $U = \frac{dY}{dT}$: and again, as one of the equations of the general integral (7.) was obtained from the other by differentiation with regard to α, it follows that both α and U must vary continuously; and that $\frac{dU}{dT}$ must not pass through infinity; in other words, if the velocity of the piston vary it must vary continuously. Neither Y nor U must be discontinuous with regard to T. Hence there must be no discontinuity of pressure within the limits of the wave at its genesis: and if discontinuity should afterwards occur in the wave during its transmission, our equations will cease to be applicable for that part of the wave where the discontinuity has occurred. For the wave in any one position may be supposed to generate its next position; and a piston or diaphragm may at any time be supposed to act the part of the generating wave. What is necessary for the diaphragm to observe as a law of genesis must be necessary for the wave considered as the generator of its next position; and therefore the part of the wave (if any) where discontinuity occurs will be beyond the reach of our equations.

14. It has been shown that the density ϱ, which at the time t is at the distance y from the plane of reference, was generated at the time T when its distance from the same plane was Y. Hence it has been transmitted through the space $y - Y$ in the time $t - T$, and consequently the velocity of its transmission (as appears from the first equation of (12.)) is

$$\sqrt{\mu} + U.$$

15. The wave as a whole is included between two points of it for each of which $U = 0$, and consequently for each of those points the velocity of transmission is $\sqrt{\mu}$. Hence the wave as a whole is transmitted with this uniform velocity. But all the parts of the wave, with the exception of its front and rear, are transmitted with velocities greater than this,—with velocities dependent on their respective densities. Hence every part of the wave, with the exception of its rear, is perpetually gaining on the front, and the result is a *constant change of type*,—the more condensed parts hurrying towards the front, with velocities greater as their densities are greater. This cannot go on perpetually without its happening at length that a *bore* (or tendency to a discontinuity of pressure) will be formed in front; which will force its way, in violation of our equations, faster than at the rate of $\sqrt{\mu}$ feet per second; and consequently in experiments, made on sound at long distances from the origin of the sound-wave, we should expect the actual velocity observed to be greater than $\sqrt{\mu}$, especially if the sound be a violent one, generated with extreme force (see art. 17).

16. We have seen that the velocity of transmission of the density ϱ is $\sqrt{\mu} + U$. Now the velocity of the particles where the density is ϱ is u, which we have shown to be equal to U. In a certain sense we may consider the velocity u to be a wind-velocity in that part of the medium, and then we have an indefinitely small disturbance at that point transmitted in that wind with the velocity $\sqrt{\mu}$ imposed upon the wind. In other

words, transmission-velocity is superimposed on particle-velocity, and in this sense transmission-velocity is everywhere the same, and equal to $\sqrt{\mu}$. A wave passes by every particle with this velocity, whatever be the particular and varying density of the medium where the particle is situated.

17. Since a wave's front cannot move faster than with the velocity $\sqrt{\mu}$, if the generating piston move faster than with this velocity, it will generate a bore; and from this we infer that a bore always moves with a velocity greater than $\sqrt{\mu}$; for wherever a bore may be situated at *any time*, we may suppose it to be *just* generated by a piston. If we write $\sqrt{\mu}$ for U, we find $p=\mu\varrho=\mu\varrho_0\varepsilon=\varepsilon p_0$. Consequently if the piston press upon the resisting air with a pressure exceeding ε atmospheres, a bore will be instantly formed.

I have defined a bore to be *a tendency to discontinuity of pressure*; and it has been shown that as a wave progresses such a tendency necessarily arises. As, however, discontinuity of pressure is a physical impossibility, it is certain Nature has a way of avoiding its actual occurrence. To examine in what way she does this, let us suppose a discontinuity to have actually occurred at the point A, in a wave which is moving forwards. Imagine a film of fluid at A forming a section at right angles to the tube. Then on the back of this film there is a certain pressure which is discontinuous with respect to the pressure on its front. To restore continuity of pressure, the film at A will rush forward with a *sudden* increase of velocity, the pressure in the front of the film not being sufficient to preserve continuity of velocity. In so doing the film will play the part of a piston generating a bit of wave in front, and a small regressive wave behind. The result will be a prolongation of the wave's front, thereby increasing the original length of the wave, and producing simultaneously a feeble regressive wave of a negative character.

Now all this supposes the discontinuity to have actually occurred, which, as has been said, is a physical impossibility. For actual occurrence we must therefore substitute a tendency to occur, and modify the preceding reasoning thus:—

Nature so contrives, that as the discontinuity is in its initial stage of beginning to take place, its actual occurrence is prevented by a *gradual* (not *sudden*) prolongation of the wave's front, and by the *constant* casting off, from its front in a retrogressive direction, of a long continuous wave of a negative character, which will be of greater or less intensity according as the tendency to discontinuity is more or less intense in the original wave.

The casting off of this long wave will probably manifest itself audibly as a continuous *hiss* or *rushing* sound.

Hence a sound-wave, from the moment that a tendency to discontinuity begins in its front, has the property of constantly prolonging itself in front, and by this means its front travels faster than at the rate $\sqrt{\mu}$. Those sounds also will travel most rapidly whose genesis was most violent; and gentlest sounds travel with velocities not much differing from $\sqrt{\mu}$. I should expect, therefore, that in circumstances where the human voice can be heard at a sufficiently great distance, the *command* to fire a gun, if instantly obeyed, and the *report* of the gun, might be heard at a long distance in an inverse order; i. e. *first* the report of the gun, and *then* the word "fire*." In a slight degree, therefore, the experimental velocity of sound will depend on its intensity, and the violence of its genesis. I consider this article as tending to account for the *discrepancy* between the *calculated* and *observed* velocities of sound (which most experimentalists have remarked and wondered at), when allowance is made (as will be done in a future part of this paper) for change of temperature.

18. It seems reasonable to suppose that the audible character of a wave is in some way dependent on its type; and consequently, if this be the case, the sound undergoes a perpetual modification as the distance of transmission increases. One modification of the sound-wave is, as we have seen, the formation of a bore in front; but there is another which cannot but have some influence on its audible properties, as it corresponds to a remarkable change of type; and this takes place when the greater densities begin to overtake the less.

Now when one degree of density overtakes another, the values of y corresponding to those two densities are equal; and hence at the time t the equation

$$y = Y + (\sqrt{\mu} + U)(t - T) \quad \cdots \cdots \cdots \quad (12.)$$

will give two equal values of y for two consecutive values of T. Hence differentiating it with regard to T, remembering that t is constant, or the same for both, as is also y, we have

$$0 = U - (\sqrt{\mu} + U) + (t - T)\frac{dU}{dT},$$

or

$$t = T + \frac{\mu}{\frac{dU}{dT}}. \quad \cdots \cdots \cdots \cdots \cdots \quad (13.)$$

The right-hand member of this equation is of course a continuous expression, and therefore its least or minimum value will be the value of t when the modification of type, of which we are speaking, *first begins* to take place; and because of the continuity of (13.), this modification once begun will gradually spread itself over the fore-part of the wave. Now t will be a minimum when

$$\left(\frac{dU}{dT}\right)^2 = \mu \cdot \frac{d^2U}{dT^2}.$$

From this equation we may find T, the time of genesis of that part of the wave where this modification begins. Then (13.) will give t, the actual time when the modification begins; and (12.) will give the place in the tube where it begins.

19. It is perhaps impossible to say what is the audible characteristic corresponding to the wave-modification just investigated; but whatever it be, we perceive from (13.) that those sound-waves soonest begin to be affected by it for which $\frac{dU}{dT}$ is largest; *i. e.* those

---

* See Supplement to Appendix of PARRY's Voyage in 1819–20, Art. "Abstract of Experiments to determine the Velocity of Sound."

whose genesis is most violent. And we may also consider it as proved that those sounds will retain their original characteristics the longest which are the most gently generated.

It is also quite evident from (13.) if the same cause generate sound-waves in different tubes filled with different gases, the wave will be soonest affected by the above modification in that tube which contains the gas for which μ is least.

We come now to speak of

2. *The Wave of Rarefaction.*

20. We shall obtain the equations for this kind of wave by writing $-U$ for $+U$ in the equations of art. 12, which is equivalent to supposing a negative wave generated on the $+y$ side of the piston. Hence the equations of a negative wave are

$$y = Y + (\sqrt{\mu} - U)(t - T),$$

$$x = \sqrt{\mu} \varepsilon^{-\frac{U}{\sqrt{\mu}}}(t - T),$$

$$\varrho = \varrho_0 \varepsilon^{-\frac{U}{\sqrt{\mu}}}.$$

21. Reasoning in the same manner as in art. 14, it appears that the velocity with which the density ϱ is transmitted is

$$\sqrt{\mu} - U.$$

From this it appears that, speaking generally, the velocity of transmission of every part of a negative wave is less than of every part of a positive wave. The exceptions to this statement are the front and rear, which in both kinds of waves move with the same velocity $\sqrt{\mu}$, because for those points $U = 0$. It is evident also that the most rarefied parts of a wave will be transmitted the most slowly, and will consequently drop continually towards the rear. Hence in this species of wave, as in the former, a constant change of type takes place; and in the end also a *negative or rarefied bore* will be formed in the *rear* of the wave.

By a process of reasoning analogous to that of art. 17, we infer that a negative sound-wave, from the moment that a tendency to discontinuity begins in its rear, has the property of constantly shortening its rear, and by this means its rear travels faster than at the rate $\sqrt{\mu}$; and also as it progresses it is constantly casting off from its rear in a regressive direction a long continuous wave of a negative character. Art. 18 also admits of easy modification to this kind of wave.

22. The velocity of transmission of a negative wave being $\sqrt{\mu} - U$, and the last term of this expression admitting of arbitrary increase, it is evident that $U = \sqrt{\mu}$ is a critical value, and that the part of the wave corresponding to that value of U is stationary. The corresponding value of ϱ is $\frac{\varrho_0}{\varepsilon}$.

Every part of the wave where the density exceeds this travels forwards; but the parts where the density is less than this are regressive; hence a wave, as a whole, in which ϱ

begins at ϱ_0, and after twice passing through $\frac{\varrho_0}{\varepsilon}$ ends at ϱ_0, will have two stationary points in its type, viz. those where $\varrho = \frac{\varrho_0}{\varepsilon}$. Between these points the wave will be stationary though constantly changing type; beyond them progressive.

23. But instead of supposing the piston to generate such a wave as this, let us suppose it to begin from the velocity zero, and according to any proposed law (continuous of course) increase its velocity till it becomes infinite; and let us consider the state of the medium at this moment.

Denote by A and B the places of the piston where its velocity became respectively $\sqrt{\mu}$ and infinite. Then whatever was the law of motion from A to B, and whether AB be great or small, provided it remains of finite length, the density at A will remain unchanged and equal to $\frac{\varrho_0}{\varepsilon}$, and the velocity of every particle as it passes by A will be equal to $\sqrt{\mu}$. The mass of air also which will rush through the section of the tube at A will be $\frac{S\varrho_0 \sqrt{\mu}}{\varepsilon}$; and this, be it observed, cannot be made either more or less by causing the piston to move in a different manner from A to B. It is also equally independent of the law of the piston's motion before it reached A. Hence the mass of air that flows through the section at A is altogether independent of the law of the piston's motion throughout its whole course.

24. Now let us inquire what quantity of air rushes through any other section of the tube. In every part where there is motion the same relation between density and velocity obtains, viz. $\varrho = \varrho_0 \varepsilon^{-\frac{u}{\sqrt{\mu}}}$; and consequently the quantity which rushes through any section is at the rate of

$$S \varrho_0 u \varepsilon^{-\frac{u}{\sqrt{\mu}}} \text{ per second.}$$

It is obvious this admits of a maximum value, which in the usual manner we find to be

$$\frac{S\varrho_0 \sqrt{\mu}}{\varepsilon},$$

at which value $u = \sqrt{\mu}$ and $\varrho = \frac{\varrho_0}{\varepsilon}$.

25. Hence one part of the tube cannot supply air to another part faster than at this rate; and consequently the greatest possible mass of air passes through the section at A: and it may be stated as a general property of motion through a tube, that a gas cannot be conveyed through a tube faster than at the rate of $\frac{S\sqrt{\mu}}{\varepsilon}$ cubic feet per second of gas of the density ϱ_0.

Hence the escaping powers of different gases through equal tubes are proportional to the velocities with which they respectively transmit sound.

26. Since this result is independent of the law of velocity of the air, both before and after passing the section A, we are entitled to say that air cannot rush through a pipe of finite length, even into a vacuum, faster than at the rate of $\frac{S\sqrt{\mu}}{\varepsilon}$ cubic feet per

second.   The length of the pipe seems to be a matter of perfect indifference, and may be nothing more than a hole through a partition of finite thickness.

27. Since one part of a tube cannot supply air to, nor convey air away from, another part, A, faster than at the maximum rate, it is easy to see that if the pipe be supposed of finite length, which conveys air into a vacuum, the velocity in every part of the pipe will soon be the same throughout, and equal to $\sqrt{\mu}$, and density everywhere equal to $\frac{\varrho_0}{\epsilon}$.

From this it would appear that the rate of discharge into a vacuum, which has generally been supposed to be that which is due to the height of the homogeneous atmosphere, is in reality that which is due to the $\left(\frac{1}{2\epsilon^2}\right)$th part only; that is, to little more than the fifteenth part of it; but this requires correction for change of temperature.

28. If the generating piston move forward and then backward, so as to generate a positive wave followed continuously by a negative wave, they will not separate; for, as we have seen, they are each transmitted, as wholes, at the same rate $\sqrt{\mu}$.   But the main body of the positive wave will gradually advance in the type towards its front, and that of the negative wave fall back towards its rear; and consequently for the purposes of audibility the central part of the compound wave, between the front of the positive and the rear of the negative wave, will become so attenuated that it may be considered of little audible effect, after the waves have been in existence a sufficient length of time to allow the formation of bores.   The compound wave will therefore have a tendency to produce the audible effect of two separate waves, separated by an interval of space nearly equal to its whole length.   If therefore the length of such a compound wave be sufficiently great, it will ultimately produce *two* distinct sounds separated by a very brief interval of time.

29. If the generating piston move backward and then forward, so as to generate a negative wave followed continuously by a positive wave, the positive and negative bores will destroy each other as rapidly as they are formed.   This, however, supposes the positive and negative portions of the original compound wave to be equal.   If one exceed the other in quantity of motion, the result will be a little modified.   A compound wave of the kind supposed in this article will therefore be entirely devoid of bores, and the sound corresponding to it will be free from that harshness which is probably the audible character of a bore.

30. If there be a continuous succession of positive and negative waves, constituting one long compound wave, such a wave will produce a continuous even sound, called a musical note, probably owing its sweetness in some degree to the property just mentioned; and as every negative portion is succeeded by a positive portion, and every positive by a negative, the length of each portion will remain unchangeable, whatever be the distance through which the compound wave travels.   Hence the pitch of a musical note cannot change by distance of transmission.

31. Suppose a portion of the tube to be filled with air of a different kind from that which fills the first part. Let p_0' ϱ_0' μ' be the quantities for this air which correspond to p_0 ϱ_0 μ of the former; and to prevent the two airs or gases from mixing, let them be supposed to be separated by an impenetrable film without weight and inertia. Then as there is equilibrium in the tube before the wave is generated, we have

$$p_0 = p_0'.$$

Let now a wave be generated in the first medium and transmitted towards the second; then when it has reached the common boundary of the media, the velocities of the particles in contact with the film on both sides will always be equal. Let U' be this velocity at any moment, and U the velocity which the film would have had at that moment, if the second medium had been the same as the first. Then $U-U'$ is the velocity lost by the particles of the first medium by the resistance due to their contact with the film. In other words, this velocity has been impressed on the particles of the first medium by the resistance of the film, in the reflex direction. This gives rise to a reflex wave in the first medium, which we may consider superimposed on the wind of the original wave. And consequently if p be the pressure at the film due to the original wave, the pressure when this reflex wave has been superimposed, i. e. the actual pressure at the film, is $= p \varepsilon^{\frac{U-U'}{\sqrt{\mu}}}$, which $= p_0 \varepsilon^{\frac{2U-U'}{\sqrt{\mu}}}$, $\because p = p_0 \varepsilon^{\frac{U}{\sqrt{\mu}}}$.

But if we now turn to the other side of the film, the velocity U' has been impressed upon the particles of the second medium in contact with the film; and hence the pressure of those particles on the film

$$= p_0' \varepsilon^{\frac{U'}{\sqrt{\mu'}}} = p_0 \varepsilon^{\frac{U'}{\sqrt{\mu'}}};$$

and consequently, as the pressures on the two sides are equal, we have

$$\frac{U'}{\sqrt{\mu'}} = \frac{2U - U'}{\sqrt{\mu}}.$$

Hence the velocities of the particles at the film, for the *incident*, *reflected* and *refracted* waves, are respectively proportional to

$$\sqrt{\mu} + \sqrt{\mu'}, \ \sqrt{\mu} - \sqrt{\mu'}, \ \text{and } 2\sqrt{\mu'}.$$

There is nothing new in these formulæ, except that they are here deduced without supposing the motions small.

II. WAVE MOTION WHEN CHANGE OF TEMPERATURE IS NOT NEGLECTED.

32. The heat developed by that change of temperature which is produced by the sudden alteration of density due to the passage of a wave, is probably taken account of by using the following equation as that which connects pressure and density,

$$\frac{p}{p_0} = \left(\frac{\varrho}{\varrho_0}\right)^k;$$

k being the ratio of the specific heat of the gas under a constant pressure, to its specific heat under a constant volume. The dynamical equation takes for this case the following form to be used instead of that in art. 1,

$$\left(\frac{dy}{dx}\right)^{k+1} \cdot \frac{d^2y}{dt^2} = k\mu \cdot \frac{d^2y}{dx^2}.$$

This equation being integrated as explained in art. 1, gives

$$\left. \begin{aligned} y &= \alpha x + \left(C \mp \frac{2\sqrt{k\mu}}{k-1}\alpha^{-\frac{k-1}{2}}\right)t + \varphi(\alpha) \\ 0 &= \alpha x \pm \sqrt{k\mu}\,\alpha^{-\frac{k-1}{2}}t + \alpha\varphi'(\alpha) \end{aligned} \right\} \quad \dots \dots \dots \quad (14.)$$

33. From these we obtain

$$u = \frac{dy}{dt} = C \mp \frac{2\sqrt{k\mu}}{k-1}\alpha^{-\frac{k-1}{2}},$$

$$= C \mp \frac{2\sqrt{k\mu}}{k-1}\left(\frac{\varrho}{\varrho_0}\right)^{\frac{k-1}{2}}.$$

For the same reasons as before we shall suppose $u=0$ and $\varrho=\varrho_0$ to be simultaneous equations; which gives

$$C = \pm\frac{2\sqrt{k\mu}}{k-1},$$

and

$$\therefore \quad \left(\frac{\varrho}{\varrho_0}\right)^{\frac{k-1}{2}} = 1 \mp \frac{(k-1)u}{2\sqrt{k\mu}}. \quad \dots \dots \dots \quad (15.)$$

This equation gives the relation between density and velocity; from which that between pressure and velocity is easily found.

34. The general integral (14.) may be expressed in terms of the original genesis precisely in the same manner as was employed in art. 4; and the result is

$$y = Y + \left(\frac{k+1}{2}U \mp \sqrt{k\mu}\right)(t-T). \quad \dots \dots \dots \quad (16.)$$

$$x = \mp\sqrt{k\mu}\left(1 \mp \frac{k-1}{2\sqrt{k\mu}}U\right)^{\frac{k+1}{k-1}}(t-T). \quad \dots \dots \quad (17.)$$

$$u = U, \text{ and } p = p_0\left(\frac{\varrho}{\varrho_0}\right)^{k}. \quad \dots \dots \dots \quad (18.)$$

These equations, with (15.), are those from which the properties of the motion are to be deduced. The degree of modification of former results required by these formulæ will be in most cases sufficiently evident, and need not therefore to be particularly pointed out.

35. The result of art. 10 takes the following form—

$$\varrho_1^{\frac{k-1}{2}} + \varrho_2^{\frac{k-1}{2}} = 2\varrho_0^{\frac{k-1}{2}};$$

and that of art. 11 the following—

$$(p_1-p_2)S=Sp_0\left\{\left(1+\frac{k-1}{2\sqrt{k\mu}}\,U\right)^{\frac{2k}{k-1}}-\left(1-\frac{k-1}{2\sqrt{k\mu}}\,U\right)^{\frac{2k}{k-1}}\right\}.$$

36. From (16.) it appears that the velocity of transmission of the front and rear of either a positive or negative wave is $\sqrt{k\mu}$; but the velocity of transmission of that part of the wave of which the density is ϱ, is, for a positive wave,

$$\sqrt{k\mu}+\frac{k+1}{2}\,U\,;$$

and for a negative wave,

$$\sqrt{k\mu}-\frac{k+1}{2}\,U.$$

The part of these expressions to which the bore is due is the term $\frac{k+1}{2}\,U$; and as k is known to be greater than unity, this is greater than U; and consequently change of temperature hastens the formation of a bore, and also renders the property of art. 16 inapplicable here.

37. As in the case of a negative wave the equation (15.) involves a negative term, it is manifestly possible for the piston, in generating a negative wave, to move so quickly as to leave a vacuum behind it.   The least velocity with which this can happen is

$$\frac{2\sqrt{k\mu}}{k-1},$$

which for common air is about 5722 feet per second.   But it is necessary to notice, that in this and similar extreme results, we are hardly justified in supposing k to be constant up to such high velocities.

38. The expression $S\varrho u$ is a maximum (see art. 24) when

$$u=\frac{2\sqrt{k\mu}}{k+1},$$

which in the case of common air is equal to about 904 feet per second; and the corresponding density is

$$\varrho=\left(\frac{2}{k+1}\right)^{\frac{2}{k-1}}\varrho_0\,;$$

or, for common air, about $\frac{2}{5}\varrho_0$.

Hence no gas can rush through a pipe faster than at the rate of

$$\sqrt{k\mu}\left(\frac{2}{k+1}\right)^{\frac{k+1}{k-1}}S$$

cubic feet per second.

39. The change of temperature due to the transmission of a wave through an elastic medium has been taken account of, by assuming a law different from that of BOYLE, to connect pressure with density (art. 32).

If we generalize the law by assuming

$$p=\varphi\left(\frac{\varrho_0}{\varrho}\right),$$

the dynamical equation takes the form

$$\frac{d^2y}{dt^2} = -\frac{1}{\varrho_0}\cdot\varphi'\left(\frac{dy}{dx}\right)\cdot\frac{d^2y}{dx^2}.$$

If now we assume

$$(\mathrm{F}'\alpha)^2 = -\frac{\varphi'\alpha}{\varrho_0},$$

the integral of the dynamical equation will be

$$\begin{cases} y = \alpha x + (\mathrm{C}\pm\mathrm{F}\alpha)t + f\alpha, \\ x = \mp\mathrm{F}'\alpha\cdot t - f'\alpha, \end{cases}$$

with $\alpha=\frac{\varrho_0}{\varrho}$, and $u = \mathrm{C}\pm\mathrm{F}\alpha = \mathrm{C}\pm\mathrm{F}\left(\frac{\varrho_0}{\varrho}\right)$.

40. These equations are true of any motion which can be generated by a piston moving subject to the laws of continuity. See art. 13. The last shows that the relation between velocity and density is independent of the law of genesis of the motion. The medium may be, as a whole, in motion with the uniform velocity $\mathrm{C}\pm\mathrm{F}(1)$, and the motion of the particles caused by the motion of a piston will be superimposed on this. For convenience we shall suppose the medium as a whole at rest, and $\therefore \mathrm{C}\pm\mathrm{F}(1)=0$.

If there be a point, or any number of points, within that part of the medium which is in motion for which $\varrho=\varrho_0$, for all such points $\alpha=1$, and the equation

$$x = \mp\mathrm{F}'(1)\cdot t - f'(1),$$

which is always true for all such points, shows that at those points x changes its value at the rate of $\mathrm{F}'(1)$ feet per second, i. e. the front of the wave travels at the rate of

$$\left\{-\frac{\varphi'(1)}{\varrho_0}\right\}^{\frac{1}{2}} \text{ feet per second,}$$

which is constant, and depends not at all on the law of genesis, but only on the assumed relation between pressure and density, and not on the *general* value of even that, but only on its limiting value when $\varrho=\varrho_0$. Now many different forms of the function φ may give the same limiting value; and consequently all the media corresponding to these various forms of φ will transmit a wave, as a whole, with the same velocity. Hence if the relation between pressure and density be given, the wave-velocity may be instantly deduced from the expression $\left\{-\frac{\varphi'(1)}{\varrho_0}\right\}^{\frac{1}{2}}$, or from its equal,

$$\left\{\frac{dp}{d\varrho}\right\}_0^{\frac{1}{2}};$$

using the subscript 0 to signify that after the differentiation has been performed ϱ_0 is to be written for ϱ.

41. Since $u = \mathrm{C}\pm\mathrm{F}\left(\frac{\varrho_0}{\varrho}\right)$, by differentiation we obtain

$$\frac{du}{d\varrho} = \mp\frac{1}{\varrho}\left(\frac{dp}{d\varrho}\right).$$

42. And if c denote the velocity of transmission of the density ϱ, then we have

$$c = C \pm F\left(\frac{\varrho_0}{\varrho}\right) \mp \frac{\varrho_0}{\varrho} F'\left(\frac{\varrho_0}{\varrho}\right) = u \mp \left(\frac{dp}{d\varrho}\right)^{\frac{1}{2}};$$

consequently

$$\frac{dc}{d\varrho} = \mp \frac{1}{\varrho} \cdot \frac{d}{d\varrho}\left(\varrho^2 \frac{dp}{d\varrho}\right)^{\frac{1}{2}}.$$

Now the former of these equations shows that unless the term $\left(\frac{dp}{d\varrho}\right)^{\frac{1}{2}}$ be constant, the property of the superposition of wave-transmission on particle-velocity, proved in art. 16, does not hold good. But if it be constant, then $p = \mu\varrho + \mu'$; which is the general relation between pressure and density when that principle of superposition holds good. Hence, as mentioned in art. 36, the development of heat puts an end to this property in all known gases.

43. In the case of negative waves we may institute a method of reasoning similar to that employed in arts. 23 *et seq.*, and arrive at analogous results. We shall also find that, taking $c = \left(\frac{dp}{d\varrho}\right)^{\frac{1}{2}} - u$ for this case, the maximum value of $S\varrho u$ will occur in that section of the tube where $c = 0$; from which it follows that at that section

$$u = \left(\frac{dp}{d\varrho}\right)^{\frac{1}{2}};$$

which is always possible and finite. Hence may be determined the limit to the quantity of a gas that can pass through a pipe in a given time, even into a vacuum.

44. The expression in art. 42 for $\frac{dc}{d\varrho}$ shows that c is in general a function of ϱ, so that in general there will be a constant change of type. In one case, however, there will be no change of type. This will take place when $\frac{dc}{d\varrho} = 0$, that is when $\varrho^2 \cdot \frac{dp}{d\varrho}$ is constant. Assume for this case

$$\varrho^2 \frac{dp}{d\varrho} = B;$$

$$\therefore p = A - \frac{B}{\varrho}.$$

This equation expresses the nature of the medium which is distinguished by the property, that it transmits waves without change of type. And if we pass from this to the dynamical equation, we find

$$\frac{d^2y}{dt^2} = \frac{B}{\varrho_0^2} \cdot \frac{d^2y}{dx^2}.$$

Now it has been usual to reduce the equation (3.) to this form for the purposes of approximation; but the process appears to be allowable only so far as the equation $p = A - \frac{B}{\varrho}$ may be taken to be a physical approximation to BOYLE's law $p = \mu\varrho$. To me it does not seem to be an allowable approximation; and consequently I do not consider the solution of the dynamical equation, which has been obtained by this means, to be applicable to the problem of sound at all.   Many *analytically* approximate forms might be invented and used for BOYLE's law, and each would have its peculiar physical attributive effects on the sound-wave; and we might thus, by adopting first one and then another of these analytically approximate laws, invent *ad libitum* an inexhaustible list of properties of the sound-wave which have no real existence where BOYLE's law is strictly true.   From which therefore it would seem to be a necessary consequence, that an equation between p and ϱ must not only be *analytically* but also *physically* approximative, in order that the results deduced from it may be accepted as real approximations to the true laws of nature.

45. By means of the expressions for $\frac{du}{d\varrho}$ and $\frac{dc}{d\varrho}$, we may not only discover the properties of motion in a tube without having recourse to the usual equations, when the relation between ϱ and p is known, but we may also solve many inverse problems.

Also, if the tube be filled with a medium of such a nature that the relation between p and ϱ changes continuously from point to point, or is different in different parts, yet if $\left(\frac{dp}{d\varrho}\right)_0$ has the same value everywhere, waves will travel through the tube with a uniform velocity.

If the nature of the medium should vary slowly and continuously, the velocity of the wave-transmission would be known, from the equations given above, by integration.

46. If, through the partial radiation of heat, or from any other cause, the dynamical equation should take the form

$$\frac{d^2y}{dt^2}=f\left(\frac{dy}{dx},\ \frac{dy}{dt}\right).\frac{d^2y}{dx^2},$$

we must integrate it as before by assuming

$$\frac{dy}{dt}=F\left(\frac{dy}{dx}\right);$$

which gives

$$(F'\alpha)^2=f(\alpha,\ F\alpha).$$

This equation being integrated will furnish the form of F; and then the integral of the proposed dynamical equation will be

$$\begin{cases} y=\alpha x+(C\pm F\alpha)t+\varphi\alpha, \\ x=\mp F'\alpha.t-\varphi'\alpha, \end{cases}$$

which does not present any new difficulty.

29

On the Theory of Sound

JOHN JAMES WATERSTON

John James Waterston (1811–1883), the Scottish engineer and amateur scientist, a somewhat neglected figure in the history of 19th century science, is now recognized to have made an outstanding contribution to the molecular theory of fluids in a memoir he submitted to the Royal Society of London in 1845. This contained the first theoretically well founded calculation of the root-mean-square velocity of molecules. It also calculated the velocity of sound on the basis of the theory that sound propagation is essentially a molecular phenomenon. The paper was not accepted and lay ignored in the archives of the Society until exhumed by Lord Rayleigh in 1892. It was then published with an introduction by Rayleigh (see p. 407 of this volume).

Waterston developed his ideas on sound propagation further in a paper which appeared in the *Philosophical Magazine* in 1858. This is presented here in its entirety. Though from the standpoint of modern physical acoustics and statistical mechanics it contains errors and unsuccessful hypotheses, it still entitles Waterston to be regarded as the pioneer in the field of molecular acoustics.

THE

LONDON, EDINBURGH AND DUBLIN

PHILOSOPHICAL MAGAZINE

AND

29 JOURNAL OF SCIENCE.

SUPPLEMENT TO VOL. XVI. FOURTH SERIES.

LV. *On the Theory of Sound.* By J. J. WATERSTON, *Esq.**

IN a paper on the Integral of Gravitation that appeared in the Philosophical Magazine of May last, I endeavoured to draw attention to the principle of physical causation supplied to us by the mechanical theory of heat,—how it leads us to study natural phænomena in their dynamical sequence, and suggests the arranging of problems of molecular statics so that they may present the aspect of motor transition to the mathematical inquirer. It may perhaps be of advantage, as testing the general argument, to discuss the question, *Can the theory of the propagation of sound developed by Newton and Laplace be viewed, consistent with the modern idea of heat, as a correct exposition of the actual mode in which pulses are transmitted through air?* It appears from recent memoirs, that this opinion is still generally entertained by physicists; and the theory of Laplace has in some instances been considered so perfect as to afford the means of deducing the specific heat of air from the velocity of sound, and this by zealous professors of the mechanical theory of heat.

The ideas upon which Laplace's theory was constructed are to be found in the second chapter of the second book of the *Mécanique Céleste.* The following extracts show how very remote they are from those now universally prevalent:—

Page 105, line 8. "Doubtless it is necessary to admit that between the molecules of air there is a repulsive force which is only sensible at imperceptible distances: the difficulty consists in deducing from it the laws of elastic fluids. This may be done by the following considerations."

P. 105, line 4 from bottom. "I suppose then that these molecules retain the heat by their attraction, and that their mutual repulsion is due to the repulsion of the molecules of the heat which I assume to extend to an insensible sphere of activity."

P. 111, line 9. "But what ought we to understand by the temperature *u*, and what is its measure? It appears natural to

* Communicated by the Author.

take for this measure only the density of caloric produced in a space by the radiation of neighbouring bodies."

P. 111, line 19. "A supposition that it appears very natural to admit is, that the action of the caloric of a molecule of gas on the caloric of another molecule, &c."

P. 113, line 6. "We have supposed, in that which precedes, that the caloric of a molecule was retained on it by the attraction of the molecule, which experienced no sensible action except by the repulsive force exerted on this caloric by the caloric of the surrounding molecules."

P. 117, line 16. "It is then extremely probable that the attractive force of the caloric of one molecule by another molecule is insensible in the state of gas."

Professor Thomson, who has written largely on the dynamical theory of heat (Philosophical Magazine; Edinb. Phil. Trans.; London Phil. Trans.), has the following in a memoir "On the Thermal Effects of Fluids in Motion" (Phil. Trans. 1854, p. 361):—

"In the notes (Prof. Thomson's notes) to Mr. Joule's paper on the Air-engine (Phil. Trans. 1852, p. 82), it was shown that if Mayer's hypothesis be true, we must have approximately—

Specific heat of air with constant pressure   ·2374
Specific heat of air with constant volume   ·1684

because observations on the velocity of sound, with *Laplace's theory*, demonstrate that $k = 1·410$ within $\frac{1}{700}$th of its own value. Now the experiments at present communicated to the Royal Society prove a very remarkable approximation to the truth in that hypothesis[*]; and we may therefore use these values as very close approximations to the specific heats of air."

The ratio k was found by MM. Gay-Lussac and Welter to be 1·37, and by MM. Clement and Desormes 1·35. The same experiments repeated by the author of the article 'Hygrometry' in the *Encyc. Brit.*, impressed him with the conviction that the *initial* ratio was exactly $\frac{4}{3}$.

Here, therefore, we have an instance of a zealous adherent and expounder of the dynamical theory of heat showing such confi-

---

[*] At p. 341, under the heading "Theoretical deductions from these Experiments," Prof. Thomson arrives at the conclusion that air and carbonic acid "evolve more heat than the amount mechanically equivalent to the work of compression," thus representing the experiments as proving an *exception* to Mayer's hypothesis. That such cooling effects must take place in consequence of the *deviation* from Mariotte's law discovered by Regnault, if Mayer's hypothesis and the dynamic theory of heat hold good, seems obvious enough (see Phil. Mag. vol. xiv. p. 279). So that they are actually a delicate test of the accuracy of that hypothesis; and taken as supplementary to Mr. Joule's previous experimental inquiries, seem completely to establish it as a theory proved by induction.

dence in Laplace's views, such conviction of their truth to nature, that he considers them entitled to supersede the result of direct observation; and no attempt is made to reconcile the statical theory of caloric with the *vis viva* theory of heat; nor is a necessity for doing so apparently even felt as a preliminary step to the adopting of one of the deductions of the former as an ally to the deductions of the latter.

As Mr. Joule's name is conjoined with that of Professor Thomson in this memoir, it is perhaps necessary to assume that he also approves of this manner of deducing the ratio of specific heats.

Another instance of the commingling of statical and dynamical theories is to be found in a memoir by M. Masson, " On the Correlation of the Physical Properties of Bodies," which appeared in a recent number of the *Annales de Chimie*. M. Masson also adopts the mechanical theory of heat (see § III. chap. 2). Yet in the beginning of the same chapter he writes, " Laplace has discovered the true mathematical expression of the velocity of sound in gas." Further, M. Masson states that he has found, by experimenting according to the method employed by MM. Gay-Lussac and Welter, that the value of k is actually what is required to make Laplace's formula agree exactly with the best observations on the velocity of sound. No details are given, although the amount of compression in such experiments is of some importance, as the ratio increases with this amount; and the formula has only to do with the initial ratio, or when the change of density is infinitely small. M. Masson has also computed the velocity of sound in a considerable number of composite gases and vapours from the pitch of the sound given by an organ pipe while immersed in them; and thence, employing Laplace's theorem, obtains the respective values of k, which are thus found to range from 1·26 to 1·42 in the composite gases, and from 1·06 to 1·27 in the vapours.

Thus all authorities, both of the statical and dynamical school, seem to agree that Newton and Laplace's theory of sound is a perfect representation of nature, and that its success is as complete as the theory of gravitation.

Newton, in several parts of the *Principia*, calls upon us to keep in mind that his principles are mathematical, not philosophical. At the beginning of Book 3 he thus expresses himself:—" In the preceding books I have laid down the principles of philosophy; *principles not philosophical, but mathematical, such to wit as we may build our reasonings upon in philosophical inquiries.*" Again, after proving in Prop. 23, Book 2, that " particles flying each other with forces that are reciprocally proportional to the distances of their centres, compose an elastic

2 I 2

fluid whose density is as the compression," he adds, in the scholium, " *But whether elastic fluids do really consist of particles so repelling each other, is a physical question. We have here demonstrated mathematically the property of fluids consisting of particles of this kind, that hence philosophers may take occasion to discuss that question.*"

In Prop. 47 of Book 2, Newton shows that " if pulses are propagated through a fluid, the several particles of the fluid going and returning with the shortest reciprocal motion are always accelerated or retarded according to the law of the oscillating pendulum." It is assumed that the elastic force is proportional to the density; and in the direction of the pulse the fluid is supposed to be divided into *physical lineolæ, which are expansible and contractile,* and exhibit a force that resists compression inversely as their breadth. The mathematical reasoning defines the law by which the breadth of these lineolæ must vary while they go and return,—and hence the law of the difference in the breadth of two adjacent lineolæ, and consequently the law of the accelerative force operating on each corpuscle, which is thus found to be the same as a body moving in a cycloid is subject to under the influence of gravity.

Newton's fundamental hypothesis is, that the particles of air in the direction of the pulse are *successively* agitated with like motions; that both the dynamic condition and the static force of repulsion, which is determined by the length of the line that separates two adjacent particles (called a lineola), is transferred onwards in the direction of the pulse from one particle to the next adjacent in regular succession.

The demonstration takes account of three orders of magnitudes :—1, the breadth of a pulse (L) ; 2, the breadth of an oscillation of a particle $(2l)$; 3, the length of a lineola (λ), each considered as infinitesimal with respect to the preceding.

If the motion of a particle forward and backward in the line $2l$ corresponds to that of a cycloidal pendulum, *i. e.* if the relation between the accelerative force (acting in the line of motion), the acquired velocity, and the time is the same in the line $2l$ as in the complete cycloid, the force in this line must vary simply as the distance (y) from its middle point. The value thus assigned to the force implies that δ, the difference between the lengths of two adjacent lineolæ, should vary also in this proportion. If a semicircle is described on the diameter $2l$, y is the cosine of an arc, of which x being the sine, we have $\delta \doteq y \doteq dx$; so that the differential of the lineola ought to be equal to the differential of the sine, and hence the absolute magnitude of the deviation of the length of a lineola from its mean length ought to be proportional to the sine.

Thus if the motion of a particle is oscillatory, like a complete cycloidal pendulum, the required sequence of force demands the above specific sequence of change in the distance of the particles. Again, if the motion of each particle is oscillatory, the required sequence in its velocity (viz. that it should vary as x the sine) demands also a specific sequence of change in the distance of the particles; and this sequence is precisely the same as what is required by the sequence of force.

To obtain a clear idea of the proof of this (which is a problem of pure mathematics), we may suppose with the same radius l another semicircle to be described, placed also in the line of the pulse, and removed to the distance λ from the preceding semicircle. Let a third also be drawn, removed the same distance λ from the second. We have further to suppose these semicircles divided into as many parts (aa_1, a_1a_2, a_2a_3, &c.; bb_1, b_1b_2, b_2b_3, &c.; cc_1, c_1c_2, c_2c_3, &c.), beginning at where the line of pulse intersects them as λ enters into L (or $\dfrac{L}{\lambda}$ number of parts). The length of each of these parts or *steps* is thus $\dfrac{\lambda}{L} 2l\pi = s$ (being infinitesimal with regard to λ).

Having made this construction, we have next to consider that the motion of each particle to be oscillatory must be such that, at the instant when particle A has traversed the versed sine of aa_n, the particle B (next in *advance* of A) being one *step* behind in its motion, has traversed only the versed sine of bb_{n-1}, and particle C the versed sine of cc_{n-2}. If B had traversed as many steps as A, the distance λ that separates them would not alter; but since it is a step behind, AB is at this point less than λ by the difference between vers aa_n and vers bb_{n-1}, or vers $aa_n -$ vers aa_{n-1}, which equals $s . \sin aa_{n-\frac{1}{2}}$. In the same way, C being a step behind B, their distance is less than λ by $s . \sin aa_{n-1\frac{1}{2}}$. Thus we have $\mathrm{BC} - \mathrm{BA} = s (\sin aa_n - \sin aa_{n-1}) = s \cos aa_n \dfrac{s}{l}$. [Here s, being an absolute magnitude, has to be divided by the absolute radius l to represent the differential of arc.] At the beginning of the vibration $n = 1$, and $\cos aa_n =$ radius; hence with B at the initial point b, C at c_{-1} (a step back on the returning half of the previous oscillation), and A at a_1 (the points on the circle being supposed projected on the diameter), the difference

$$\mathrm{BC} - \mathrm{BA} = \frac{s^2}{l} = 4l\left(\frac{\pi\lambda}{L}\right)^2.$$

This initial amount determines the accelerative force acting at the beginning of the motion of each particle, which is obtained by comparing it with the reciprocal of λ, which represents the whole static force of repulsion between two particles at the distance λ. This force having to support the weight of $\dfrac{H}{\lambda}$ particles (H being the height of a uniform atmosphere), $\dfrac{1}{\lambda}$ represents the force $\dfrac{H}{\lambda} g$, viz. a force that in one second is capable of communicating a velocity of $\dfrac{H}{\lambda} g$ feet per second.

To obtain the value of the initial force acting on particle B when it is at b, we have the following proportion:

$$\frac{1}{\lambda} : \frac{1}{\lambda} - \frac{1}{\lambda - \dfrac{s^2}{l}} :: \frac{H}{\lambda} g : \frac{s^2}{l\lambda} \frac{H}{\lambda} g = Hg4l\left(\frac{\pi}{L}\right)^2.$$

The time (τ) taken by a particle to traverse $2l$, with this force diminishing as the distance from the centre of the semicircle, is the same as the time required for one oscillation of a pendulum whose length is l, if subject to an influence of gravity equal to this force, and is the same as the time taken by the pulse to travel through $\frac{1}{2}$L. By the law of the pendulum, τ is equal to π multiplied by the square root of quotient of length of pendulum by force of gravity, hence

$$\tau = \pi \sqrt{\frac{l}{\pi^2} \frac{L^2}{Hg4l}} = \tfrac{1}{2} L \sqrt{\frac{1}{Hg}} ;$$

and the velocity of the pulse per second is $\sqrt{Hg}$.

This supposes Mariotte's law maintained; but the experiments made at the suggestion of Laplace, proved that for rapid compressions the elastic force increases in a higher ratio than the density; and for small increments of density, the correspondent increments of elastic force are very nearly $\frac{4}{3}$ those computed on the hypothesis that Mariotte's law is maintained; that is, as if the repulsive force of a particle supported the weight of $\dfrac{4}{3} \dfrac{H}{\lambda}$ particles: hence Newton's theorem for the velocity of sound in elastic fluids is strictly represented by $\sqrt{\frac{4}{3}Hg}$.

The numerical results from this formula, taking Regnault's data, compared with observations made at low temperatures (so that the influence of aqueous vapour should be avoided), show a difference at most of about $\frac{1}{36}$, or as if $\frac{4}{3}$ should be augmented $\frac{1}{18}$th part. Thus no objection to Newton's theory can well be made on the ground of its not according with observation, as no theory can be expected to embrace all the circumstances that may affect the result; *e. g.* the repulsive force may not emanate from the centre of the particles, so that the size of the particles may influence the result. Also the repulsive action is necessarily assumed to be limited to adjacent particles, not extending through the interstices of these to the particles beyond (for such is the extraordinary and improbable hypothesis required to deduce Mariotte's law from a static repulsive force). This may be supposed subject to modification during vibratory action.

But the hypothesis upon which the mathematical demonstration rests is open to three grounds of objection:—1. It does not take account of the condition of the front of a pulse when the particles from a condition of rest enter into the cycle of motion defined by the theory. 2. The force of repulsion between two adjacent particles required by the theory is extravagantly large. 3. The other physical properties of gases are not deducible from the hypothesis.

To these may be added, that the dynamical theory of heat has suggested another hypothesis which is free from these objections, and which therefore claims a preference according to Newton's first " rule of reasoning in philosophy," viz. " We are to admit no more causes of natural things than such as are both true and sufficient to explain their appearances. To this purpose the philosophers say that Nature does nothing in vain, and *more is in vain when less will serve*; for Nature is pleased with simplicity, and affects not the pomp of superfluous causes."

1. The theory does not take account of the condition of the front of the pulse, or rather of the front of the first of the series of pulses of which a sound consists. This is apparent if we consider that a particle is represented by the theory as at rest at each extremity of its oscillation, and at those points the accelerative force is at its maximum, and is derived from the difference between the lengths of the lineolæ that issue from the particle in front and in rear. The front lineola cannot differ from the mean length so long as the front particle is at rest unaffected by the advancing pulse. The rear lineola is less than the mean length by a certain small amount α. If the front particle were in action in a pulse cycle, the length of the front lineola would be greater than the mean length of a lineola by the same amount α, so that the accelerative force at each extremity of the oscillation of a particle is represented by 2α; and unless it were so, the condition required to sustain the beautiful relation of velocity and propelling force would be wanting. But at the front of the first pulse the lineola does not differ from the mean length, so that the accelerative force is represented by α, and this is only one-half the amount required by the theory to begin the oscillation. In truth, the demonstration only applies to a pulse having similar pulses operating on both sides.

2. The force of repulsion between two adjacent particles re-

quired by the theory is extravagantly large.   The recent advances in the theory of heat have, in a measure, compelled us to realize the dynamic value of natural forces.   To compute the absolute value of the repulsive force acting between two adjacent molecules of air, we have to consider that it has to support the gravity of the number of molecules in the height of a uniform atmosphere $\left[\dfrac{H}{\lambda}\right]$; it must therefore exceed the force of gravity of one molecule in this ratio.   Now the force of gravity in one second can communicate a velocity of 32 feet per second, so that the force of repulsion between two adjacent molecules of air must be capable in one second of communicating a velocity of $32\,\dfrac{H}{\lambda}$ feet.   The absolute value of λ, the distance between two adjacent molecules of air, we can now with great probability deduce from the phænomena of capillarity (Phil. Mag. vol. xv. p. 1).   At the boiling-point of water the number of molecules of steam in a cubic inch is the same as the number of molecules of air in the same volume.   At 86° the number of layers of aqueous molecules in a cubic inch is 215 millions (Phil. Mag. vol. xv. p. 11).   Hence at ordinary temperatures the distance between two adjacent molecules of air must be about $\frac{1}{17}$th of a millionth of an inch, and the value of $\left(32\,\dfrac{H}{\lambda}\right)$, the velocity communicable in a second, is 160 thousand times the velocity of light.   Can we for a moment believe that such a force has any real existence, that it is other than a mathematical fiction?

3. The other physical properties are not deducible from the hypothesis of a static force of repulsion.   The deductive power of Newton's theory is confessedly limited to Mariotte's law and the velocity of sound.   Laplace, by the invention of calorific atmospheres, is allowed to have added to these Dalton and Gay-Lussac's theory of expansion; but it is a question whether the reciprocal action between heat-atmospheres and molecules, which he expresses by mathematical symbols, can be realized by the mind.   In judging of this, we must not forget the chapter of the *Mécanique Céleste*, in which the author speculates upon what the laws of motion would have been if force had been as a function of the velocity, instead of as the simple velocity*.   What is to be expected from a superstructure resting upon such a foundation as this reveals?   Nevertheless, granting that

---

* The author of the article 'Virtual Velocities' in the Penny Cyclopædia has the following remark upon this chapter of the *Mécanique Céleste*:— " We have never met with any one who could give us an intelligible account of the meaning of this investigation."

Mariotte's law, Dalton and Gay-Lussac's law, and the velocity of sound are represented by the statical hypothesis, we have still Dalton and Graham's law of diffusion and diffusive velocity; Gay-Lussac's law of volumes; Dulong and Petit's law of specific heat, extended to the more simple gaseous bodies by Haycraft and the French physicists; the law of latent heat partially discovered by Gay-Lussac and Welter's experiments; also the diminution of temperature in ascending the atmosphere,—all as yet undeduced from any statical theory of elastic fluids. It may be that additions to the mathematical hypotheses of Laplace will be attempted with the view of extending their capacity, as indeed there seems to be no limit to this artificial and barren system of procedure, which is as far removed from the simplicity of nature as the hideous epicycles of Ptolemy.

There is another mode by which pulses may be conceived to be transmitted, which admits of being set forth in a popular way.

Suppose we range a number of ivory balls in a straight line upon a billiard table, and strike the first of the row upon the second, the initial velocity will be carried forward from one ball to the next adjacent, and so will make its appearance in the last —supposing perfect elasticity and no resistance in rolling— undiminished as if the motion of the first ball had continued, and the impulse had been carried by it alone, and not transferred by impact through others. These balls, confined to one line, are supposed to be in motion among themselves, so that those adjacent alternately strike against each other in opposite directions; the end ones being reflected from the cushions, and then back again after striking the next adjacent ball, the *vis viva* in one direction being at all times equal to the *vis viva* in the opposite.

If we now suppose one cushion to move forward with comparatively slow velocity, each time the adjacent ball strikes it it will be reflected with a velocity greater than that with which it impinged. This increment of velocity it transfers to the next ball, and so on; and the velocity with which the impulse is transferred along the line is equal to the common velocity with which the balls move. We may suppose the line extended indefinitely, and the motion of the cushion to be alternately forwards and backwards. While the adjacent ball impinges many times during each advance and retreat—during the former carrying forward a succession of small increments of *vis viva*, during the latter a succession of small decrements of *vis viva*—a pulse is formed, the intensity and duration of which depends on the motion of the cushion, but the velocity of propagation upon the motions of the balls, upon their common velocity.

Now suppose these balls reduced excessively in their dimensions, and to be *perfectly* elastic as well as the cushion, and we shall have obtained an idea of how pulses may be conveyed in a manner quite different from that depending upon a statical repulsive force between adjacent particles. Instead of such stupendous force, we have to substitute molecules simply moving with the velocity of a cannon-ball, and assume the atmosphere to maintain its elasticity by its particles striking against each other with such velocity, which, viewed cosmically, by no means exceeds what is moderate, and even highly probable if heat is molecular motion.

Such a theory of elastic fluids was started by Mr. Herapath so far back as 1821 in the pages of the Annals of Philosophy, and has more recently suggested itself to M. Krönig, as we find noticed by M. Clausius in his memoir " On the Nature of the Motion we call Heat " (Phil. Mag. vol. xiv. p. 108).

The following is an extract from Mr. Herapath's memoir, p. 278, vol. i. Annals of Philosophy, April 1821 :—" . . . . . . if gases, instead of having their particles endued with repulsive forces, subject to so curious a limitation as Newton proposed, were made up of particles or atoms mutually impinging on one another, and the sides of the vessel containing them, such a constitution of aëriform bodies would not only be more simple than repulsive powers, but, as far as I could perceive, would be consistent with phænomena in other respects, and would admit of an easy application of the theory of heat by intestine motion. Such bodies, I easily saw, possessed several of the properties of gases : for instance, they would expand, and if the particles be vastly small, contract almost indefinitely ; their elastic force would increase by an increase of motion or temperature, and diminish by a diminution ; they would conceive heat rapidly and conduct it slowly ; would generate heat by sudden compression, and destroy it by sudden rarefaction ; and any two having ever so small a communication, would quickly and equally intermix."

At p. 341, in paragraph beginning " These impulses," &c., and in those which succeed, ending " have from one another," we have a very clear announcement of the mode by which a static force of pressure is counterbalanced by a dynamic force of elastic impact.

At p. 345, Prop. 8. " The same things remaining, the elasticity of a gas under a variable temperature and compression is proportional to its numeratom (number of atoms in constant volume) and the *square of its temperature* conjointly ; or the elasticity varies as the *square of the temperature* directly, and the simple of the space inversely.

Mr. Herapath unfortunately assumed heat or temperature to be represented by the simple ratio of the velocity instead of the square of the velocity—being in this apparently led astray· by the definition of motion generally received—and thus was baffled in his attempts to reconcile his theory with observation. If we make this change in Mr. Herapath's definition of heat or temperature, viz. that it is proportional to the *vis viva* or square velocity of the moving particle, not to the momentum or simple ratio of the velocity, we can without much difficulty deduce, not only the primary laws of elastic fluids, but also the other physical properties of gases enumerated above in the third objection to Newton's hypothesis. In the Archives of the Royal Society for 1845–46, there is a paper " On the Physics of Media that consists of perfectly Elastic Molecules in a state of Motion," which contains the synthetical reasoning upon which the demonstration of these matters rests. The velocity of sound is therein deduced to be equal to the velocity acquired in falling through three-fourths of a uniform atmosphere. This theory does not take account of the size of the molecules. It assumes that no time is lost at the impact; and that if the impacts produce rotatory motion, the *vis viva* thus invested bears a constant ratio to the rectilineal *vis viva*, so as not to require separate consideration. It also does not take account of the probable internal motion of composite molecules; yet the results so closely accord with observation in every part of the subject, as to leave no doubt that Mr. Herapath's idea of the physical constitution of gases approximates closely to the truth. M. Krönig appears to have entered upon the subject in an independent manner, and arrived at the same result; M. Clausius, too, as we learn from his paper " On the Nature of the Motion we call Heat " (Phil. Mag. vol. xiv. p. 108).

The physics of such media is a study that must be ungenial, perhaps repulsive, to mathematicians brought up in the statical school. The fundamental hypothesis does not permit us at once to transfer the subject to the domain of pure mathematics, as Newton's hypothesis converted physical astronomy into a purely mathematical study. The *mode* of action by which certain phænomena make their appearance must be realized at each step in conformity with the conservation of force: the causal relation must be ever present to the mind, or no true progress can be made. On the other hand, the mathematics required is simple, and almost every one of the applications of the theory admits of popular illustration. An attempt of this kind I have given above in reference to the conveyance of sonorous impulses, introducing a dynamic theory of sound. The strict demonstration of the velocity is given in the memoir above referred to. A few further illustrations may here be added to show in brief the capabilities of the theory; but it is difficult to limit these to one branch, all the physical properties of elastic fluids being so interwoven with each other; and it is an admirable instance of the simplicity of nature, that the cluster of elegant quantitative relations which the physics of gases present, should flow from the constitution assigned, which indeed is the simplest that it is possible to imagine.

The velocity of sound is not affected by the height of the barometer, but it is sensibly influenced by a change of temperature. This latter is to be looked for; since the velocity of the particles of air increases with the temperature, the velocity with which they convey pulses must increase in the same proportion: but it is not so obvious that the height of the barometer or weight of the atmosphere should have no effect either to accelerate or retard.

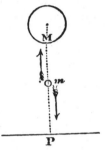

Let m be an elastic ball traversing the vertical P M backwards and forwards from the sphere, M, to the plane, P, the surfaces of m, M, and P being perfectly elastic. The condition of permanence in the mean distance of M from P requires that the impacts of m upon M should have the effect of changing the velocity of M downwards into the same velocity upwards. Gravity affects M in the interval of time that elapses while m descends from M to P and ascends from P to M; during half this time gravity is employed in destroying the upward motion of M, and during the second half in producing the velocity downwards with which it encounters m on its return,—m and M thus meeting each other, and separating after impact with the same velocity, but with directions reversed.

The relation between the distance $MP(=\lambda)$, the velocity of $m(=v)$, the weight of $M\left(=\dfrac{H}{\lambda}m\right)$, and of m, is very simple, and enables us to compute the absolute value of v.

The time taken by m to traverse MP is $\dfrac{\lambda}{v}$ part of a second; and in this time gravity communicates to M the velocity $\dfrac{\lambda}{v}g$.

From the law of elastic collision, two bodies impinging and reflected back in the direction they came with unaltered velocities, must have their velocities inversely proportioned to their masses, so that

$$M : m :: v : \frac{vm}{M} = \frac{\lambda}{H}v = \frac{\lambda g}{v} \text{ as above.}$$

Thus we have $v^2 = Hg$, or the velocity is that produced by gravity in a body falling through $\frac{1}{2}H$. [Strictly, the square velocity of air-molecules must be six times this, because the above calculus only takes account of the action in one of the six rectangular directions of space.]

To trace the influence of the barometer or weight of the uniform atmosphere, we may suppose the weight of M doubled and λ reduced one-half: this leaves H and v unaltered; so that if the density of air increases as the weight of the uniform atmosphere, the velocity of sound is unaffected by the barometer. If with M constant the density represented by $\frac{1}{\lambda}$ diminishes, H must increase in the same ratio, and thence $v^2 \doteq \lambda$, or the volume under constant pressure as the square molecular velocity,—which conforms to Dalton and Gay-Lussac's law, if $v^2 \doteq$ temperature from zero of gaseous tension.

If we view these relations in another elastic fluid, where the weight of the molecule is twice that of air, M being unaltered, and the number of molecules in a unit volume also the same as with air, we have H inversely as m, or one-half the height of a uniform atmosphere of air, and v^2 reduced in the same proportion; also the velocity of sound reduced inversely as the square root of the molecular weight or specific gravity of the gas.

To explain the increase of temperature that arises from suddenly condensing air, we may imagine an elastic ball traversing a vertical between two horizontal plates and striking alternately against them. Those plates being also considered as perfectly elastic, the velocity of the ball will continue uniform without its motion being impaired. If we now suppose the distance between the plates to be gradually diminished by one of them assuming a velocity incomparably less than that of the ball, the ball will, each time it strikes this advancing plane, receive an increment to its velocity, and thus to its *vis viva*.

Let v represent the velocity of the ball, δ the distance between the planes, $\frac{v}{n}$ the velocity of the plane. The number of impacts upon the advancing plate in a unit of time is $\frac{v}{2\delta}$. The velocity after one impact has increased from v to $v + \frac{v}{n}$, and the square velocity from v^2 to $v^2 + \frac{2v^2}{n}$, the increment being $\frac{2}{n}$ of the square velocity; at the same time the decrement of space is $\frac{v}{n}$ (the space moved over by the plate in a unit of time) divided by $\frac{v}{2\delta}$; this gives $\dfrac{2\delta}{n}$; and the ratio of this to δ is also $\dfrac{2}{n}$: thus the increment of *vis viva* is equal to the decrement of distance, irrespective of the velocities and distance, and is equal to the mechanical force exerted by the plate. In an elastic medium, the increment of absolute temperature is equal to one-third the decrement of volume, and the increment of temperature is the equivalent of the force expended in the act of compression. We thus gain a knowledge of the mechanical equivalent of heat, and further deduce the amount of *vis viva* in a gas to be equal to the work performed by its pressure acting through three times its volume.

The diminution of temperature found when ascending the atmosphere may be illustrated by supposing a series of elastic balls, *a, b, c, d,* &c., to be arranged in a vertical, and moving in the vertical so that those adjacent shall alternately strike against each other at the extremity of their up-and-down motion without any transference of *vis viva,* which requires that they should encounter each other with equal velocities, e. g. *b* in its up-motion striking *c,* and in its down-motion striking *a.* Now we have to mark, that, between the upper and lower impact, *b* receives an accession of *vis viva* from the force of gravity which is proportional to the vertical distance traversed; so that comparing the *vis viva* of *b* with any other of the series, such as *z,* we shall find that the higher ball *z* has less *vis viva* than *b,* and the difference is equal to gravity acting through *bz.* Thus in the atmosphere we might expect the decrease of temperature to be uniform if its constitution agrees with this hypothesis.

The strict demonstration represents this to be the case, and that the gradient of temperature is $1°$ in 319 feet; also that the absolute height of the atmosphere is six times the height of a uniform atmosphere, the density in a stratum as the fifth power of the depth of that stratum below the summit, and the elastic force or height of barometer as the sixth power of that depth.

Here we find that the diminishing temperature, in ascending the atmosphere, is represented as the natural condition of vertical equilibrium; and the question occurs, may not the increasing temperature found in descending through the earth's crust be also its natural condition of vertical equilibrium? This may be cited as one of many instances of the suggestive power of the *vis viva* theory, marking it as specially the natural introduction to the dynamic theory of heat, and as likely to promote a beneficial change in the application of mathematics to molecular physics generally. Upon this account it seems to merit the attention of the educational authorities in the higher depart-ments of scientific instruction. The subject, however, is so remote from practical application, that there is little hope of any impression being made in such quarters for many years. In the mean time it seems the duty of those who have profited by it, to do what lies in their power to proclaim its merits and acknowledge the value of the idea first struck out by Mr. Herapath, and perhaps saved from oblivion by the Philosophical Magazine of that period.

Edinburgh, Nov. 6, 1858.

30

Sensations of Tone

HERMANN VON HELMHOLTZ

Hermann von Helmholtz (1821–1894) was probably the most distinguished German physicist of the 19th century. Trained as an army surgeon he early made important discoveries in physiology and from 1849–1872 held chairs in physiology in German universities. In the meantime his intensive study of physics led him to combine the two fields in research in vision and audition. In 1872 he became professor of physics at the university of Berlin and investigated many fields of theoretical physics. A scientists of very wide interests, he devoted much attention to mathematics, philosophy, and the arts.

The three extracts from Helmholtz's work presented here are taken from his famous book, *Sensations of Tone*, in which he endeavored to provide a physical and physiological basis for hearing and in particular for the perception of musical sounds. The Introduction sets forth the plan of the work and gives an insight into the author's wide ranging interests in science and the arts. The second extract describes Helmholtz's method of analyzing sound waves by means of resonators, which he invented and which are known by his name. The third extract discusses the analysis of musical tones by the ear.

Excerpts reprinted from an English translation of "Die Lehre von den Tunernpfindungen als Physiologische Grundlage für die Theorie der Musik," 3rd edition (1895) (Dover Publications, Inc., New York, 1954)

30

INTRODUCTION.

— ◆ —

In the present work an attempt will be made to connect the boundaries of two sciences, which, although drawn towards each other by many natural affinities, have hitherto remained practically distinct—I mean the boundaries of *physical and physiological acoustics* on the one side, and of *musical science and esthetics* on the other. The class of readers addressed will, consequently, have had very different cultivation, and will be affected by very different interests. It will therefore not be superfluous for the author at the outset distinctly to state his intention in undertaking the work, and the aim he has sought to attain. The horizons of physics, philosophy, and art have of late been too widely separated, and, as a consequence, the language, the methods, and the aims of any one of these studies present a certain amount of difficulty for the student of any other ¶ of them; and possibly this is the principal cause why the problem here undertaken has not been long ago more thoroughly considered and advanced towards its solution.

It is true that acoustics constantly employs conceptions and names borrowed from the theory of harmony, and speaks of the ' scale,' ' intervals,' ' consonances,' and so forth; and similarly, manuals of Thorough Bass generally begin with a physical chapter which speaks of 'the numbers of vibrations,' and fixes their ' ratios ' for the different intervals; but, up to the present time, this apparent *connection* of acoustics and music has been wholly external, and may be regarded rather as an expression given to the feeling that such a connection must exist, than as its actual formulation. Physical knowledge may indeed have been useful for musical instrument makers, but for the development and foundation of the theory of harmony ¶ it has hitherto been totally barren. And yet the essential facts within the field here to be explained and turned to account, have been known from the earliest times. Even Pythagoras (fl. circa B.C. 540–510) knew that when strings of different lengths but of the same make, and subjected to the same tension, were used to give the perfect consonances of the Octave, Fifth, or Fourth, their lengths must be in the ratios of 1 to 2, 2 to 3, or 3 to 4 respectively, and if, as is probable, his knowledge was partly derived from the Egyptian priests, it is impossible to conjecture in what remote antiquity this law was first known. Later physics has extended the law of Pythagoras by passing from the lengths of strings to the number of vibra-tions, and thus making it applicable to the tones of all musical instruments, and the numerical relations 4 to 5 and 5 to 6 have been added to the above for the less perfect consonances of the major and minor Thirds, but I am not aware that any real step was ever made towards answering the question : *What have musical consonances to do with the ratios of the first six numbers ?* Musicians, as well as philosophers and physicists, have generally contented themselves with saying in effect that human minds were in some unknown manner so constituted as to discover the numerical relations of musical vibrations, and to have a peculiar pleasure in contemplating simple ratios which are readily comprehensible.

Meanwhile musical esthetics has made unmistakable advances in those points which depend for their solution rather on psychological feeling than on the action of the senses, by introducing the conception of movement in
¶ the examination of musical works of art. E. Hanslick, in his book ' on the Beautiful in Music ' (*Ueber das musikalisch Schöne*), triumphantly attacked the false standpoint of exaggerated sentimentality, from which it was fashionable to theorise on music, and referred the critic to the simple elements of melodic movement. The esthetic relations for the structure of musical compositions, and the characteristic differences of individual forms of composition, are explained more fully in Vischer's ' Esthetics ' (*Aesthetik*). In the inorganic world the kind of motion we see, reveals the kind of moving force in action, and in the last resort the only method of recognising and measuring the elementary powers of nature consists in determining the motions they generate, and this is also the case for the motions of bodies or of voices which take place under the influence of human feelings. Hence
¶ the properties of musical movements which possess a graceful, dallying, or a heavy, forced, a dull, or a powerful, a quiet, or excited character, and so on, evidently chiefly depend on psychological action. In the same way questions relating to the equilibrium of the separate parts of a musical composition, to their development from one another and their connection as one clearly intelligible whole, bear a close analogy to similar questions in architecture. But all such investigations, however fertile they may have been, cannot have been otherwise than imperfect and uncertain, so long as they were without their proper origin and foundation, that is, so long as there was no scientific foundation for their elementary rules relating to the construction of scales, chords, keys and modes, in short, to all that is usually contained in works on ' Thorough Bass.' In this elementary region
¶ we have to deal not merely with unfettered artistic inventions, but with the natural power of immediate sensation. Music stands in a much closer connection with pure sensation than any of the other arts. The latter rather deal with what the senses apprehend, that is with the images of outward objects, collected by psychical processes from immediate sensation. *Poetry* aims most distinctly of all at merely exciting the formation of images, by addressing itself especially to imagination and memory, and it is only by subordinate auxiliaries of a more musical kind, such as rhythm, and imitations of sounds, that it appeals to the immediate sensation of hearing. Hence its effects depend mainly on psychical action. The *plastic arts*, although they make use of the sensation of sight, address the eye almost in the same way as *poetry* addresses the ear. Their main purpose s to excite in us the image of an external object of determinate form and colour. The spectator is essentially intended to interest himself in this

image, and enjoy its beauty; not to dwell upon the means by which it was created. It must at least be allowed that the pleasure of a connoisseur or virtuoso in the constructive art shewn in a statue or a picture, is not an essential element of artistic enjoyment.

It is only in painting that we find colour as an element which is directly appreciated by sensation, without any intervening act of the intellect. On the contrary, in *music*, the sensations of tone are the material of the art. So far as these sensations are excited in music, we do not create out of them any images of external objects or actions. Again, when in hearing a concert we recognise one tone as due to a violin and another to a clarinet, our artistic enjoyment does not depend upon our conception of a violin or clarinet, but solely on our hearing of the tones they produce, whereas the ¶ artistic enjoyment resulting from viewing a marble statue does not depend on the white light which it reflects into the eye, but upon the mental image of the beautiful human form which it calls up. In this sense it is clear that music has a more immediate connection with pure sensation than any other of the fine arts, and, consequently, that the theory of the sensations of hearing is destined to play a much more important part in musical esthetics, than, for example, the theory of *chiaroscuro* or of perspective in painting. Those theories are certainly useful to the artist, as means for attaining the most perfect representation of nature, but they have no part in the artistic effect of his work. In music, on the other hand, no such perfect representation of nature is aimed at; tones and the sensations of tone exist for themselves alone, and produce their effects independently of anything behind ¶ them.

This theory of the sensations of hearing belongs to natural science, and comes in the first place under *physiological acoustics*. Hitherto it is the *physical* part of the *theory of sound* that has been almost exclusively treated at length, that is, the investigations refer exclusively to the motions produced by solid, liquid, or gaseous bodies when they occasion the sounds which the ear appreciates. This *physical acoustics* is essentially nothing but a section of the theory of the motions of elastic bodies. It is physically indifferent whether observations are made on stretched strings, by means of spirals of brass wire, (which vibrate so slowly that the eye can easily follow their motions, and, consequently, do not excite any sensation of sound,) or by means of a violin string, (where the eye can scarcely perceive the vibrations ¶ which the ear readily appreciates). The laws of vibratory motion are precisely the same in both cases; its rapidity or slowness does not affect the laws themselves in the slightest degree, although it compels the observer to apply different methods of observation, the eye for one and the ear for the other. In physical acoustics, therefore, the phenomena of hearing are taken into consideration solely because the ear is the most convenient and handy means of observing the more rapid elastic vibrations, and the physicist is compelled to study the peculiarities of the natural instrument which he is employing, in order to control the correctness of its indications. In this way, although physical acoustics as hitherto pursued, has, undoubtedly, collected many observations and much knowledge concerning the action of the ear, which, therefore, belong to *physiological acoustics*, these results were not the principal object of its investigations; they were merely secondary and isolated facts. The only justification for devoting a separate chapter to acoustics in the theory of the motions of elastic bodies, to which it essentially belongs, is, that the application of the ear as an instrument of research influenced the nature of the experiments and the methods of observation.

But in addition to a *physical* there is a *physiological theory of acoustics*, the aim of which is to investigate the processes that take place within the ear itself. The section of this science which treats of the conduction of the motions to which sound is due, from the entrance of the external ear to the expansions of the nerves in the labyrinth of the inner ear, has received much attention, especially in Germany, since ground was broken by ¶ Johannes Mueller. At the same time it must be confessed that not many results have as yet been established with certainty. But these attempts attacked only a portion of the problem, and left the rest untouched. Investigations into the processes of each of our organs of sense, have in general three different parts. First we have to discover how the agent reaches the nerves to be excited, as light for the eye and sound for the ear. This may be called the *physical* part of the corresponding physiological investigation. Secondly we have to investigate the various modes in which the nerves themselves are excited, giving rise to their various *sensations*, and finally the laws according to which these sensations result in mental images of determinate external objects, that is, in *perceptions*. Hence we have secondly a specially *physiological* investigation for sensations, and ¶ thirdly, a specially *psychological* investigation for perceptions. Now whilst the physical side of the theory of hearing has been already frequently attacked, the results obtained for its *physiological* and *psychological* sections are few, imperfect, and accidental. Yet it is precisely the physiological part in especial—the theory of the sensations of hearing—to which the theory of music has to look for the foundation of its structure.

In the present work, then, I have endeavoured in the first place to collect and arrange such materials for a theory of the *sensations of hearing* as already existed, or as I was able to add from my own personal investigations. Of course such a first attempt must necessarily be somewhat imperfect, and be limited to the elements and the most interesting divisions of the subject discussed. It is in this light that I wish these studies to be regarded. ¶ Although in the propositions thus collected there is little of entirely new discoveries, and although even such apparently new facts and observations as they contain are, for the most part, more properly speaking the immediate consequences of my having more completely carried out known theories and methods of investigation to their legitimate consequences, and of my having more thoroughly exhausted their results than had heretofore been attempted, yet I cannot but think that the facts frequently receive new importance and new illumination, by being regarded from a fresh point of view and in a fresh connection.

The First Part of the following investigation is essentially physical and physiological. It contains a general investigation of the phenomenon of harmonic *upper partial tones*. The nature of this phenomenon is established, and its relation to *quality of tone* is proved. A series of qualities of tone are analysed in respect to their harmonic upper partial tones, and it results

327

that these upper partial tones are not, as was hitherto thought, isolated phenomena of small importance, but that, with very few exceptions, they determine the qualities of tone of almost all instruments, and are of the greatest importance for those qualities of tone which are best adapted for musical purposes. The question of how the ear is able to perceive these harmonic upper partial tones then leads to an hypothesis respecting the mode in which the auditory nerves are excited, which is well fitted to reduce all the facts and laws in this department to a relatively simple mechanical conception.

The Second Part treats of the disturbances produced by the simultaneous production of two tones, namely the *combinational tones* and *beats*. The physiologico-physical investigation shews that two tones can be simul- ¶ taneously heard by the ear without mutual disturbance, when and only when they stand to each other in the perfectly determinate and well-known relations of intervals which form musical consonance. We are thus immediately introduced into the field of music proper, and are led to discover the physiological reason for that enigmatical numerical relation announced by Pythagoras. The magnitude of the consonant intervals is independent of the quality of tone, but the harmoniousness of the consonances, and the distinctness of their separation from dissonances, depend on the quality of tone. The conclusions of physiological theory here agree precisely with the musical rules for the formation of chords; they even go more into particulars than it was possible for the latter to do, and have, as I believe, the authority of the best composers in their favour. ¶

In these first two Parts of the book, no attention is paid to esthetic considerations. Natural phenomena obeying a blind necessity, are alone treated. The Third Part treats of the construction of *musical scales* and *notes*. Here we come at once upon esthetic ground, and the differences of national and individual tastes begin to appear. Modern music has especially developed the principle of *tonality*, which connects all the tones in a piece of music by their relationship to one chief tone, called the tonic. On admitting this principle, the results of the preceding investigations furnish a method of constructing our modern musical scales and modes, from which all arbitrary assumption is excluded.

I was unwilling to separate the physiological investigation from its musical consequences, because the correctness of these consequences must ¶ be to the physiologist a verification of the correctness of the physical and physiological views advanced, and the reader, who takes up my book for its musical conclusions alone, cannot form a perfectly clear view of the meaning and bearing of these consequences, unless he has endeavoured to get at least some conception of their foundations in natural science. But in order to facilitate the use of the book by readers who have no special knowledge of physics and mathematics, I have transferred to appendices, at the end of the book, all special instructions for performing the more complicated experiments, and also all mathematical investigations. These appendices are therefore especially intended for the physicist, and contain the proofs of my assertions.* In this way I hope to have consulted the interests of both classes of readers.

* [The additional Appendix XX. by the Translator is intended especially for the use of musical students.—*Translator.*]

It is of course impossible for any one to understand the investigations thoroughly, who does not take the trouble of becoming acquainted by personal observation with at least the fundamental phenomena mentioned. Fortunately with the assistance of common musical instruments it is easy for any one to become acquainted with harmonic upper partial tones, combinational tones, beats, and the like.* Personal observation is better than the exactest description, especially when, as here, the subject of investigation is an analysis of sensations themselves, which are always extremely difficult to describe to those who have not experienced them.

In my somewhat unusual attempt to pass from natural philosophy into the theory of the arts, I hope that I have kept the regions of physiology
¶ and esthetics sufficiently distinct. But I can scarcely disguise from myself, that although my researches are confined to the lowest grade of musical grammar, they may probably appear too mechanical and unworthy of the dignity of art, to those theoreticians who are accustomed to summon the enthusiastic feelings called forth by the highest works of art to the scientific investigation of its basis. To these I would simply remark in conclusion, that the following investigation really deals only with the analysis of actually existing sensations—that the physical methods of observation employed are almost solely meant to facilitate and assure the work of this analysis and check its completeness—and that this analysis of the sensations would suffice to furnish all the results required for musical theory, even independently of my physiological hypothesis concerning the mechanism of
¶ hearing, already mentioned (p. 5a), but that I was unwilling to omit that hypothesis because it is so well suited to furnish an extremely simple connection between all the very various and very complicated phenomena which present themselves in the course of this investigation.†

* [But the use of the *Harmonical*, described in App. XX. sect. F. No. 1, and invented for the purpose of illustrating the theories of this work, is recommended as greatly superior for students and teachers to any other instrument. —*Translator.*]

† Readers unaccustomed to mathematical and physical considerations will find an abridged account of the essential contents of this book in Sedley Taylor, *Sound and Music,* London, Macmillan, 1873. Such readers will also find a clear exposition of the physical relations of sound in J. Tyndall, *On Sound,* a course of eight lectures, London, 1867, (the last or fourth edition 1883) Longmans, Green, & Co. A German translation of this work, entitled *Der Schall,* edited by R. Helmholtz and G. Wiedemann, was published at Brunswick in 1874.

*** [The marks ¶ in the outer margin of each page, separate the page into 4 sections, referred to as *a, b, c, d,* placed after the number of the page. If any section is in double columns, the letter of the second column is accented, as p. 13*d'*.]

CHAPTER III.

ANALYSIS OF MUSICAL TONES BY SYMPATHETIC RESONANCE.

WE proceed to shew that the simple partial tones contained in a composite mass of musical tones, produce peculiar mechanical effects in nature, altogether independent of the human ear and its sensations, and also altogether independent of merely theoretical considerations. These effects consequently give a peculiar objective significance to this peculiar method of analysing vibrational forms.

Such an effect occurs in the phenomenon of *sympathetic resonance*. This phenomenon is always found in those bodies which when once set in motion by any impulse, continue to perform a long series of vibrations before they come to rest. When these bodies are struck gently, but periodically, although each blow may be separately quite insufficient to produce a sensible motion in the vibratory body, yet, provided the periodic time of the gentle blows is precisely the same as the periodic time of the body's own vibrations, very large and powerful oscillations may result. But if the periodic time of the regular blows is different from the periodic time of the oscillations, the resulting motion will be weak or quite insensible.

Periodic impulses of this kind generally proceed from another body which is already vibrating regularly, and in this case the swings of the latter in the course of a little time, call into action the swings of the former. Under these circumstances we have the process called *sympathetic oscillation* or *sympathetic resonance*. The essence of the mechanical effect is independent of the rate of motion, which may be fast enough to excite the sensation of sound, or slow enough not to produce anything of the kind. Musicians are well acquainted with *sympathetic resonance*. When, for example, the strings of two violins are in exact unison, and one string is bowed, the other will begin to vibrate. But the nature of the process is best seen in instances where the vibrations are slow enough for the eye to follow the whole of their successive phases.

Thus, for example, it is known that the largest church-bells may be set in motion by a man, or even a boy, who pulls the ropes attached to them at proper and regular intervals, even when their weight of metal is so great that the strongest man could scarcely move them sensibly, if he did not apply his strength in determinate periodical intervals. When such a bell is once set in motion, it continues, like a struck pendulum, to oscillate for some time, until it gradually returns to rest, even if it is left quite by itself, and no force is employed to arrest its motion. The motion diminishes gradually, as we know, because the friction on the axis and the resistance of the air at every swing destroy a portion of the existing moving force.

As the bell swings backwards and forwards, the lever and rope fixed to its axis rise and fall. If when the lever falls a boy clings to the lower end of the bell-rope, his weight will act so as to increase the rapidity of the existing motion. This increase of velocity may be very small, and yet it will produce a corresponding increase in the extent of the bell's swings, which again will continue for a while, until destroyed by the friction and resistance of the air. But if the boy clung to the bell-rope at a wrong time, while it was ascending, for instance, the weight of his body would act in opposition to the motion of the bell, and the extent of swing would decrease. Now, if the boy continued to cling to the rope at each swing so long as it was falling, and then let it ascend freely, at every swing the motion of the bell would be only increased in speed, and its swings would gradually become greater and greater, until by their increase the motion imparted on every oscillation of the bell to the walls of the belfry, and the external air would become so great as exactly to be covered by the power exerted by the boy at each swing.

The success of this process depends, therefore, essentially on the boy's applying his force only at those moments when it will increase the motion of the bell. That is, he must employ his strength periodically, and the periodic time must be equal to that of the bell's swing, or he will not be successful. He would just as easily bring the swinging bell to rest, if he clung to the rope only during its ascent, and thus let his weight be raised by the bell.

A similar experiment which can be tried at any instant is the following. Construct a pendulum by hanging a heavy body (such as a ring) to the lower end of a thread, holding the upper end in the hand. On setting the ring into gentle pendular vibration, it will be found that this motion can be gradually and considerably increased by watching the moment when the pendulum has reached its greatest ¶ departure from the vertical, and then giving the hand a very small motion in the opposite direction. Thus, when the pendulum is furthest to the right, move the hand very slightly to the left ; and when the pendulum is furthest to the left, move the hand to the right. The pendulum may be also set in motion from a state of rest by giving the hand similar very slight motions having the same periodic time as the pendulum's own swings. The displacements of the hand may be so small under these circumstances, that they can scarcely be perceived with the closest attention, a circumstance to which is due the superstitious application of this little apparatus as a divining rod. If namely the observer, without thinking of his hand, follows the swings of the pendulum with his eye, the hand readily follows the eye, and involuntarily moves a little backwards or forwards, precisely in the same time as the pendulum, after this has accidentally begun to move. These involuntary motions of the hand are usually overlooked, at least when the observer is not accustomed to exact observations on such unobtrusive influences. By this ¶ means any existing vibration of the pendulum is increased and kept up, and any accidental motion of the ring is readily converted into pendular vibrations, which seem to arise spontaneously without any co-operation of the observer, and are hence attributed to the influence of hidden metals, running streams, and so on.

If on the other hand the motion of the hand is intentionally made in the contrary direction, the pendulum soon comes to rest.

The explanation of the process is very simple. When the upper end of the thread is fastened to an immovable support, the pendulum, once struck, continues to swing for a long time, and the extent of its swings diminishes very slowly. We can suppose the extent of the swings to be measured by the angle which the thread makes with the vertical on its greatest deflection from it. If the attached body at the point of greatest deflection lies to the right, and we move the hand to the left, we manifestly increase the angle between the string and the vertical, and con- ¶ sequently also augment the extent of the swing. By moving the upper end of the string in the opposite direction we should decrease the extent of the swing.

In this case there is no necessity for moving the hand in the same periodic time as the pendulum swings. We might move the hand backwards and forwards only at every third or fifth or other swing of the pendulum, and we should still produce large swings. Thus, when the pendulum is to the right, move the hand to the left, and keep it still, till the pendulum has swung to the left, then again to the right, and then once more to the left, and then return the hand to its first position, afterwards wait till the pendulum has swung to the right, then to the left, and again to the right, and then recommence the first motion of the hand. In this way three complete vibrations, or double excursions of the pendulum, will correspond to one left and right motion of the hand. In the same way one left and right motion of the hand may be made to correspond with seven or more swings of the pendulum. The meaning of this process is always that the motion of the

331

hand must in each case be made at such a time and in such a direction as to be opposed to the deflection of the pendulum and consequently to increase it.

By a slight alteration of the process we can easily make two, four, six, &c., swings of the pendulum correspond to one left and right motion of the hand; for a sudden motion of the hand at the instant of the pendulum's passage through the vertical has no influence on the size of the swings. Hence when the pendulum lies to the right move the hand to the left, and so increase its velocity, let it swing to the left, watch for the moment of its passing the vertical line, and at that instant return the hand to its original position, allow it to reach the right, and then again the left and once more the right extremity of its arc, and then recommence the first motion of the hand.

We are able then to communicate violent motion to the pendulum by very small periodical vibrations of the hand, having their periodic time exactly as great, ¶ or else two, three, four, &c., times as great as that of the pendular oscillation. We have here considered that the motion of the hand is backwards. This is not necessary. It may take place continuously in any other way we please. When it moves continuously there will be generally portions of time during which it will increase the pendulum's motion, and others perhaps in which it will diminish the same. In order to create strong vibrations in the pendulum, then, it will be necessary that the increments of motion should be permanently predominant, and should not be neutralised by the sum of the decrements.

Now if a determinate periodic motion were assigned to the hand, and we wished to discover whether it would produce considerable vibrations in the pendulum, we could not always predict the result without calculation. Theoretical mechanics would, however, prescribe the following process to be pursued : *Analyse the periodic motion of the hand into a sum of simple pendular vibrations of the hand*—exactly in the same way as was laid down in the last chapter for the periodic motions of ¶ the particles of air,—*then, if the periodic time of one of these vibrations is equal to the periodic time of the pendulum's own oscillations, the pendulum will be set into violent motion, but not otherwise.* We might compound small pendular motions of the hand out of vibrations of other periodic times, as much as we liked, but we should fail to produce any lasting strong swings of the pendulum. Hence the analysis of the motion of the hand into pendular swings has a real meaning in nature, producing determinate mechanical effects, and for the present purpose no other analysis of the motion of the hand into any other partial motions can be substituted for it.

In the above examples the pendulum could be set into sympathetic vibration, when the hand moved periodically at the same rate as the pendulum; in this case the longest partial vibration of the hand, corresponding to the prime tone of a resonant vibration, was, so to speak, in unison with the pendulum. When three swings of the pendulum went to one backwards and forwards motion of the hand, ¶ it was the third partial swing of the hand, answering as it were to the Twelfth of its prime tone, which set the pendulum in motion. And so on.

The same process that we have thus become acquainted with for swings of long periodic time, holds precisely for swings of so short a period as sonorous vibrations. Any elastic body which is so fastened as to admit of continuing its vibrations for some length of time when once set in motion, can also be made to vibrate sympathetically, when it receives periodic agitations of comparatively small amounts, having a periodic time corresponding to that of its own tone.

Gently touch one of the keys of a pianoforte without striking the string, so as to raise the damper only, and then sing a note of the corresponding pitch forcibly directing the voice against the strings of the instrument. On ceasing to sing, the note will be echoed back from the piano. It is easy to discover that this echo is caused by the string which is in unison with the note, for directly the hand is removed from the key, and the damper is allowed to fall, the echo ceases. The sympathetic vibration of the string is still better shown by putting little paper riders upon it, which are jerked off as soon as the string vibrates. The more exactly the singer hits the pitch of the string, the more strongly it vibrates. A very little deviation from the exact pitch fails in exciting sympathetic vibration.

In this experiment the sounding board of the instrument is first struck by the vibrations of the air excited by the human voice. The sounding board is well known to consist of a broad flexible wooden plate, which, owing to its extensive surface, is better adapted to convey the agitation of the strings to the air, and of the air to the strings, than the small surface over which string and air are themselves directly in contact. The sounding board first communicates the agitations which it receives from the air excited by the singer, to the points where the string is fastened. The magnitude of any single such agitation is of course infinitesimally small. A very large number of such effects must necessarily be aggregated, before any sensible motion of the string can be caused. And such a continuous addition of effects really takes place, if, as in the preceding experiments with ¶ the bell and the pendulum, the periodic time of the small agitations which are communicated to the extremities of the string by the air, through the intervention of the sounding board, exactly corresponds to the periodic time of the string's own vibrations. When this is the case, a long series of such vibrations will really set the string into motion which is very violent in comparison with the exciting cause.

In place of the human voice we might of course use any other musical instrument. Provided only that it can produce the tone of the pianoforte string accurately and sustain it powerfully, it will bring the latter into sympathetic vibration. In place of a pianoforte, again, we can employ any other stringed instrument having a sounding board, as a violin, guitar, harp, &c., and also stretched membranes, bells, elastic tongues or plates, &c., provided only that the latter are so fastened as to admit of their giving a tone of sensible duration when once made to sound.

When the pitch of the original sounding body is not exactly that of the sym- ¶ pathising body, or that which is meant to vibrate in sympathy with it, the latter will nevertheless often make sensible sympathetic vibrations, which will diminish in amplitude as the difference of pitch increases. But in this respect different sounding bodies shew great differences, according to the length of time for which they continue to sound after having been set in action before communicating their whole motion to the air.

Bodies of small mass, which readily communicate their motion to the air, and quickly cease to sound, as, for example, stretched membranes, or violin strings, are readily set in sympathetic vibration, because the motion of the air is conversely readily transferred to them, and they are also sensibly moved by sufficiently strong agitations of the air, even when the latter have not precisely the same periodic time as the natural tone of the sympathising bodies. The limits of pitch capable of exciting sympathetic vibration are consequently a little wider in this case. By the comparatively greater influence of the motion of the air upon light elastic ¶ bodies of this kind which offer but little resistance, their natural periodic time can be slightly altered, and adapted to that of the exciting tone. Massive elastic bodies, on the other hand, which are not readily movable, and are slow in communicating their sonorous vibrations to the air, such as bells and plates, and continue to sound for a long time, are also more difficult to move by the air. A much longer addition of effects is required for this purpose, and consequently it is also necessary to hit the pitch of their own tone with much greater nicety, in order to make them vibrate sympathetically. Still it is well known that bell-shaped glasses can be put into violent motion by singing their proper tone into them ; indeed it is related that singers with very powerful and pure voices, have sometimes been able to crack them by the agitation thus caused. The principal difficulty in this experiment is in hitting the pitch with sufficient precision, and retaining the tone at that exact pitch for a sufficient length of time.

Tuning-forks are the most difficult bodies to set in sympathetic vibration. To

333

effect this they may be fastened on sounding boxes which have been exactly tuned to their tone, as shewn in fig. 13. If we have two such forks of exactly the same pitch, and excite one by a violin bow, the other will begin to vibrate in sympathy, even if placed at the further end of the same room, and it will continue to sound, after the first has been damped. The astonishing nature of such a case of sympathetic vibration will appear, if we merely compare the heavy and powerful mass of steel set in motion, with the light yielding mass of air which produces the effect by such ¶ small motive powers that they could not stir the lightest spring which was not in tune with the fork. With such forks the time required to set them in full swing by sympathetic action, is also of sensible duration, and the

FIG. 13.

slightest disagreement in pitch is sufficient to produce a sensible diminution in the sympathetic effect. By sticking a piece of wax to one prong of the second fork, sufficient to make it vibrate once in a second less than the first—a difference of pitch scarcely sensible to the finest ear—the sympathetic vibration will be wholly destroyed.

After having thus described the phenomenon of sympathetic vibration in general, we proceed to investigate the influence exerted in sympathetic resonance by the different forms of wave of a musical tone.

¶ First, it must be observed that most elastic bodies which have been set into sustained vibration by a gentle force acting periodically, are (with a few exceptions

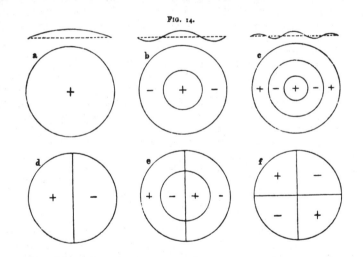

FIG. 14.

to be considered hereafter) always made to swing in pendular vibrations. But they are in general capable of executing several kinds of such vibration, with different periodic times and with a different distribution over the various parts of the vibrating body. Hence to the different lengths of the periodic times correspond different simple tones producible on such an elastic body. These are its so-called *proper tones*. It is, however, only exceptionally, as in strings and the narrower kinds of organ pipes, that these proper tones correspond in pitch with the harmonic upper partial tones of a musical tone already mentioned. They are for the most part inharmonic in relation to the prime tone.

In many cases the vibrations and their mode of distribution over the vibrating bodies can be rendered visible by strewing a little fine sand over the latter. Take, for example, a membrane (as a bladder or piece of thin india-rubber) stretched over a circular ring. In fig. 14 are shewn the various forms which a membrane can assume when it vibrates. The diameters and circles on the surface of the membrane, mark those points which remain at rest during the vibration, and are known as *nodal lines*. By these the surface is divided into a number of compartments which bend alternately up and down, in such a way that while those marked (+) rise, those marked (−) fall. Over the figures a, b, c, are shewn the forms of a section of the membrane during vibration. Only those forms of motion are drawn which correspond with the deepest and most easily producible tones of the membrane. The number of circles and diameters can be increased at pleasure by ¶ taking a sufficiently thin membrane, and stretching it with sufficient regularity, and in this case the tones would continually sharpen in pitch. By strewing sand on the membrane the figures are easily rendered visible, for as soon as it begins to vibrate the particles of sand collect on the nodal lines.

In the same way it is possible to render visible the nodal lines and forms of vibration of oval and square membranes, and of differently-shaped plane elastic plates, bars, and so on. These form a series of very interesting phenomena discovered by Chladni, but to pursue them would lead us too far from our proper subject. It will suffice to give a few details respecting the simplest case, that of a circular membrane.

In the time required by the membrane to execute 100 vibrations of the form a, fig. 14 (p. 40c), the number of vibrations executed by the other forms is as follows :—

Form of Vibration	Pitch Number	Cents *	Notes nearly
a without nodal lines	100	0	c
b with one circle	229·6	1439	d′ +
c with two circles	359·9	2217	b′♭ +
d with one diameter	159	805	a♭
e with one diameter and one circle . . .	292	1858	g′ −
f with two diameters	214	1317	c′♯ +

The prime tone has been here arbitrarily assumed as c, in order to note the intervals of the higher tones. Those simple tones produced by the membrane which are slightly higher than those of the note written, are marked (+); those lower, by (−). In this case there is no commensurable ratio between the prime tone and the other tones, that is, none expressible in whole numbers.

Strew a very thin membrane of this kind with sand, and sound its prime tone strongly in its neighbourhood; the sand will be driven by the vibrations towards ¶ the edge, where it collects. On producing another of the tones of the membrane, the sand collects in the corresponding nodal lines, and we are thus easily able to determine to which of its tones the membrane has responded. A singer who knows how to hit the tones of the membrane correctly, can thus easily make the

* [*Cents* are hundredths of an equal Semitone, and are exceedingly valuable as measures of any, especially unusual, musical intervals. They are fully explained, and the method of calculating them from the Interval Ratios is given in App. XX. sect. C. Here it need only be said that the number of hundreds of cents is the number of *equal*, that is, pianoforte Semitones in the interval, and these may be counted on the keys of any piano, while the units and tens shew the number of hundredths of a Semitone in excess. Wherever *cents* are spoken of in the text, (as in this table), they must be considered as additions by the translator. In the present case, they give the intervals exactly, and not roughly as in the column of notes. Thus, 1439 cents is sharper than 14 Semitones above c, that is, sharper than d′ by 39 hundredths of a Semitone, or about ⅖ of a Semitone, and 1858 is flatter than 19 Semitones above c, that is, flatter than g′ by 42 hundredths of a Semitone, or nearly ½ a Semitone. —*Translator.*]

send arrange itself at pleasure in one order or the other, by singing the corresponding tones powerfully at a distance. But in general the simpler figures of the deeper tones are more easily generated than the complicated figures of the upper tones. It is easiest of all to set the membrane in general motion by sounding its prime tone, and hence such membranes have been much used in acoustics to prove the existence of some determinate tone in some determinate spot of the surrounding air. It is most suitable for this purpose to connect the membrane with an inclosed mass of air. A, fig. 15, is a glass bottle, having an open mouth a, and in place of its bottom b, a stretched membrane, consisting of wet pig's bladder, allowed to dry after it has been stretched and fastened. At c is attached a ¶ single fibre of a silk cocoon, bearing a drop of sealing-wax, and hanging down like a pendulum against the membrane.

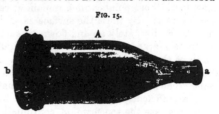

FIG. 15.

As soon as the membrane vibrates, the little pendulum is violently agitated. Such a pendulum is very convenient as long as we have no reason to apprehend any confusion of the prime tone of the membrane with any other of its proper tones. There is no scattering of sand, and the apparatus is therefore always in order. But to decide with certainty what tones are really agitating the membrane, we must after all place the bottle with its mouth downwards and strew sand on the membrane. However, when the bottle is of the right size, and the membrane uniformly stretched and fastened, it is only the prime tone of the membrane (slightly altered by that of the sympathetically vibrating mass of air in the bottle) which is easily excited. This prime tone can be made deeper by increasing the size of the membrane, or the volume of the bottle, or by diminishing the tension of the membrane ¶ or size of the orifice of the bottle.

A stretched membrane of this kind, whether it is or is not attached to the bottom of a bottle, will not only be set in vibration by musical tones of the same pitch as its own proper tone, but also by such musical tones as contain the proper tone of the membrane among its upper partial tones. Generally, given a number of interlacing waves, to discover whether the membrane will vibrate sympathetically, we must suppose the motion of the air at the given place to be mathematically analysed into a sum of pendular vibrations. If there is one such vibration among them, of which the periodic time is the same as that of any one of the proper tones of the membrane, the corresponding vibrational form of the membrane will be superinduced. But if there are none such, or none sufficiently powerful, the membrane will remain at rest.

In this case, then, we also find that the analysis of the motion of the air into pendular vibrations, and the existence of certain vibrations of this kind, are decisive for the sympathetic vibration of the membrane, and for this purpose no other ¶ similar analysis of the motion of the air can be substituted for its analysis into pendular vibrations. The pendular vibrations into which the composite motion of the air can be analysed, here shew themselves capable of producing mechanical effects in external nature, independently of the ear, and independently of mathematical theory. Hence the statement is confirmed, that the theoretical view which first led mathematicians to this method of analysing compound vibrations, is founded in the nature of the thing itself.

As an example take the following description of a single experiment :—

A bottle of the shape shewn in fig. 15 above was covered with a thin vulcanised india-rubber membrane, of which the vibrating surface was 49 millimetres (1·93 inches)* in diameter, the bottle being 140 millimetres (5·51 inches) high, and

---

* [As 10 inches are exactly 254 millimetres and 100 metres, that is, 100,000 millimetres are 3937 inches, it is easy to form little tables for the calculation of one set of measures from the other. Roughly we may assume 25 mm. to be 1 inch. But whenever dimensions are

having an opening at the brass mouth of 13 millimetres ('51 inches) in diameter. When blown it gave $f'\sharp$, and the sand heaped itself in a circle near the edge of the membrane. The same circle resulted from my giving the same tone $f'\sharp$ on an harmonium, or its deeper Octave $f\sharp$, or the deeper Twelfth B. Both $F\sharp$ and D gave the same circle, but more weakly. Now the $f'\sharp$ of the membrane is the prime tone of the harmonium tone $f'\sharp$, the second partial tone of $f\sharp$, the third of B, the fourth of $F\sharp$ and fifth of D.* All these notes on being sounded set the membrane in the motion due to its deepest tone. A second smaller circle, 19 millimetres ('75 inches) in diameter was produced on the membrane by b' and the same more faintly by b, and there was a trace of it for the deeper Twelfth e, that is, for simple tones of which vibrational numbers were $\frac{1}{2}$ and $\frac{1}{3}$ that of b'.†

Stretched membranes of this kind are very convenient for these and similar experiments on the partials of compound tones. They have the great advantage of being independent of the ear, but they are not very sensitive for the fainter simple tones. Their sensitiveness is far inferior to that of the *res'onätors* which I have introduced. These are hollow spheres of glass or metal, or tubes, with two openings as shewn in figs. 16 a and 16 b. One opening (a) has sharp edges, the other (b) is funnel-shaped, and adapted for insertion into the ear. This smaller end I usually coat with melted sealing wax, and when the wax has cooled down enough not to hurt the finger on being touched, but is still soft, I press the opening into the entrance of my ear. The sealing wax thus moulds itself to the shape of the inner surface of this opening, and when I subsequently use the resonator, it fits easily and is air-tight. ¶ Such an instrument is very like the resonance bottle already described, fig. 15 (p. 42a), for which the observer's own tympanic membrane has been made to replace the former artificial membrane.

Fig. 16 a.

b

Fig. 16 b.

B

a

b

The mass of air in a resonator, together with that in the aural passage, and with the tympanic membrane or drumskin itself, forms an elastic system which is capable of vibrating in a peculiar manner, and, in especial, the prime tone of the sphere, which is much deeper than any other of its proper tones, can be set into very powerful sympathetic vibration, and then the ear, which is in immediate connection with the air inside the sphere, perceives this augmented tone by direct action. If we stop one ear (which is best done by a plug of sealing wax moulded into the ¶ form of the entrance of the ear),‡ and apply a resonator to the other, most of the tones produced in the surrounding air will be considerably damped; but if the proper tone of the resonator is sounded, it brays into the ear most powerfully.

given in the text in mm. (that is, millimetres), they will be reduced to inches and decimals of an inch.—*Translator.*]

* [As the instrument was tempered, we should have, approximately, for $f\sharp$ the partials $f\sharp$, $f'\sharp$, &c.; for B the partials B, b, $f'\sharp$, &c.; for $F\sharp$ the partials $F\sharp$, $f\sharp$, $c\sharp$, $f'\sharp$, &c.; and for D the partials D, d, a, d', $f'\sharp$, &c. To prevent confusion I have reduced the *upper* partials of the text to ordinary partials, as suggested in p. 23b', note.—*Translator.*]

† [Here the partials of b are b, b', &c., and of e are e, e', b', &c., so that both b and e contain b'.—*Translator.*]

‡ [For ordinary purposes this is quite enough, indeed it is generally unnecessary to stop the other ear at all. But for such experiments as Mr. Bosanquet had to make on beats (see App. XX. section L. art. 4, b) he was obliged to use a jar as the resonator, conduct the sound from it through first a glass and then an elastic tube to a semicircular metal tube which reached from ear to ear, to each end of which a tube coated with india-rubber, could be screwed into the ear. By this means, when proper care was taken, all sound but that coming from the resonance jar was perfectly excluded.—*Translator.*]

Hence any one, even if he has no ear for music or is quite unpractised in detecting musical sounds, is put in a condition to pick the required simple tone, even if comparatively faint, from out of a great number of others. The proper tone of the resonator may even be sometimes heard cropping up in the whistling of the wind, the rattling of carriage wheels, the splashing of water. For these purposes such resonators are incomparably more sensitive than tuned membranes. When the simple tone to be observed is faint in comparison with those which accompany it, it is of advantage to alternately apply and withdraw the resonator. We thus easily feel whether the proper tone of the resonator begins to sound when the instrument is applied, whereas a uniform continuous tone is not so readily perceived.

A properly tuned series of such resonators is therefore an important instrument for experiments in which individual faint tones have to be distinctly heard, although accompanied by others which are strong, as in observations on the combinational ¶ and upper partial tones, and a series of other phenomena to be hereafter described relating to chords. By their means such researches can be carried out even by ears quite untrained in musical observation, whereas it had been previously impossible to conduct them except by trained musical ears, and much strained attention properly assisted. These tones were consequently accessible to the observation of only a very few individuals; and indeed a large number of physicists and even musicians had never succeeded in distinguishing them. And again even the trained ear is now able, with the assistance of resonators, to carry the analysis of a mass of musical tones much further than before. Without their help, indeed, I should scarcely have succeeded in making the observations hereafter described, with so much precision and certainty, as I have been enabled to attain at present.[*]

It must be carefully noted that the ear does not hear the required tone with augmented force, unless that tone attains a considerable intensity within the mass ¶ of air inclosed in the resonator. Now the mathematical theory of the motion of the air shews that, so long as the amplitude of the vibrations is sufficiently small, the inclosed air will execute pendular oscillations of the same periodic time as those in the external air, and none other, and that only those pendular oscillations whose periodic time corresponds with that of the proper tone of the resonator, have any considerable strength; the intensity of the rest diminishing as the difference of their pitch from that of the proper tone increases. All this is independent of the connection of the ear and resonator, except in so far as its tympanic membrane forms one of the inclosing walls of the mass of air. Theoretically this apparatus does not differ from the bottle with an elastic membrane, in fig. 15 (p. 42a), but its sensitiveness is amazingly increased by using the drumskin of the ear for the closing membrane of the bottle, and thus bringing it in direct connection with the auditory nerves themselves. Hence we cannot obtain a powerful tone in the resonator except when an analysis of the motion of the external air into ¶ pendular vibrations, would shew that one of them has the same periodic time as the proper tone of the resonator. Here again no other analysis but that into pendular vibrations would give a correct result.

It is easy for an observer to convince himself of the above-named properties of resonators. Apply one to the ear, and let a piece of harmonised music, in which the proper tone of the resonator frequently occurs, be executed by any instruments. As often as this tone is struck, the ear to which the instrument is held, will hear it violently contrast with all the other tones of the chord.

This proper tone will also often be heard, but more weakly, when deeper musical tones occur, and on investigation we find that in such cases tones have been struck which include the proper tone of the resonator among their upper partial tones. Such deeper musical tones are called the *harmonic under tones* of the resonator. They are musical tones whose periodic time is exactly 2, 3, 4, 5, and so on, times as great as that of the resonator. Thus if the proper tone of

---

[*] See Appendix II. for the measures and different forms of these Resonators.

the resonator is c'', it will be heard when a musical instrument sounds c', f, c, $A\flat$, F, D, C, and so on.* In this case the resonator is made to sound in sympathy with one of the harmonic upper partial tones of the compound musical tone which is vibrating in the external air. It must, however, be noted that by no means all the harmonic upper partial tones occur in the compound tones of every instrument, and that they have very different degrees of intensity in different instruments. In the musical tones of violins, pianofortes, and harmoniums, the first five or six are generally very distinctly present. A more detailed account of the upper partial tones of strings will be given in the next chapter. On the harmonium the unevenly numbered partial tones (1, 3, 5, &c.) are generally stronger than the evenly numbered ones (2, 4, 6, &c.). In the same way, the upper partial tones are clearly heard by means of the resonators in the singing tones of the human voice, but differ in strength for the different vowels, as will be shewn hereafter.   ¶

Among the bodies capable of strong sympathetic vibration must be reckoned stretched strings which are connected with a sounding board, as on the pianoforte.

The principal mark of distinction between strings and the other bodies which vibrate sympathetically, is that different vibrating forms of strings give simple tones corresponding to the *harmonic* upper partial tones of the prime tone, whereas the secondary simple tones of membranes, bells, rods, &c., are *in*harmonic with the prime tone, and the masses of air in resonators have generally only very high upper partial tones, also chiefly *in*harmonic with the prime tone, and not capable of being much reinforced by the resonator.

The vibrations of strings may be studied either on elastic chords loosely stretched, and not sonorous, but swinging so slowly that their motion may be followed with the hand and eye, or else on sonorous strings, as those of the pianoforte, guitar, monochord, or violin. Strings of the first kind are best made of thin ¶ spirals of brass wire, six to ten feet in length. They should be gently stretched, and both ends should be fastened. A string of this construction is capable of making very large excursions with great regularity, which are easily seen by a large audience. The swings are excited by moving the string regularly backwards and forwards by the finger near to one of its extremities.

A string may be first made to vibrate as in fig. 17, a (p. 46b), so that its appearance when displaced from its position of rest is always that of a simple half wave. The string in this case gives a single simple tone, the deepest it can produce, and no other harmonic secondary tones are audible.

But the string may also during its motion assume the forms fig. 17, b, c, d. In this case the form of the string is that of two, three, or four half waves of a simple wave-curve. In the vibrational form b the string produces only the upper Octave of its prime tone, in the form c the Twelfth, and in the form d the second Octave. The dotted lines shew the position of the string at the end of half its ¶ periodic time. In b the point β remains at rest, in c two points γ_1 and γ_2 remain at rest, in d three points δ_1, δ_2, δ_3. These points are called *nodes*. In a swinging spiral wire the nodes are readily seen, and for a resonant string they are shewn by little paper riders, which are jerked off from the vibrating parts and remain sitting on the nodes. When, then, the string is divided by a node into two swinging sections, it produces a simple tone having a pitch number double that of the prime

* [The c'' occurs as the 2nd, 3rd, 4th, 5th, 6th, 7th, 8th partials of these notes, the 7th being rather flat. The partials are in fact :—

c'	c''						
f	f'	c''					
c	c'	f'	c''				
$A\flat$	$a\flat$	$e\flat$	$a'\flat$	c''			
F	f	c'	f'	a'	c''		
D	d	a	d'	$f'\sharp$	a	c''	
C	c	f	c'	e'	f'	$b'\flat$	c''. —*Translator.*]

tone. For three sections the pitch number is tripled, for four sections quadrupled, and so on.

To bring a spiral wire into these different forms of vibration, we move it periodically with the finger near one extremity, adopting the period of its slowest swings for a, twice that rate for b, three times for c, and four times for d. Or else we just gently touch one of the nodes nearest the extremity with the finger, and pluck the string half-way between this node and the nearest end. Hence when γ_1 in c, or δ_1 in d, is kept at rest by the finger, we pluck the string at ϵ. The other nodes then appear when the vibration commences.

FIG. 17.

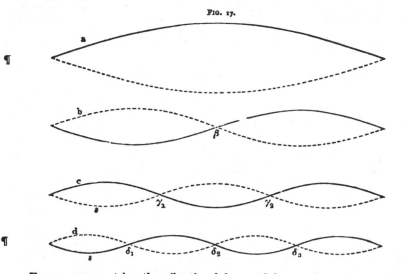

For a sonorous string the vibrational forms of fig. 17 above are most purely produced by applying to its sounding board the handle of a tuning-fork which has been struck and gives the simple tone corresponding to the form required. If only a determinate number of nodes are desired, and it is indifferent whether the individual points of the string do or do not execute simple vibrations, it is sufficient to touch the string very gently at one of the nodes and either pluck the string or rub it with a violin bow. By touching the string with the finger all those simple vibrations are damped which have no node at that point, and only those remain which allow the string to be at rest in that place.

The number of nodes in long thin strings may be considerable. They cease to be formed when the sections which lie between the nodes are too short and stiff to ¶ be capable of sonorous vibration. Very fine strings consequently give a greater number of higher tones than thicker ones. On the violin and the lower pianoforte strings it is not very difficult to produce tones with 10 sections; but with extremely fine wires tones with 16 or 20 sections can be made to sound. [Also compare p. 78d.]

The forms of vibration here spoken of are those in which each point of the string performs pendular oscillations. Hence these motions excite in the ear the sensation of only a single simple tone. In all other vibrational forms of the strings, the oscillations are not simply pendular, but take place according to a different and more complicated law. This is always the case when the string is plucked in the usual way with the finger (as for guitar, harp, zither) or is struck with a hammer (as on the pianoforte), or is rubbed with a violin bow. The resulting motions may then be regarded as compounded of many simple vibrations, which, when taken separately, correspond to those in fig. 17. The multiplicity of such composite forms of motion is infinitely great, the string may indeed be considered as capable of assuming any given form (provided we confine ourselves in all cases to very small deviations from the position of rest), because, according to what was said in Chapter II., any given form of wave can be compounded out of a number of simple waves such as those indicated in fig. 17, a, b, c, d. A plucked, struck, or bowed string therefore allows a great number of harmonic upper partial tones to be heard at the same time as the prime tone, and generally the number increases with the thinness of the string. The peculiar tinkling sound of very fine metallic strings, is clearly due to these very high secondary tones. It is easy to distinguish the upper simple tones up to the sixteenth by means of resonators. Beyond the sixteenth they are too close to each other to be distinctly separable by this means.

Hence when a string is sympathetically excited by a musical tone in its neighbourhood, answering to the pitch of the prime tone of the string, a whole series of different simple vibrational forms will generally be at the same time generated in the string. For when the prime of the musical tone corresponds to the prime of the string all the harmonic upper partials of the first correspond to those of the ¶ second, and are hence capable of exciting the corresponding vibrational forms in the string. Generally the string will be brought into as many forms of sympathetic vibration by the motion of the air, as the analysis of that motion shews that it possesses simple vibrational forms, having a periodic time equal to that of some vibrational form, that the string is capable of assuming. But as a general rule when there is one such simple vibrational form in the air, there are several such, and it will often be difficult to determine by which one, out of the many possible simple tones which would produce the effect, the string has been excited. Consequently the usual unweighted strings are not so convenient for the determination of the pitch of any simple tones which exist in a composite mass of air, as the membranes or the inclosed air of resonators.

To make experiments with the pianoforte on the sympathetic vibrations of strings, select a flat instrument, raise its lid so as to expose the strings, then press down the key of the string (for c' suppose) which you wish to put into sympathetic ¶ vibration, but so slowly that the hammer does not strike, and place a little chip of wood across this c' string. You will find the chip put in motion, or even thrown off, when certain other strings are struck. The motion of the chip is greatest when one of the *under tones* of c' (p. 44d) is struck, as c, F, C, $A\flat$, $F_{,}$, $D_{,,}$, or $C_{,}$. Some, but much less, motion also occurs when one of the upper partial tones of c' is struck, as c'', g'', or c''', but in this last case the chip will not move if it has been placed over one of the corresponding nodes of the string. Thus if it is laid across the middle of the string it will be still for c'' and c''', but will move for g''. Placed at one third the length of the string from its extremity, it will not stir for g'', but will move for c'' or c'''. Finally the string c' will also be put in motion when an under tone of one of its upper partial tones is struck; for example, the note f, of which the third partial tone c'' is identical with the second partial tone of c'. In this case also the chip remains at rest when put on to the middle of the string c', which is its node for c''. In the same way the string c' will move, with the formation of ¶ two nodes, for g', g, or $e\flat$, all which notes have g'' as an upper partial tone, which is also the third partial of c'.*

Observe that on the pianoforte, where one end of the strings is commonly concealed, the position of the nodes is easily found by pressing the string gently on both sides and striking the key. If the finger is at a node the corresponding upper partial tone will be heard purely and distinctly, otherwise the tone of the string is dull and bad.

As long as only one upper partial tone of the string c' is excited, the corresponding nodes can be discovered, and hence the particular form of its vibration determined. But this is no longer possible by the above mechanical method when

---

* [These experiments can of course not be conducted on the usual upright cottage piano. But the experimenter can at least *hear* the tone of c', if c, F, C, &c., are struck and immediately damped, or if c'', g'', c''' are struck and damped. And this sounding of c', although unstruck, is itself a very interesting phenomenon. But of course, as it depends on the ear, it does not establish the results of the text.—*Translator*.]

two upper partial tones are excited, such as c'' and g'', as would be the case if both these notes were struck at once on the pianoforte, because the whole string of c' would then be in motion.

Although the relations for strings appear more complicated to the eye, their sympathetic vibration is subject to the same law as that which holds for resonators, membranes, and other elastic bodies. The sympathetic vibration is always determined by the analysis of whatever sonorous motions exist, into simple pendular vibrations. If the periodic time of one of these simple vibrations corresponds to the periodic time of one of the proper tones of the elastic body, that body, whether it be a string, a membrane, or a mass of air, will be put into strong sympathetic vibration.

These facts give a real objective value to the analysis of sonorous motion into simple pendular vibration, and no such value would attach to any other analysis. ¶ Every individual single system of waves formed by pendular vibrations exists as an independent mechanical unit, expands, and sets in motion other elastic bodies having the corresponding proper tone, perfectly undisturbed by any other simple tones of other pitches which may be expanding at the same time, and which may proceed either from the same or any other source of sound. Each single simple tone, then, can, as we have seen, be separated from the composite mass of tones, by mechanical means, namely by bodies which will vibrate sympathetically with it. Hence every individual partial tone exists in the compound musical tone produced by a single musical instrument, just as truly, and in the same sense, as the different colours of the rainbow exist in the white light proceeding from the sun or any other luminous body. Light is also only a vibrational motion of a peculiar elastic medium, the luminous ether, just as sound is a vibrational motion of the air. In a beam of white light there is a species of motion which *may* be represented as the sum of many oscillatory motions of various periodic times, each of ¶ which corresponds to one particular colour of the solar spectrum. But of course each particle of ether at any particular moment has only *one* determinate velocity, and only *one* determinate departure from its mean position, just like each particle of air in a space traversed by many systems of sonorous waves. The really existing motion of any particle of ether is of course only one and individual; and our theoretical treatment of it as compound, is in a certain sense arbitrary. But the undulatory motion of light can also be analysed into the waves corresponding to the separate colours, by external mechanical means, such as by refraction in a prism, or by transmission through fine gratings, and each individual simple wave of light corresponding to a simple colour, exists mechanically by itself, independently of any other colour.

We must therefore not hold it to be an illusion of the ear, or to be mere imagination, when in the musical tone of a single note emanating from a musical instrument, we distinguish many partial tones, as I have found musicians inclined ¶ to think, even when they have heard those partial tones quite distinctly with their own ears. If we admitted this, we should have also to look upon the colours of the spectrum which are separated from white light, as a mere illusion of the eye. The real outward existence of partial tones in nature can be established at any moment by a sympathetically vibrating membrane which casts up the sand strewn upon it.

Finally I would observe that, as respects the conditions of sympathetic vibration, I have been obliged to refer frequently to the mechanical theory of the motion of air. Since in the theory of sound we have to deal with well-known mechanical forces, as the pressure of the air, and with motions of material particles, and not with any hypothetical explanation, theoretical mechanics have an unassailable authority in this department of science. Of course those readers who are unacquainted with mathematics, must accept the results on faith. An experimental way of examining the problems in question will be described in the next chapter, in which the laws of the analysis of musical tones by the ear have to be established. The experimental proof there given for the ear, can also be carried out in precisely the same way for membranes and masses of air which vibrate sympathetically, and the identity of the laws in both cases will result from those investigations.*

CHAPTER IV.

ON THE ANALYSIS OF MUSICAL TONES BY THE EAR.

It was frequently mentioned in the preceding chapter that musical tones could be resolved by the ear alone, unassisted by any peculiar apparatus, into a series of partial tones corresponding to the simple pendular vibrations in a mass of air, that ¶ is, into the same constituents as those into which the motion of the air is resolved by the sympathetic vibration of elastic bodies. We proceed to shew the correctness of this assertion.

Any one who endeavours for the first time to distinguish the upper partial tones of a musical tone, generally finds considerable difficulty in merely hearing them.

The analysis of our sensations when it cannot be attached to corresponding differences in external objects, meets with peculiar difficulties, the nature and significance of which will have to be considered hereafter. The attention of the observer has generally to be drawn to the phenomenon he has to observe, by peculiar aids properly selected, until he knows precisely what to look for ; after he has once succeeded, he will be able to throw aside such crutches. Similar difficulties meet us in the observation of the upper partials of a musical tone. I shall first give a description of such processes as will most easily put an untrained ¶ observer into a position to recognise upper partial tones, and I will remark in passing that a musically trained ear will not necessarily hear upper partial tones with greater ease and certainty than an untrained ear. Success depends rather upon a peculiar power of mental abstraction or a peculiar mastery over attention, than upon musical training. But a musically trained observer has an essential advantage over one not so trained in his power of figuring to himself how the simple tones sought for, ought to sound, whereas the untrained observer has continually to hear these tones sounded by other means in order to keep their effect fresh in his mind.

First we must note, that the unevenly numbered partials, as the Fifths, Thirds, Sevenths, &c., of the prime tones, are usually easier to hear than the even ones, which are Octaves either of the prime tone or of some of the upper partials which lie near it, just as in a chord we more readily distinguish whether it contains Fifths and Thirds than whether it has Octaves. The second, fourth, and eighth ¶ partials are higher Octaves of the prime, the sixth partial an Octave above the third partial, that is, the Twelfth of the prime ; and some practice is required for distinguishing these. Among the uneven partials which are more easily distinguished, the first place must be assigned, from its usual loudness, to the third partial, the Twelfth of the prime, or the Fifth of its first higher Octave. Then follows the fifth partial as the major Third of the prime, and, generally very faint, the seventh partial as the minor Seventh† of the second higher Octave of the prime, as will be seen by their following expression in musical notation, for the compound tone *c*.

* Optical means for rendering visible weak sympathetic motions of sonorous masses of air, are described in App. II. These means are valuable for demonstrating the facts to hearers unaccustomed to the observing and distinguishing musical tones.

† [Or more correctly *sub*-minor Seventh ; as the real minor Seventh, formed by taking two Fifths down and then two Octaves up, is sharper by 27 cents, or in the ratio of 63 : 64. —*Translator.*]

	1	2		3	4	5	6	7	8
	c	c'		g'	c''	e''	g''	$b''\flat$	c'''
[Cents.	0	1200		1902	2400	2786	3102	3369	3600] *

In commencing to observe upper partial tones, it is advisable just before producing the musical tone itself which you wish to analyse, to sound the note you wish to distinguish in it, very gently, and if possible in the same quality of tone as the compound itself. The pianoforte and harmonium are well adapted for these experiments, because they both have upper partial tones of considerable power.

¶ First gently strike on a piano the note g', as marked above, and after letting the digital† rise so as to damp the string, strike the note c, of which g' is the third partial, with great force, and keep your attention directed to the pitch of the g' which you had just heard, and you will hear it again in the compound tone of c. Similarly, first stroke the fifth partial e'' gently, and then c strongly. These upper partial tones are often more distinct as the sound dies away, because they appear to lose force more slowly than the prime. The seventh and ninth partials $b''\flat$ and d''' are mostly weak, or quite absent on modern pianos. If the same experiments are tried with an harmonium in one of its louder stops, the seventh partial will generally be well heard, and sometimes even the ninth.

To the objection which is sometimes made that the observer only imagines he hears the partial tone in the compound, because he had just heard it by itself, I need only remark at present that if e'' is first heard as a partial tone of c on a good piano, tuned in equal temperament, and then e'' is struck on the instrument ¶ itself, it is quite easy to perceive that the latter is a little sharper. This follows from the method of tuning. But if there is a difference in pitch between the two tones, one is certainly not a continuation of the mental effect produced by the other. Other facts which completely refute the above conception, will be subsequently adduced.

A still more suitable process than that just described for the piano, can be adopted on any stringed instrument, as the piano, monochord, or violin. It consists in first producing the tone we wish to hear, as an harmonic, [p. 25d, note] by touching the corresponding node of the string when it is struck or rubbed. The resemblance of the tone first heard to the corresponding partial of the compound is then much greater, and the ear discovers it more readily. It is usual to place a divided scale by the string of a monochord, to facilitate the discovery of the nodes. Those for the third partial, as shewn in Chap. III. (p. 45d), divide the string into three equal parts, those for the fifth into five, and so on. On the piano and violin ¶ the position of these points is easily found experimentally, by touching the string gently with the finger in the neighbourhood of the node, which has been approximatively determined by the eye, then striking or bowing the string, and moving the finger about till the required harmonic comes out strongly and purely. By then sounding the string, at one time with the finger on the node, and at another without, we obtain the required upper partial at one time as an harmonic, and at another in the compound tone of the whole string, and thus learn to recognise the existence of the first as part of the second, with comparative ease. Using thin strings which have loud upper partials, I have thus been able to recognise the

---

* [The cents, (see p. 41d, note) reckoned from the lowest note, are assigned on the supposition that the harmonics are perfect, as on the Harmonical, not tempered as on the pianoforte. See also diagram, p. 22c.—*Translator.*]

† [The keys played by the fingers on a piano or organ, are best called *digitals* or finger-keys, on the analogy of *pedals* and foot-keys on the organ. The word *key* having another musical sense, namely, the scale in which a piece of music is written, will without prefix be confined to this meaning.—*Translator.*]

partials separately, up to the sixteenth. Those which lie still higher are too near to each other in pitch for the ear to separate them readily.

In such experiments I recommend the following process. Touch the node of the string on the pianoforte or monochord with a camel's-hair pencil, strike the note, and immediately remove the pencil from the string. If the pencil has been pressed tightly on the string, we either continue to hear the required partial as an harmonic, or else in addition hear the prime tone gently sounding with it. On repeating the excitement of the string, and continuing to press more and more lightly with the camel's-hair pencil, and at last removing the pencil entirely, the prime tone of the string will be heard more and more distinctly with the harmonic till we have finally the full natural musical tone of the string. By this means we obtain a series of gradual transitional stages between the isolated partial and the compound tone, in which the first is readily retained by the ear. By applying this last process I have generally succeeded in making perfectly untrained ears ¶ recognise the existence of upper partial tones.

It is at first more difficult to hear the upper partials on most wind instruments and in the human voice, than on stringed instruments, harmoniums, and the more penetrating stops of an organ, because it is then not so easy first to produce the upper partial softly in the same quality of tone. But still a little practice suffices to lead the ear to the required partial tone, by previously touching it on the piano. The partial tones of the human voice are comparatively most difficult to distinguish for reasons which will be given subsequently. Nevertheless they were distinguished even by Rameau * without the assistance of any apparatus. The process is as follows :—

Get a powerful bass voice to sing $e\flat$ to the vowel O, in *sore* [more like *aw* in *saw* than *o* in *so*], gently touch $b\flat$ on the piano, which is the Twelfth, or third partial tone of the note $e\flat$, and let its sound die away while you are listening to it attentively. The note $b\flat$ on the piano will appear really not to die away, ¶ but to keep on sounding, even when its string is damped by removing the finger from the digital, because the ear unconsciously passes from the tone of the piano to the partial tone of the same pitch produced by the singer, and takes the latter for a continuation of the former. But when the finger is removed from the key, and the damper has fallen, it is of course impossible that the tone of the string should have continued sounding. To make the experiment for g'' the fifth partial, or major Third of the second Octave above $e\flat$, the voice should sing to the vowel A in *father*.

The resonators described in the last chapter furnish an excellent means for this purpose, and can be used for the tones of any musical instrument. On applying to the ear the resonator corresponding to any given upper partial of the compound c, such as g', this g' is rendered much more powerful when c is sounded. Now hearing and distinguishing g' in this case by no means proves that the ear alone and without this apparatus would hear g' as part of the compound c. But ¶ the increase of the loudness of g' caused by the resonator may be used to direct the attention of the ear to the tone it is required to distinguish. On gradually removing the resonator from the ear, the force of g' will decrease. But the attention once directed to it by this means, remains more readily fixed upon it, and the observer continues to hear this tone in the natural and unchanged compound tone of the given note, even with his unassisted ear. The sole office of the resonators in this case is to direct the attention of the ear to the required tone.

By frequently instituting similar experiments for perceiving the upper partial tones, the observer comes to discover them more and more easily, till he is finally able to dispense with any aids. But a certain amount of undisturbed concentration is always necessary for analysing musical tones by the ear alone, and hence the use of resonators is quite indispensable for an accurate comparison of different

* *Nouveau Système de Musique théorique.* Paris : 1726. Préface.

qualities of tones, especially in respect to the weaker upper partials. At least, I must confess, that my own attempts to discover the upper partial tones in the human voice, and to determine their differences for different vowels, were most unsatisfactory until I applied the resonators.

We now proceed to prove that the human ear really does analyse musical tones according to the law of simple vibrations. Since it is not possible to institute an exact comparison of the strength of our sensations for different simple tones, we must confine ourselves to proving that when an analysis of a composite tone into simple vibrations, effected by theoretic calculation or by sympathetic resonance, shews that certain upper partial tones are absent, the ear also does not perceive them.

The tones of strings are again best adapted for conducting this proof, because they admit of many alterations in their quality of tone, according to the manner ¶ and the spot in which they are excited, and also because the theoretic or experimental analysis is most easily and completely performed for this case. Thomas Young * first shewed that when a string is plucked or struck, or, as we may add, bowed at any point in its length which is the node of any of its so-called harmonics, those simple vibrational forms of the string which have a node in that point are not contained in the compound vibrational form. Hence, if we attack the string at its middle point, all the simple vibrations due to the evenly numbered partials, each of which has a note at that point, will be absent. This gives the sound of the string a peculiarly hollow or nasal twang. If we excite the string at $\frac{1}{3}$ of its length, the vibrations corresponding to the third, sixth, and ninth partials will be absent; if at $\frac{1}{4}$, then those corresponding to the fourth, eighth, and twelfth partials will fail; and so on.†

This result of mathematical theory is confirmed, in the first place, by analysing the compound tone of the string by sympathetic resonance, either by the ¶ resonators or by other strings. The experiments may be easily made on the pianoforte. Press down the digitals for the notes c and c', without allowing the hammer to strike, so as merely to free them from their dampers, and then pluck the string c with the nail till it sounds. On damping the c string the higher c' will echo the sound, except in the particular case when the c string has been plucked exactly at its middle point, which is the point where it would have to be touched in order to give its first harmonic when struck by the hammer.

If we touch the c string at $\frac{1}{3}$ or $\frac{2}{3}$ its length, and strike it with the hammer, we obtain the harmonic g'; and if the damper of the g' is raised, this string echoes the sound. But if we pluck the c string with the nail, at either $\frac{1}{3}$ or $\frac{2}{3}$ its length, g' is not echoed, as it will be if the c string is plucked at any other spot.

In the same way observations with the resonators shew that when the c string is plucked at its middle the Octave c' is missing, and when at $\frac{1}{3}$ or $\frac{2}{3}$ its length the Twelfth g' is absent. The analysis of the sound of a string by the sympathetic ¶ resonance of strings or resonators, consequently fully confirms Thomas Young's law.

But for the vibration of strings we have a more direct means of analysis than that furnished by sympathetic resonance. If we, namely, touch a vibrating string gently for a moment with the finger or a camel's-hair pencil, we damp all those simple vibrations which have no node at the point touched. Those vibrations, however, which have a node there are not damped, and hence will continue to sound without the others. Consequently, if a string has been made to speak in any way whatever, and we wish to know whether there exists among its simple vibrations one corresponding to the Twelfth of the prime tone, we need only touch one of the nodes of this vibrational form at $\frac{1}{3}$ or $\frac{2}{3}$ the length of the string, in order to reduce to silence all simple tones which have no such node, and leave the Twelfth sounding, if it were there. If neither it, nor any of the sixth, ninth,

* London. *Philosophical Transactions*, 1800, vol. i. p. 137.
† See Appendix III.

twelfth, &c., of the partial tones were present, giving corresponding harmonics, the string will be reduced to absolute silence by this contact of the finger.

Press down one of the digitals of a piano, in order to free a string from its damper. Pluck the string at its middle point, and immediately touch it there. The string will be completely silenced, shewing that plucking it in its middle excited none of the evenly numbered partials of its compound tone. Pluck it at $\frac{1}{3}$ or $\frac{2}{3}$ its length, and immediately touch it in the same place; the string will be silent, proving the absence of the third partial tone. Pluck the string anywhere else than in the points named, and the second partial will be heard when the middle is touched, the third when the string is touched at $\frac{1}{3}$ or $\frac{2}{3}$ of its length.

The agreement of this kind of proof with the results from sympathetic resonance, is well adapted for the experimental establishment of the proposition based in the last chapter solely upon the results of mathematical theory, namely, that sympathetic vibration occurs or not, according as the corresponding simple ¶ vibrations are or are not contained in the compound motion. In the last described method of analysing the tone of a string, we are quite independent of the theory of sympathetic vibration, and the simple vibrations of strings are exactly characterised and recognisable by their nodes. If the compound tones admitted of being analysed by sympathetic resonance according to any other vibrational forms except those of simple vibration, this agreement could not exist.

If, after having thus experimentally proved the correctness of Thomas Young's law, we try to analyse the tones of strings by the unassisted ear, we shall continue to find complete agreement.* If we pluck or strike a string in one of its nodes, all those upper partial tones of the compound tone of the string to which the node belongs, disappear for the ear also, but they are heard if the string is plucked at any other place. Thus, if the string c be plucked at $\frac{1}{3}$ its length, the partial tone g' cannot be heard, but if the string be plucked at only a little distance from this point the partial tone g' is distinctly audible. Hence the ear analyses the sound ¶ of a string into precisely the same constituents as are found by sympathetic resonance, that is, into simple tones, according to Ohm's definition of this conception. These experiments are also well adapted to shew that it is no mere play of imagination when we hear upper partial tones, as some people believe on hearing them for the first time, for those tones are not heard when they do not exist.

The following modification of this process is also very well adapted to make the upper partial tones of strings audible. First, strike alternately in rhythmical sequence, the third and fourth partial tone of the string alone, by damping it in the corresponding nodes, and request the listener to observe the simple melody thus produced. Then strike the undamped string alternately and in the same rhythmical sequence, in these nodes, and thus reproduce the same melody in the upper partials, which the listener will then easily recognise. Of course, in order to hear the third partial, we must strike the string in the node of the fourth, and conversely.

The compound tone of a plucked string is also a remarkably striking example ¶ of the power of the ear to analyse into a long series of partial tones, a motion which the eye and the imagination are able to conceive in a much simpler manner. A string, which is pulled aside by a sharp point, or the finger nail, assumes the form fig. 18, A (p. 54a), before it is released. It then passes through the series of forms, fig. 18, B, C, D, E, F, till it reaches G, which is the inversion of A, and then returns, through the same, to A again. Hence it alternates between the forms A and G. All these forms, as is clear, are composed of three straight lines, and on expressing the velocity of the individual points of the strings by vibrational curves, these would have the same form. Now the string scarcely imparts any perceptible portion of its own motion directly to the air. Scarcely any audible tone results when both ends of a string are fastened to immovable supports, as metal bridges, which are again fastened to the walls of a room. The sound of

* See Brandt in Poggendorff's *Annalen der Physik*, vol. cxii. p. 324, where this fact is proved.

the string reaches the air through that one of its extremities which rests upon a bridge standing on an elastic sounding board. Hence the sound of the string essentially depends on the motion of this

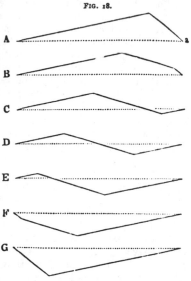

FIG. 18.

extremity, through the pressure which it exerts on the sounding board. The magnitude of this pressure, as it alters periodically with the time, is shewn in fig. 19, where the height of the line h h corresponds to the amount of pressure exerted on the bridge by that extremity of the string when the string is at rest. Along h h suppose lengths to be set off corresponding to consecutive intervals of time, the vertical heights of the broken line above or below h h represent the corresponding augmentations or diminutions of pressure at those times. The pressure of the string on the sounding board consequently alternates, as the figure shews, between a higher and a lower value. For some time the greater pressure remains unaltered; then the lower suddenly ensues, and likewise remains for a time unaltered. The letters a to g in fig. 19 correspond to the times at which the string assumes the forms A to G in fig. 18. It is this alteration between a greater and a smaller pressure which produces the sound in the air. We cannot but feel astonished that a motion produced by means so simple and so easy to comprehend, should be analysed by the ear into such a complicated sum of simple tones. For the eye and the understanding the action of the string on the sounding board can be figured with extreme simplicity. What has the simple broken line of fig. 19 to do with wave-curves, which, in the course of one of their periods, shew

FIG. 19.

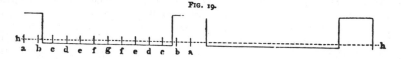

3, 4, 5, up to 16, and more, crests and troughs? This is one of the most striking examples of the different ways in which eye and ear comprehend a periodic motion.

There is no sonorous body whose motions under varied conditions can be so completely calculated theoretically and contrasted with observation as a string. The following are examples in which theory can be compared with analysis by ear :—

I have discovered a means of exciting simple pendular vibrations in the air. A tuning-fork when struck gives no harmonic upper partial tones, or, at most, traces of them when it is brought into such excessively strong vibration that it no longer exactly follows the law of the pendulum.* On the other hand, tuning-forks have some very high inharmonic secondary tones, which produce that peculiar sharp

* [On all ordinary tuning-forks between a and d'' in pitch, I have been able to hear the second partial or Octave of the prime. In some low forks this Octave is so powerful that on pressing the handle of the fork against the table, the prime quite disappears and the Octave only is heard, and this has often proved a source of embarrassment in tuning the forks, or in counting beats to determine pitch numbers. But the prime can always be heard when the fork is held to the ear or over a properly tuned resonance jar, as described in this paragraph. I tune such jars by pouring water in or out until the resonance is strongest, and then I register the height of the water and pitch of the fork for future use on a slip of paper gummed to the side of the jar. I have found that it is not at all necessary to

tinkling of the fork at the moment of being struck, and generally become rapidly inaudible. If the tuning-fork is held in the fingers, it imparts very little of its tone to the air, and cannot be heard unless it is held close to the ear. Instead of holding it in the fingers, we may screw it into a thick board, on the under side of which some pieces of india-rubber tubing have been fastened. When this is laid upon a table, the india-rubber tubes on which it is supported convey no sound to the table, and the tone of the tuning-fork is so weak that it may be considered inaudible. Now if the prongs of the fork be brought near a resonance chamber * of a bottle-form of such a size and shape that, when we blow over its mouth, the air it contains gives a tone of the same pitch as the fork's, the air within this chamber vibrates sympathetically, and the tone of the fork is thus conducted with great strength to the outer air. Now the higher secondary tones of such resonance chambers are also inharmonic to the prime tone, and in general the secondary tones of the chambers correspond neither with the harmonic nor the inharmonic ¶ secondary tones of the forks; this can be determined in each particular case by producing the secondary tones of the bottle by stronger blowing, and discovering those of the forks with the help of strings set into sympathetic vibration, as will be presently described. If, then, only one of the tones of the fork, namely the prime tone, corresponds with one of the tones of the chamber, this alone will be reinforced by sympathetic vibration, and this alone will be communicated to the external air, and thus conducted to the observer's ear. The examination of the motion of the air by resonators shews that in this case, provided the tuning-fork be not set into too violent motion, no tone but the prime is present, and in such case the unassisted ear hears only a single simple tone, namely the common prime of the tuning-fork and of the chamber, without any accompanying upper partial tones.

The tone of a tuning-fork can also be purified from secondary tones by placing its handle upon a string and moving it so near to the bridge that one of the proper tones of the section of string lying between the fork and the bridge is the same as ¶ that of the tuning-fork. The string then begins to vibrate strongly, and conducts the tone of the tuning-fork with great power to the sounding board and surrounding air, whereas the tone is scarcely, if at all, heard as long as the above-named section is not in unison with the tone of the fork. In this way it is easy to find the lengths of string which correspond to the prime and upper partial tones of the fork, and accurately determine the pitch of the latter. If this experiment is conducted with ordinary strings which are uniform throughout their length, we shield the ear from the inharmonic secondary tones of the fork, but not from the harmonic upper partials, which are sometimes faintly present when the fork is made to vibrate strongly. Hence to conduct this experiment in such a way as to create purely pendular vibrations of the air, it is best to weight one point of the string, if only so much as by letting a drop of melting sealing-wax fall upon it. This causes the upper proper tones of the string itself to be inharmonic to the prime tone, and hence there is a distinct interval between the points where the fork must be placed ¶ to bring out the prime tone and its audible Octave, if it exists.

In most other cases the mathematical analysis of the motions of sound is not nearly far enough advanced to determine with certainty what upper partials will be present and what intensity they will possess. In circular plates and stretched membranes which are struck, it is theoretically possible to do so, but their inhar-

---

put the fork into excessively strong vibration in order to make the Octave sensible. Thus, taking a fork of 232 and another of 468 vibrations, after striking them both, and letting the deeper fork spend most of its energy until I could not see the vibrations with the eye at all, the beats were heard distinctly, when I pressed both on to a table, and continued to be heard even after the forks themselves were separately inaudible. See also Prof. Helmholtz's experiments on a fork of 64 vibrations at the close

of Chap. VII., and Prof. Preyer's in App. XX. sect. L. art. 4, c. The conditions according to Koenig that tuning-forks should have no upper partials are given in App. XX. sect. L. art. 2, a.—*Translator*.]

* Either a bottle of a proper size, which can readily be more accurately tuned by pouring oil or water into it, or a tube of pasteboard quite closed at one end, and having a small round opening at the other. See the proper sizes of such resonance chambers in App. IV.

monic secondary tones are so numerous and so nearly of the same pitch that most observers would probably fail to separate them satisfactorily. On elastic rods, however, the secondary tones are very distant from each other, and are inharmonic, so that they can be readily distinguished from each other by the ear. The following are the proper tones of a rod which is free at both ends; the vibrational number of the prime tone taken to be c, is reckoned as 1 :—

	Pitch Number	Cents *	Notation
Prime tone	1·0000	0	c
Second proper tone	2·7576	1200 + 556	f + 0·2
Third proper tone	5·4041	2400 + 521	f' + 0·1
Fourth proper tone	13·3444	3600 + 886	a''' − 0·1

¶ The notation is adapted to the equal temperament, and the appended fractions are parts of the interval of a complete tone.

Where we are unable to execute the theoretical analysis of the motion, we can, at any rate, by means of resonators and other sympathetically vibrating bodies, analyse any individual musical tone that is produced, and then compare this analysis, which is determined by the laws of sympathetic vibration, with that effected by the unassisted ear. The latter is naturally much less sensitive than one armed with a resonator; so that it is frequently impossible for the unarmed ear to recognise amongst a number of other stronger simple tones those which the resonator itself can only faintly indicate. On the other hand, so far as my experience goes, there is complete agreement to this extent: the ear recognises without resonators the simple tones which the resonators greatly reinforce, and perceives no upper partial tone which the resonator does not indicate. To verify this conclusion, I performed numerous experiments, both with the human voice and the harmonium, and they all confirmed it.†

¶ By the above experiments the proposition enunciated and defended by G. S. Ohm must be regarded as proved, viz. that *the human ear perceives pendular vibrations alone as simple tones, and resolves all other periodic motions of the air into a series of pendular vibrations, hearing the series of simple tones which correspond with these simple vibrations.*

¶ * [For cents see note p. 41d. As a Tone is 200 ct., 0·1 Tone = 20 ct., these would give for the Author's notation f + 40 ct., f' + 20 ct., a''' − 10 ct., whereas the column of cents shews that they are more accurately f + 56 ct., f' + 21 ct., a''' − 14 ct. For convenience, the cents for Octaves are separated, thus 1200 + 556 in place of 1756, but this separation is quite unnecessary. The cents again shew the intervals of the inharmonic partial tones without any assumption as to the value of the prime. By a misprint in all the German editions, followed in the first English edition, the second proper tone was made f − 0·2 in place of f + 0·2.—*Translator*.]

† [In my ' Notes of Observations on Musical Beats,' *Proceedings of the Royal Society,* May 1880, vol. xxx. p. 531, largely cited in App. XX. sect. B. No. 7, I showed that I was able to determine the pitch numbers of deep reed tones, by the beats (Chap. VIII.) that their upper partials made with the primes of a set of Scheibler's tuning-forks. The correctness of the process was proved by the fact that the results obtained from different partials of the same reed tone, which were made to beat with different forks, gave the same pitch numbers for the primes, within one or two hundredths of a vibration in a second. I not only employed such low partials as 3, 4, 5 for one tone, and 4, 5, 6 for others, but I determined the pitch number 31·47, by partials 7, 8, 9, 10, 11, 12, 13, and the pitch number 15·94 by partials 25 and 27. The objective reality of these extremely high upper partials, and their independence of resonators or resonance jars, was therefore conclusively shewn. On the Harmonical the beats of the 16th partial of C 66, with c'''', when slightly flattened by pressing the note lightly down, are very clear.—*Translator*.

31

Singing and Sensitive Flames

JOHN TYNDALL

John Tyndall (1820–1893) was not only a distinguished experimental physicist, known for his investigations in radiation, magnetism, acoustics, and geophysics, but also one of the most brilliant lecturers on science the world has ever known. For most of his active career he was professor of natural philosophy at the Royal Institution of Great Britain, where his lectures and lecture demonstrations gained him world wide recognition.

The following extracts are taken from Tyndall's book *Sound*. This book was based on lectures which Tyndall gave at the Royal Institution. The passages reproduced concern the nature of singing flames and the use of the sensitive flame as a detector of sound waves, particularly those inaudible to the human ear (ultrasonics). It must be recalled that Tyndall worked before the days of electroacoustics and the modern microphone.

Reprinted from "Sound," J. Tyndall (Appleton) 217–237 (1867)

31

LECTURE VI.

SOUNDING FLAMES—INFLUENCE OF THE TUBE SURROUNDING THE FLAME—
INFLUENCE OF SIZE OF FLAME—HARMONIC NOTES OF FLAMES—EFFECT OF
UNISONANT NOTES ON SINGING FLAMES—ACTION OF SOUND ON NAKED
FLAMES—EXPERIMENTS WITH FISH-TAIL AND BAT'S-WING BURNERS—EX-
PERIMENTS ON TALL FLAMES—EXTRAORDINARY DELICACY OF FLAMES AS
ACOUSTIC REAGENTS — THE VOWEL FLAME—ACTION OF CONVERSATIONAL
TONES UPON FLAMES—ACTION OF MUSICAL SOUNDS ON UNIGNITED JETS
OF GAS—CONSTITUTION OF WATER JETS—ACTION OF MUSICAL SOUNDS ON
WATER JETS—A LIQUID VEIN MAY COMPETE IN POINT OF DELICACY WITH
THE EAR.

FRICTION is always rhythmic. When we pass a resined bow across a string, the tension of the string secures the perfect rhythm of the friction. When we pass the wetted finger round the edge of a glass, the breaking up of the friction into rhythmic pulses expresses itself in music. Savart's experiments prove the friction of a liquid against the sides of an orifice through which it passes to be competent to produce musical sounds. We have here the means of repeating his experiment. The tube A B, fig. 110, is filled with water, its extremity, B, being closed by a plate of brass, which is pierced by a circular orifice of a diameter equal to the thickness of the plate. Removing a little peg which stops the orifice, the water issues from it, and as it sinks in the tube a musical note of great sweetness issues from the liquid column. This note is due to the intermittent flow of the liquid through the orifice, by which the whole column above it is thrown into vibration. The tendency to this effect shows itself when tea is poured from a teapot, in the circular ripples that cover the falling liquid. The same intermittence is observed in the black dense smoke which rolls in rhythmic rings from the funnel of a steamer. The unpleasant noise of unoiled machinery is also a declaration of the fact that the friction is not uniform, but is due to the alternate 'bite' and release of the rubbing surfaces.

FIG. 110.

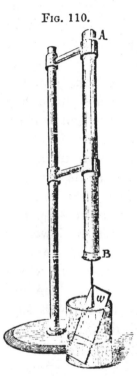

Where gases are concerned friction is of the same intermittent character. A rifle bullet sings in its passage through the air; while to the rubbing of the wind against the boles and branches of the trees are to be ascribed the 'waterfall tones' of an agitated pine-wood. Pass a steadily burning candle rapidly through the air; an indented band of light, declaring intermittence, is the consequence, while the almost musical sound which accompanies the appearance of this band is the audible expression of the rhythm. On the other hand, if you blow gently against a candle flame, the fluttering noise announces a rhythmic action. We have already learned what can be done when a pipe is associated with such a flutter; we have learned that the pipe selects a special pulse from the flutter, and raises it by resonance to a musical sound. In a similar manner the noise of a flame may be turned to account. The blowpipe flame of our laboratory, for example, when enclosed within an appropriate tube, has its flutter raised to a musical roar. The special pulse first selected soon reacts upon the flame so as to abolish in a great degree the other pulses, compelling the flame to vibrate in periods answering to the selected one.   And this reaction can become so powerful—the timed shock of the reflected pulses may accumulate to such an extent—as to beat the flame, even when very large, into extinction.

Nor is it necessary to produce this flutter by any extraneous means.   When a gas flame is simply enclosed within a tube, the passage of the air over it is usually sufficient to produce the necessary rhythmic action, so as to cause the flame to burst spontaneously into song.   Not all, however, are aware of the intensity to which this flame-music may rise.   I have here a ring burner with twenty-eight orifices, from which issues a gas flame.   I place over the flame this tin tube, 5 feet long, and $2\frac{1}{2}$ inches in diameter. The flame flutters at first, but it soon chastens its impulses into perfect periodicity, and a deep and clear musical note is the result.   The quickness of its pulses depends in some measure on the size of the flame, and by lowering the gas I finally stop the note which is now sounded. After a momentary interval of silence, another note, which is the octave of the last, is yielded by the flame.   The first note was the fundamental note of the tube which surrounds the flame: this is the first harmonic.   In fact, here, exactly as in the case of open organ-pipes, we have the aërial column dividing itself into vibrating segments, separated from each other by nodes.

Permit me now to try the effect of this larger tube, *a b*, fig. 111, 15 feet long, and 4 inches wide, which was formed for a totally different use.   It is supported by a steady stand *s s'*, and into it is lifted the tall burner, shown enlarged at B.   You hear the incipient flutter ; you now hear the more powerful sound.   As the flame is lifted higher the action becomes more violent, until finally a storm of music issues from the tube.   And now all has suddenly ceased ;

Fig. 111.

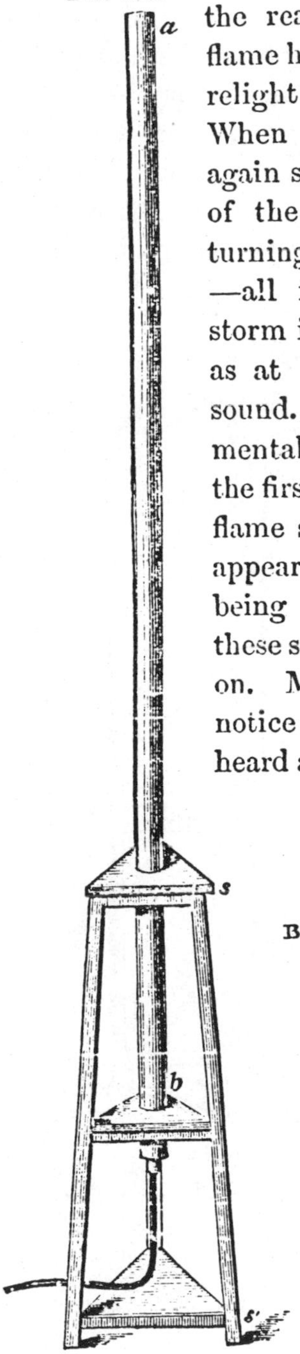

the reaction of its own pulses upon the flame has beaten it into extinction. I now relight the flame and make it very small. When raised within the tube, the flame again sings, but it is one of the harmonics of the tube that you now hear. On turning the gas fully on, the note ceases —all is silent for a moment; but the storm is brewing, and soon it bursts forth, as at first, in a kind of hurricane of sound. By lowering the flame the fundamental note is abolished, and now you hear the first harmonic of the tube. Making the flame still smaller, the first harmonic disappears, and the second is heard. Your ears being disciplined to the apprehension of these sounds, I turn the gas once more fully on. Mingling with the deepest note you notice the harmonics, as if struggling to be heard amid the general uproar of the flame.

With a large Bunsen's rose burner, the sound of this tube becomes powerful enough to shake the floor and seats, and the large audience that occupies the seats of this room, while the extinction of the flame, by the reaction of the sonorous pulses, announces itself by an explosion almost as loud as a pistol shot. It must occur to you that a chimney is a tube of this kind upon a large scale, and that the roar of a flame in a chimney is simply a rough attempt at music.

I now pass on to shorter tubes and smaller flames. Here is a series of eight of them. Placing the tubes over the flames, each of them starts into song, and you notice that as the tubes lengthen the tones deepen. The lengths of these tubes are so chosen that they yield in succession the eight notes of the gamut. Round some of them you observe a paper slider, s, fig. 112, by which the resounding tube can be lengthened or shortened. While the flame is sounding I raise the slider; the pitch instantly falls. I now lower the slider; the pitch instantly rises. These experiments prove the flame to be governed by the tube. By the reaction of the pulses, reflected back upon the flame, its flutter is rendered perfectly periodic, the length of the period being determined, as in the case of organ-pipes, by the length of the tube.

Fig. 112.

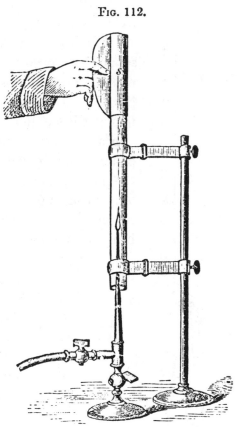

*   *   *   *   *   *   *   *   *   *   *   *   *

We have hitherto dealt with flames surrounded by resonant tubes; and none of these flames, if naked, would respond in any way to such noise or music as could be here applied. Still it is possible to make naked flames thus sympathetic. In a former lecture (p. 101), I referred to the oscillations of water in a bottle, as revealing the existence of vibrations of a definite period in the general jar of a railway train. The fish-tail flames in some of our metropolitan railway carriages are far more sensitive acoustic reagents. If you pay the requisite attention, you will find single flames here and there jumping in synchronism with certain tremors of the train. A flame, for example, having a horizontal edge, when the train is still, will, during the motion, periodically thrust forth a central tongue, and continue to jump as long as a special set of vibrations is present. It will subside when those vibrations disappear, and jump again when they are restored. When the train is at rest, the tapping of the glass shade which surrounds the flame rarely fails, when it is sensitive, to cause it to jump.

This action of sound upon a naked fish-tail flame

was first observed by Professor Leconte at a musical party in the United States. His observation is thus described:—'Soon after the music commenced, I observed that the flame exhibited pulsations which were *exactly synchronous* with the audible beats. This phenomenon was very striking to every one in the room, and especially so when the strong notes of the violoncello came in. It was exceedingly interesting to observe how perfectly even the *trills* of this instrument were reflected on the sheet of flame. *A deaf man might have seen the harmony.* As the evening advanced, and the diminished consumption of gas in the city *increased the pressure*, the phenomenon became more conspicuous. The *jumping* of the flame gradually increased, became somewhat irregular, and, finally, it began to flare continuously, emitting the characteristic sound indicating the escape of a greater amount of gas than could be properly consumed. I then ascertained by experiment, that the phenomenon *did not* take place unless the discharge of gas was so regulated that the flame approximated to the condition of *flaring*. I likewise determined by experiment, that the effects *were not* produced by jarring or shaking the floor and walls of the room by means of repeated concussions. Hence it is obvious that the pulsations of the flame *were not* owing to *indirect* vibrations propagated through the medium of the walls of the room to the burning apparatus, but must have been produced by the *direct* influence of aërial sonorous pulses on the burning jet.'*

The significant remark, that the jumping of the flame was not observed until it was near flaring, suggests the means of repeating the experiments of Dr. Leconte; while a more intimate knowledge of the conditions of success enables us to vary and exalt them in a striking

* Philosophical Magazine, March 1858, p. 235.

degree.   Before you burns a bright candle-flame : I may
shout, clap my hands, sound this whistle, strike this anvil
with a hammer, or explode a mixture of oxygen and
hydrogen.   Though sonorous waves pass in each case
through the air, the candle is absolutely insensible to the
sound ; there is no motion of the flame.

I now urge from this small blow-pipe a narrow stream
of air through the flame of the candle, producing thereby an
incipient flutter, and reducing at the same time the brightness of the flame.   When I now sound a whistle, the flame
jumps visibly.   The experiment may be so arranged that
when the whistle sounds, the flame shall be either restored
almost to its pristine brightness, or that the amount of
light it still possesses shall disappear.

The blow-pipe flame of our laboratory is totally un-

FIG. 115.

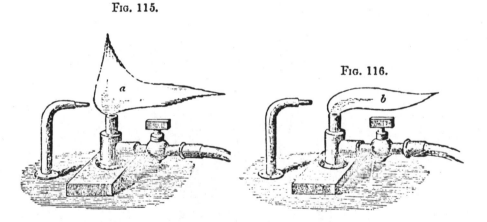

FIG. 116.

affected by the sound of the whistle as long as no air is
urged through it.   By properly tempering the force of
the blast I obtain a flame of the shape shown in fig.
115, the blast not being sufficiently powerful to urge the
whole of the flame forwards.   On sounding the whistle
the erect portion of the flame drops down, and while it
continues to sound we have a flame of the form shown in
fig. 116.

Here moreover is a fish-tail flame, which burns brightly and steadily, refusing to respond to any sound, musical or unmusical. I urge against the broad face of the flame a stream of air from a blow-pipe. The flame is cut in two by the air, and now, when the whistle is sounded, it instantly starts. A knock on the table causes the two half-flames to unite, and form, for an instant, a single flame of the ordinary shape. By a slight variation of the experiment, the two side-flames disappear when the whistle is sounded, a central luminous tongue being thrust forth in their stead.

Before you now is another thin sheet of flame, also issuing from a common fish-tail burner, fig. 117. You might

Fig. 118.

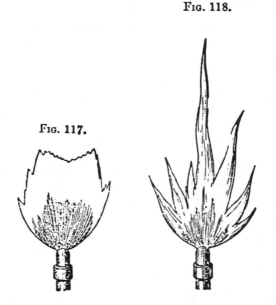

Fig. 117.

sing to it, varying the pitch of your voice, no shiver of the flame would be visible. You might employ pitch-pipes, tuning-forks, bells, and trumpets, with a like absence of all effect. A barely perceptible motion of the interior of the flame may be noticed when this shrill whistle is blown

close to it.   By turning the cock more fully on I bring the
flame to the verge of flaring.   And now, when the whistle
is blown, you see an extraordinary appearance.   The flame
thrusts out seven quivering tongues, fig. 118.   As long as
the sound continues the tongues jut forth, being violently
agitated ; the moment the sound ceases, the tongues disappear, and the flame becomes quiescent.

Passing from a fish-tail to a bat's-wing burner, we obtain
this broad, steady flame, fig. 119.   It is quite insensible to
the loudest sound which would be tolerable here.   The flame
is fed from this small gas-holder,* which places a greater
pressure at my disposal than that existing in the pipes
of the Institution.   I enlarge the flame and now a slight
flutter of its edge answers to the sound of the whistle.

Fig. 120.

Fig. 119.

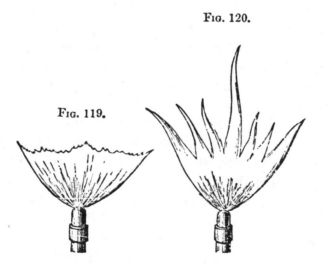

Finally I turn on gas until the flame is on the point of
roaring, as flames do when the pressure is too great.   I
now sound the whistle; the flame roars, and suddenly
assumes the form shown in fig. 120.

* A gas-bag properly weighted also answers for these experiments.

When I strike a distant anvil with a hammer, the flame instantly responds by thrusting forth its tongues.

An essential condition to entire success in these experiments disclosed itself in the following manner. I was in a room illuminated by two fish-tail flames. One of them jumped to a whistle, the other did not. The gas of the non-sensitive flame was turned off, additional pressure being thereby thrown upon the other flame; it flared, and its cock was turned so as to lower the flame. It now proved non-sensitive, however close it might be brought to the point of flaring. The narrow orifice of the half-turned cock appeared to interfere with the action of the sound. When the gas was turned fully on, the flame being lowered by opening the cock of the second burner, it became again sensitive. Up to this time a great number of burners had been tried, including some with single orifices, but, with many of them, the action was *nil.* Acting, however, upon the hint conveyed by this observation, the pipes which fed the flames were widely opened; the consequence was, that our most refractory burners were thus rendered sensitive.

The observation of Dr. Leconte is thus easily and strikingly illustrated; in our subsequent, and far more delicate experiments, the precaution just referred to is still more essential.

Mr. Barrett, late laboratory assistant in this place, first observed the shortening of a tall flame issuing from the single orifice of this old burner, when the higher notes of a circular plate were sounded; and, by the selection of more suitable burners, he afterwards succeeded in rendering the flame extremely sensitive.* Observing the precaution above adverted to, we can readily obtain, in an exalted degree, the shortening of the flame. It is

* For Mr. Barrett's own account of his experiments I refer the reader to the Philosophical Magazine for March, 1867.

now before you, being 18 inches long and smoking co-
piously. When I sound the whistle the flame falls to a
height of 9 inches, the smoke disappearing, and the flame
increasing in brightness.

A long flame may be shortened and a short one lengthened, according to circumstances, by these sonorous vibrations. Here, for example, are two flames, issuing from rough
burners formed from pewter tubing. The one flame,
fig. 121, is long, straight, and smoky; the other, fig. 122,

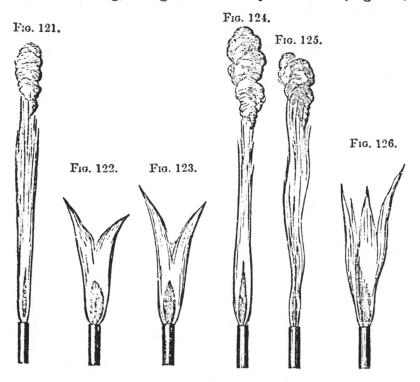

FIG. 124.

FIG. 121. FIG. 125.

FIG. 122. FIG. 123. FIG. 126.

is short, forked, and brilliant. On sounding the whistle,
the long flame becomes short, forked, and brilliant, as
in fig. 123; the forked flame becomes long and smoky, as
in fig. 124. As regards, therefore, their response to the
sound of the whistle, one of these flames is the complement
of the other.

In fig. 125 is represented another smoky flame which, when the whistle sounds, breaks up into the form shown in fig. 126.

The foregoing experiments illustrate the lengthening and shortening of flames by sonorous vibrations. They are also able to produce *rotation*. We have here several home-made burners, from which issue flat flames, each about ten inches high, and three inches across at their widest part. The burners are purposely so formed that the flames are dumpy and forked. When the whistle sounds, the plane of each flame turns ninety degrees round, and continues in its new position as long as the sound continues.

32

A Method of Recording Articulate Vibrations by Means of Photography

ELI WHITNEY BLAKE, JR.

Eli Whitney Blake, Jr. (1836–1895), American physicist, was the first Hazard Professor of Physics at Brown University, having held this chair from 1870 until his death. He was the grand nephew of Eli Whitney, the American inventor. Blake was much interested in telephoning and made contributions to the early development of the telephone. He also devoted much attention to the problem of rendering sound waves visible. We reproduce here an extract from his 1878 paper on this subject. It was of course the forerunner of much modern equipment like the phonodeik of D. C. Miller and the cathode-ray oscillograph.

Reprinted from *American Journal of Science and Arts,* series 3, 16, 54–59 (1878)

32

ART. VII.—*A method of recording Articulate Vibrations by means of Photography;* by E. W. BLAKE, Jr., Hazard Professor of Physics, Brown University.

THE extreme minuteness of the vibrations of the iron disc of the Bell telephone withdraws them from all ordinary methods of observation and measurement. A pointed wire fastened to the center of a ferrotype disc $2\frac{3}{4}$ inches in diameter, and moving on smoked glass, gave $\frac{1}{50}$ inches as the extreme amplitude of vibration under a powerful impulse of the voice, while sounds moderated to such a point as to be fairly articulate, were with difficulty detected by the movement which they communicated.

Animal membranes, possessing greater flexibility than the metal disc, seemed to promise better results, but the inertia of the attached wire, and the resistance offered by the smoked surface, become of importance, and throw doubt on the accuracy of the results obtained. Dr. Clarence J. Blake employs the human *membrana tympani* as a logograph,[*] and has obtained very beautiful and interesting tracings. I find by examination of some, which he kindly sent me, that the number of vibrations as recorded falls considerably below the ordinary pitch of the voice, being in some cases as low as 80 per second.

The logograph described by W. H. Barlow, F.R.S., in a paper read before the Royal Society,[†] serves to record the varying pressures of the expelled air taken as a whole. With a single exception the diagrams give no suggestion of the *musical* character of the sounds. The width of the line drawn by a camel's hair brush and the slow movement of the paper would mask the minute vibrations even if the apparatus were otherwise adapted to showing them.

The opeioscope, invented by Professor A. E. Dolbear, consisting of a tense membrane, to the center of which a small mirror is attached, is well adapted to proving the existence of musical vibrations in human speech, but not to determining their character.

The phonograph of Mr. Edison records on tin-foil enough of the vocal elements to reproduce intelligible articulation. The minute indentations are therefore a record of great scientific value. In the hands of Mr. Fleeming Jenkin they promise to lead to valuable results[‡] in the analysis of vocal sounds.

Dr. S. Th. Stein[§] described in 1876 a method of photograph-

[*] Archives of Opthalmology and Otology, vol. v, No. 1, 1876.
[†] Reprinted in the Popular Science Review, London, July, 1874.
[‡] Nature, May 9th, 1878. Article on Phonograph by Mr. J. Ellis.
[§] Poggendorff's Annalen, Bd. clix, S. 142.

ing the vibrations of tuning forks, strings, &c., by attaching to them plates of blackened mica punctured with small holes. A beam of sunlight passing through the hole strikes a sensitive plate moving with uniform velocity, and leaves a permanent record of the combined motions. Dr. Stein considers his method applicable to *vocal* sounds, but I cannot learn that he has ever attempted this application. My own experiments in that direction by Stein's method resulted in failure.

The object of this paper is to describe a method of obtaining photographs of minute vibrations on a magnified scale.

A plane mirror of steel, A, is supported by its axis in the metal frame B. The ends of the axis are conical, and carefully fitted into sockets in the ends of the screws C, C. On the back of the mirror is a slight projection D pierced by a small hole.

The vibrating disc, as hitherto employed, is a circular plate of ferrotype iron, $2\frac{3}{4}$ inches in diameter, screwed to the back of a telephone mouth-piece of the form invented by Professor John Peirce, and now universally used. From the center of the back of this disc a stiff steel wire projects, the end of which is bent at a right angle. This wire serves to connect the vibrating disc with the mirror by hooking into the hole in D, as represented in the figure. The mirror frame and the vibrating disc are kept in a fixed re-

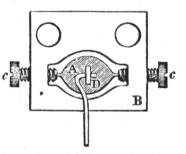

Back view of Mirror, actual size.

lation to each other by a block of hard wood, to which both are firmly screwed. The mirror is set with its axis parallel, and its reflecting surface perpendicular, to the vibrating disc.

A heliostat sends a beam of sunlight horizontally through a small circular opening. This beam passes into a dark closet and at a distance of several feet from the circular opening falls upon the mirror above described placed with its axis inclined 45° to the horizon. The rays, reflected vertically downward, pass through a lens at whose focus they form an intensely luminous image of the circular opening.

A carriage moving smoothly on four wheels travels beneath the lens at such a distance that the sensitized plate laid upon it comes at the focus for actinic rays. A uniform velocity is given to the carriage by a string fastened to it and passing over a pulley. To this string a lead weight, just sufficient to balance friction, is permanently attached, while a supplemental weight acts at the beginning of motion and is removed just before the sensitized plate reaches the spot of light above described.

The velocity attained by the carriage is determined by placing a sheet of smoked glass upon it and letting it run under a tuning fork (Ut 3—512 v. s.) provided with a pointed wire. In every case more than 200 vibrations were counted and measured, and careful comparisons made between the earlier and later ones, so as to be certain of the uniformity of the motion.

From the description it will be evident, that when the carriage alone is in motion a straight line will be photographed upon the plate. On speaking into the mouth-piece the disc is set in vibration, each movement causing change of angular position of the mirror, the reflected light moves through twice this angle, and the resulting photograph gives us the combination of its motion with that of the carriage.*

The general character of the curves obtained is shown in the accompanying figures, which are about one-half (0·56) the actual size of the originals. The reduction was accomplished by photography on the wood itself, so that the skill of the engraver was employed simply to follow the lines, which he has done with great fidelity.

The velocity of the carriage for the vowel-sounds was $21\frac{1}{2}$, for *Brown University*, 40, and for *How do you do*, 14 inches per second.

In the mathematical discussion of these curves the abscissas are measured by the known velocity of the carriage, and serve to determine the *pitch*, the ordinates represent the amplitude of vibration of the center of the disc, magnified 200 times in the photographs. The reduction of scale makes the magnifying in the wood-cuts only 112 times.

The ordinates are not strictly straight lines, but parts of the vertex of a parabola, and closely approximate to circular arcs whose radius is the focal length of the lens employed.† In the figures given, the centers of curvature these arcs is at the right hand.

With an ordinary tone of voice an amplitude of nearly an inch is obtained, implying a movement of the center of the disc of ·005 inches as determined by actual measurement.

By varying the accelerating weight and its fall, any manageable velocity may be given to the carriage. Each syllable requires for its articulation about one-fourth of a second, hence

* The carriage should run from *right* to *left*. The negative (examined from the *glass* side), and prints taken from it, then give the syllables in their proper order, and show movements of the disc from the speaker by lines going from the observer. The arrangement of my dark room compelled me to make my carriage move from *left* to *right*; hence, in the figures given, forward positions of the disc are represented by the lower portions of the curves.

† It can easily be shown that the reflected beam describes the envelope of a cone, whose apex has an angle of 90°, and whose axis is inclined 45°. The intersection of this cone with the horizontal plane gives the parabola. The lens employed transfers the apex to its own optical center. The ordinates may be made practically straight lines by placing the mirror with its axis vertical so as to reflect the beam *almost* directly back on its path, and having the sensitized plate move up and down in a vertical plane.

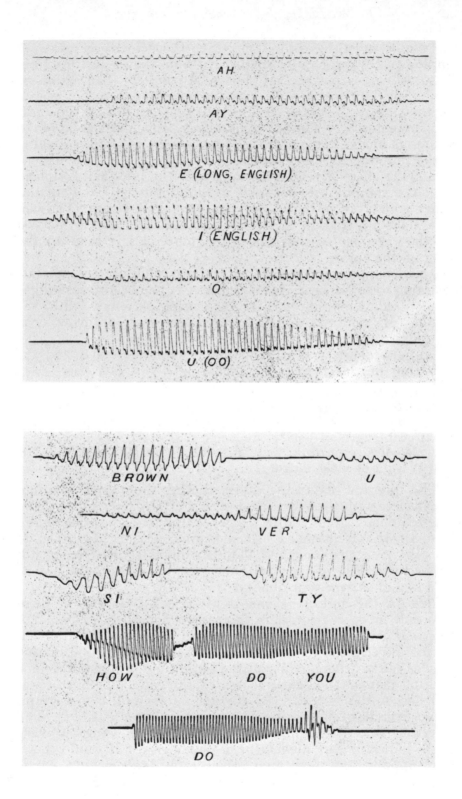

the plates must be quite long when the velocity is great. I employ plates two feet in length, and find that velocities from 16 to 40 inches per second give good results. The action of the light is however inversely as the velocity. To compensate for this, the size of the circular opening admitting the light may be increased. This, of course, causes an enlargement of the luminous image, and apparently involves an injurious widening of the line traced, but, as observed by Dr. Stein in his experiments, the effect of velocity is to narrow the line photographed, since the maximum exposure is in that diameter of the circular image which lies in the line of motion. This is a great advantage, since a variation of velocity in the vibration is marked by the widening of the line, often more clearly than by the form of the curve.

I have employed the ordinary photographic process, not attempting to obtain special sensitiveness. The brightest sunlight is required, a slight haziness interfering seriously with the result. My heliostat employs two reflectors of ordinary looking-glass, and the loss of light is considerable.

To guide those who may wish to try this method I add the following measurements :

Diameter of circular opening _____$\frac{3}{4}$ $\frac{7}{16}$ inches.*
Distance of mirror from circular opening_____28 feet.
Distance of mirror from photographic plate___11$\frac{1}{2}$ inches.
Focal length of lens_____9$\frac{1}{4}$ inches.
Size of steel mirror_____0·46x0·34 inches.
Weight of steel mirror_____ _____0·065 gram.

The question naturally arises whether the mirror may not so interfere with the vibrating disc as to destroy its articulation. The telephone gives a direct answer and banishes the doubt. The mirror was attached in the manner already described to the disc of a telephone, and the instrument showed itself still perfectly capable of 'sending,' and 'receiving,' without noticeable loss of clearness or quality.

Are *all* the audible elements of speech traceable in these records? in other words, is the record complete? I am not prepared as yet to answer this question definitely, but the following experiment leads me to doubt whether an affirmative answer can be given, while at the same time it illustrates in a striking manner the sensitiveness of the ear. The mirror was attached to the disc of a *receiving* telephone and a photograph taken from it while the instrument was talking audibly. The resulting record was almost a smooth line, showing but very slight indications of movement of the mirror. It would therefore, appear that there are distinctly audible elements, which

---

* Depending on velocity required, and on actinic intensity of the light.

are too minute to be recorded by this method. It is to be noted, however, that the width of the line traced where the vibrations are extremely small, is so great as to mask the curvature, so that the experiment just cited is not entirely fair.

The clearness and beauty of the curves obtained can hardly be appreciated without inspection of the originals. Their complexity and variety open a large field for investigation, and they seem to offer the means of analysis of articulate speech.

33

Crystal Physics—Development by Pressure of Polar Electricity in Hemihedral Crystals with Inclined Faces

PAUL-JACQUES CURIE and PIERRE CURIE

Paul-Jacques Curie (1855–1941) and Pierre Curie (1859–1906), brothers and physicists, whose names will ever be associated with the discovery of piezoelectricity, both became professors of physics in Paris. Pierre Curie gained greater fame through his work with his wife, Marie, on radioactivity. In the early part of their careers they were both interested in the physical properties of crystals and it was as a result of this that in 1880 they discovered piezoelectricity.

The following is a translation of their original paper. This was followed by a later paper in the same year, also in *Comptes Rendus*, containing further details but nothing of fundamental character. In neither paper does the term piezoelectricity appear. This nomenclature was suggested immediately after the publication of the Curie papers by the German physicist W. G. Hankel (1814–1898). Further information on the subject will be found in *Piezoelectricity* by W. G. Cady (1946, revised ed., 1964). The use of piezoelectric crystals as electroacoustic transducers revolutionized the production and reception of high-frequency sound. The discovery of piezoelectricity was thus a benchmark in the development of acoustics.

33

Translated by R. Bruce Lindsay from
*Comptes Rendus Hebdomadaires des Seances
de l'Academie des Sciences, Paris,* **91**, 294 (1880).

1. Crystals possessing one or more faces with dissimilar extremities, that is to say, hemihedral crystals with inclined faces, possess a special physical property, that of giving birth to two electric charges of opposite sign at the extremities of the axes in question when they are subjected to a change in temperature. This is the phenomenon known as *pyroelectricity*.

We have found a new way to develop polar electricity in the same crystals. This consists in subjecting them to variations of pressure along their hemihedral axes.[1]

The effects produced are entirely analogous to those produced by heat. The extremities of the axis along which the pressure is exerted receive electric charges of opposite sign. Once the crystal has returned to its normal state, if one decompresses it, the phenomenon is repeated, but with an inversion of sign: the end which was charged positively on compression becomes negative on decompression and conversely.[2]

In order to make an experiment two parallel faces are cut perpendicular to a hemihedral axis in the substance one wishes to study. The faces are covered with sheets of tin foil which are insulated from the surroundings by two plates of hard rubber. By putting the whole between the jaws of a vise, for example, one can exert pressure on the two faces that are cut, that is to say, along the hemihedral axis

---

[1] *Bulletin de la Société minéralogique,* 1880.

[2] Hemihedral crystals with inclined faces are the only pyroelectric crystals. They are also the only ones capable of acquiring polar electricity by the influence of pressure. Certain holohedral crystals, like calcspar, for example, are also charged by the application of pressure, but with electricity of one sign only. This is a surface phenomenon and entirely different. The effect of it was not perceptible under the conditions of our experiments.

itself. To verify the electric charge, we use a Thomson electrometer. One can show the voltage difference between the two ends of the crystal by connecting each piece of tin foil with two of the sectors of the instrument; the needle being given a known electric charge. One can also collect each electric charge separately. For this it suffices to ground one of the tin foil sheets, the other being in communication with the needle and the two sectors being charged with the aid of a cell.

Although we have not embarked on the study of the laws which govern the phenomenon, we are able to say that it presents characteristics identical to those of pyroelectricity as Gaugain has determined them in his beautiful work on tourmaline.

2. We have made a comparative study of the two ways of developing polar electricity on a number of nonconducting substances, hemihedral with inclined faces, comprising nearly all those known to be pyroelectric.[3]

The action of heat has been studied with the aid of the procedure indicated by M. Friedel, a procedure of great convenience.[4]

Our experiments have been carried out on blende, sodium chlorate, borax, tourmaline, quartz, calamine, topaz, right-handed tartaric acid, sugar, and Rochelle salt.

For all these crystals the effects produced by compression are in the same sense as those produced by cooling. Those associated with decompression are in the same sense as those due to heating.

There is thus an evident relation which permits us to relate the phenomena in both cases to a single cause and to unite them in the following statement.

Whatever may be the determining cause, every time a nonconducting hemihedral crystal with inclined face contracts, electric charges appear in a certain direction. Every time such a crystal expands the appearance of electricity takes place in the opposite direction.

If this way of looking at the matter is correct, the effects due to compression ought to be in the same direction as those due to heating in a substance possessing a negative coefficient of expansion along the hemihedral axis.[5]

---

[3]One can predict that many others exist among artificially crystallized substances. Bodies active with respect to polarized light, for example, furnish crystals with certain diameters having dissimilar extremities.

[4]*Bulletin de la Société Minéralogique,* 1879.

[5]This work has been done in the laboratory of mineralogy of the Faculté des Sciences (Paris).

34–38

Fundamental Papers on Acoustics

LORD RAYLEIGH

John William Strutt, Third Baron Rayleigh (1842–1919), was perhaps the most versatile of the British physical scientists of the second half of the 19th century and the first quarter of the 20th. Active both as an experimentalist and a theoretician, he turned his attention during a long and active professional career to practically every branch of physics and many phases of chemistry. Though perhaps best known as the co-discoverer (with William Ramsay) of the rare gas argon, he devoted much attention to acoustics. His book, *The Theory of Sound* (1877–1878) was a landmark in the development of the subject, and has remained a standard treatise ever since its publication. Of the 130 articles which he published in acoustics we present here five as typical of his important contributions. In each one he broke new ground in some field of sound phenomena. We have restricted ourselves to papers that appeared before 1900. Later papers belong more appropriately in the volume on physical acoustics, to appear later in this series.

34

5.

ON THE THEORY OF RESONANCE.

[*Phil. Trans.* CLXI. pp. 77—118; Read Nov. 1870.*]

Introduction.

ALTHOUGH the theory of aërial vibrations has been treated by more
than one generation of mathematicians and experimenters, comparatively
little has been done towards obtaining a clear view of what goes on in
any but the more simple cases. The extreme difficulty of anything like a
general deductive investigation of the question is no doubt one reason.
On the other hand, experimenters on this, as on other subjects, have too
often observed and measured blindly without taking sufficient care to simplify
the conditions of their experiments, so as to attack as few difficulties as
possible at a time. The result has been vast accumulations of isolated
facts and measurements which lie as a sort of dead weight on the scientific
stomach, and which must remain undigested until theory supplies a more
powerful solvent than any now at our command. The motion of the air
in cylindrical organ-pipes was successfully investigated by Bernoulli and
Euler, at least in its main features; but their treatment of the question
of the open pipe was incomplete, or even erroneous, on account of the
assumption that at the open end the air remains of invariable density
during the vibration. Although attacked by many others, this difficulty
was not finally overcome until Helmholtz†, in a paper which I shall have
repeated occasion to refer to, gave a solution of the problem under certain
restrictions, free from any arbitrary assumptions as to what takes place
at the open end. Poisson and Stokes ‡ have solved the problem of the

---

* Additions made since the paper was first sent to the Royal Society are inclosed in square
brackets [ ].
† Theorie der Luftschwingungen in Röhren mit offenen Enden. *Crelle*, 1860.
‡ *Phil. Trans.* 1868, or *Phil. Mag.* Dec. 1868.

vibrations communicated to an infinite mass of air from the surface of a sphere or circular cylinder. The solution for the sphere is very instructive, because the vibrations outside any imaginary sphere enclosing vibrating bodies of any kind may be supposed to take their rise in the surface of the sphere itself.

More important in its relation to the subject of the present paper is an investigation by Helmholtz of the air-vibrations in cavernous spaces (*Hohlraüme*), whose three dimensions are very small compared to the wave-length, and which communicate with the external atmosphere by small holes in their surfaces. If the opening be circular of area σ, and if S denote the volume, n the number of vibrations per second in the fundamental note, and a the velocity of sound,

$$n = \frac{a\sigma^{\frac{1}{2}}}{2^{\frac{1}{2}}\pi^{\frac{1}{4}}S^{\frac{1}{2}}}.$$

Helmholtz's theory is also applicable when there are more openings than one in the side of the vessel.

In the present paper I have attempted to give the theory of vibrations of this sort in a more general form. The extension to the case where the communication with the external air is no longer by a mere hole in the side, but by a neck of greater or less length, is important, not only because resonators with necks are frequently used in practice, but also by reason of the fact that the theory itself is applicable within wider limits. The mathematical reasoning is very different from that of Helmholtz, at least in form, and will I hope be found easier. In order to assist those who may wish only for clear general ideas on the subject, I have broken up the investigation as much as possible into distinct problems, the results of which may in many cases be taken for granted without the rest becoming un-intelligible. In Part I. my object has been to put what may be called the dynamical part of the subject in a clear light, deferring as much as possible special mathematical calculations. In the first place, I have con-sidered the general theory of resonance for air-spaces confined nearly all round by rigid walls, and communicating with the external air by any number of passages which may be of the nature of necks or merely holes, under the limitation that both the length of the necks and the dimensions of the vessel are very small compared to the wave-length. To prevent misapprehension, I ought to say that the theory applies only to the funda-mental note of the resonators, for the vibrations corresponding to the overtones are of an altogether different character. There are, however, cases of multiple resonance to which our theory is applicable. These occur when two or more vessels communicate with each other and with the external air by necks or otherwise; and are easily treated by Lagrange's general dynamical method, subject to a restriction as to the relative magnitudes of the wave-lengths and the dimensions of the system corresponding to that stated above for a single vessel. I am not aware whether this kind of resonance has been investigated before, either mathematically or experimentally. Lastly, I have sketched a solution of the problem of the open organ-pipe on the same general plan, which may be acceptable to those who are not acquainted with Helmholtz's most valuable paper. The method here adopted, though it leads to results essentially the same as his, is I think more calculated to give an insight into the real nature of the question, and at the same time presents fewer mathematical difficulties. For a discussion of the solution, however, I must refer to Helmholtz.

In Part II. the calculation of a certain quantity depending on the form of the necks of common resonators, and involved in the results of Part I., is entered upon. This quantity, denoted by *c*, is of the nature of a length, and is identical with what would be called in the theory of electricity the *electric conductivity* of the passage, supposed to be occupied by uniformly conducting matter. The question is accordingly similar to that of determining the electrical resistance of variously shaped conductors—an analogy of which I have not hesitated to avail myself freely both in investigation and statement. Much circumlocution is in this way avoided on account of the greater completeness of electrical phraseology. Passing over the case of mere holes, which has been already considered by Helmholtz, and need not be dwelt upon here, we come to the value of the resistance for necks in the form of circular cylinders. For the sake of simplicity each end is supposed to be in an infinite plane. In this form the mathematical problem is definite, but has not been solved rigorously. Two limits, however (a higher and a lower), are investigated, between which it is proved that the true resistance must lie. The lower corresponds to a correction to the length of the tube equal to $\frac{1}{4}\pi \times$ (radius) for each end. It is a remarkable coincidence that Helmholtz also finds the same quantity as an approximate correction to the length of an organ-pipe, although the two methods are entirely different and neither of them rigorous. His consists of an exact solution of the problem for an approximate cylinder, and mine of an approximate solution for a true cylinder; while both indicate on which side the truth must lie. The final result for a cylinder infinitely long is that the correction lies between $\cdot 785\,R$ and $\cdot 828\,R$. When the cylinder is finite, the upper limit is rather smaller. In a somewhat similar manner I have investigated limits for the resistance of a tube of revolution, which is shown to lie between

$$\int \frac{dx}{\pi y^2} \quad \text{and} \quad \int \frac{dx}{\pi y^2}\left\{1 + \tfrac{1}{2}\left(\frac{dy}{dx}\right)^2\right\},$$

where *y* denotes the radius of the tube at any point *x* along the axis. These formulæ apply whatever may be in other respects the form of the tube, but are especially valuable when it is so nearly cylindrical that dy/dx is everywhere small. The two limits are then very near each other, and either of them gives very approximately the true value. The resistance of tubes, which are either not of revolution or are not nearly straight, is afterwards approximately determined. The only experimental results bearing on the subject of this paper, and available for comparison with theory, that I have met with are some arrived at by Sondhauss* and Wertheim†. Besides those quoted by Helmholtz, I have only to mention a series of observations by Sondhauss‡ on the pitch of flasks with long necks which led him to the empirical formula

$$n = 46705 \frac{\sigma^{\frac{1}{2}}}{L^{\frac{1}{2}}S^{\frac{1}{2}}},$$

σ, L being the area and length of the neck, and S the volume of the flask. The corresponding equation derived from the theory of the present paper is

$$n = 54470 \frac{\sigma^{\frac{1}{2}}}{L^{\frac{1}{2}}S^{\frac{1}{2}}},$$

which is only applicable, however, when the necks are so long that the corrections at the ends may be neglected—a condition not likely to be fulfilled. This consideration sufficiently explains the discordance. Being anxious to give the formulæ of Parts I. and II. a fair trial, I investigated experimentally the resonance of a considerable number of vessels which were of such a form that the theoretical pitch could be calculated with tolerable accuracy. The result of the comparison is detailed in Part III., and appears on the whole very satisfactory; but it is not necessary that I should describe it more minutely here. I will only mention, as perhaps a novelty, that the experimental determination of the pitch was not made by causing the resonators to speak by a stream of air blown over their mouths. The grounds of my dissatisfaction with this method are explained in the proper place.

[Since this paper was written there has appeared another memoir by Dr Sondhauss§ on the subject of resonance. An empirical formula is obtained bearing resemblance to the results of Parts I. and II., and agreeing fairly well with observation. No attempt is made to connect it with the fundamental principles of mechanics. In the *Philosophical Magazine* for September 1870 [Art. IV. above], I have discussed the differences between Dr Sondhauss's formula and my own from the experimental side, and shall not therefore go any further into the matter on the present occasion.]

* Pogg. *Ann.* vol. LXXXI.
† *Annales de Chimie*, vol. XXXI.
‡ Pogg. *Ann.* vol. LXXIX.
§ Pogg. *Ann.* 1870.

PART I.

The class of resonators to which attention will chiefly be given in this paper are those where a mass of air confined almost all round by rigid walls communicates with the external atmosphere by one or more narrow passages. For the present it may be supposed that the boundary of the principal mass of air is part of an oval surface, nowhere contracted into anything like a narrow neck, although some cases not coming under this description will be considered later. In its general character the fundamental vibration of such an air-space is sufficiently simple, consisting of a periodical rush of air through the narrow channel (if there is only one) into and out of the confined space, which acts the part of a reservoir. The channel spoken of may be either a mere hole of any shape in the side of the vessel, or may consist of a more or less elongated tube-like passage.

If the linear dimension of the reservoir be small as compared to the wave-length of the vibration considered, or, as perhaps it ought rather to be said, the quarter wave-length, the motion is remarkably amenable to deductive treatment. Vibration in general may be considered as a periodic transformation of energy from the potential to the kinetic, and from the kinetic to the potential forms. In our case the kinetic energy is that of the air in the neighbourhood of the opening as it rushes backwards or forwards. It may be easily seen that relatively to this the energy of the motion inside the reservoir is, under the restriction specified, very small. A formal proof would require the assistance of the general equations to the motion of an elastic fluid, whose use I wish to avoid in this paper. Moreover the motion in the passage and its neighbourhood will not differ sensibly from that of an incompressible fluid, and its energy will depend only on the rate of total flow through the opening. A quarter of a period later this energy of motion will be completely converted into the potential energy of the compressed or rarefied air inside the reservoir. So soon as the mathematical expressions for the potential and kinetic energies are known, the determination of the period of vibration or resonant note of the air-space presents no difficulty.

The motion of an incompressible frictionless fluid which has been once at rest is subject to the same formal laws as those which regulate the flow of heat or electricity through uniform conductors, and depends on the properties of the potential function, to which so much attention has of late years been given. In consequence of this analogy many of the results obtained in this paper are of as much interest in the theory of electricity as in acoustics, while, on the other hand, known modes of expression in the former subject will save circumlocution in stating some of the results of the present problem.

Let h_0 be the density, and ϕ the velocity-potential of the fluid motion through an opening. The kinetic energy or *vis viva*

$$= \tfrac{1}{2} h_0 \iiint \left[\left(\frac{d\phi}{dx}\right)^2 + \left(\frac{d\phi}{dy}\right)^2 + \left(\frac{d\phi}{dz}\right)^2 \right] dx \, dy \, dz,$$

the integration extending over the volume of the fluid considered

$$= \tfrac{1}{2} h_0 \iint \phi \frac{d\phi}{dn} \, dS, \quad \text{by Green's theorem.}$$

Over the rigid boundary of the opening or passage, $d\phi/dn = 0$, so that if the portion of fluid considered be bounded by two equipotential surfaces, ϕ_1 and ϕ_2, one on each side of the opening,

$$vis \ viva = \tfrac{1}{2} h_0 (\phi_1 - \phi_2) \iint \frac{d\phi}{dn} \, dS = \tfrac{1}{2} h_0 (\phi_1 - \phi_2) \dot{X},$$

if $\dot{X}$ denote the rate of total flow through the opening.

At a sufficient distance on either side ϕ becomes constant, and the rate of total flow is proportional to the difference of its values on the two sides. We may therefore put

$$\phi_1 - \phi_2 = \frac{1}{c} \iint \frac{d\phi}{dn} \, dS = \frac{\dot{X}}{c},$$

where c is a linear quantity depending on the size and shape of the opening, and representing in the electrical interpretation the reciprocal of the *resistance* to the passage of electricity through the space in question, the specific resistance of the conducting matter being taken for unity. The same thing may be otherwise expressed by saying that c is the side of a cube, whose resistance between opposite faces is the same as that of the opening.

The expression for the *vis viva* in terms of the rate of total flow is accordingly

$$vis \ viva = \frac{h_0}{2} \frac{\dot{X}^2}{c}. \quad \dotfill (1)$$

If S be the capacity of the reservoir, the condensation at any time inside it is given by X/S, of which the mechanical value is

$$\tfrac{1}{2} h_0 a^2 \frac{X^2}{S}, \quad \dotfill (2)$$

a denoting, as throughout the paper, the velocity of sound.

The whole energy at any time, both actual and potential, is therefore

$$\frac{h_0}{2} \frac{\dot{X}^2}{c} + \frac{h_0}{2} a^2 \frac{X^2}{S}, \quad \dotfill (3)$$

and is constant. Differentiating with respect to time, we arrive at

$$\ddot{X} + \frac{a^2 c}{S} X = 0 \quad \dotfill (4)$$

as the equation to the motion, which indicates simple oscillations performed in a time

$$2\pi \div \sqrt{\left(\frac{a^2 c}{S}\right)}.$$

Hence if n denote the number of vibrations per second in the resonant note,

$$n = \frac{a}{2\pi} \sqrt{\left(\frac{c}{S}\right)}. \quad \dots\dots\dots\dots\dots\dots\dots(5)$$

The wave-length λ, which is the quantity most immediately connected with the dimensions of the resonant space, is given by

$$\lambda = \frac{a}{n} = 2\pi \sqrt{\left(\frac{S}{c}\right)}. \quad \dots\dots\dots\dots\dots(6)$$

A law of Savart, not nearly so well known as it ought to be, is in agreement with equations (5) and (6). It is an immediate consequence of the principle of dynamical similarity, of extreme generality, to the effect that *similar* vibrating bodies, whether they be gaseous, such as the air in organpipes or in the resonators here considered, or solid, such as tuning-forks, vibrate in a time which is directly as their linear dimensions. Of course the material must be the same in two cases that are to be compared, and the geometrical similarity must be complete, extending to the shape of the opening as well as to the other parts of the resonant vessel. Although the wave-length λ is a function of the size and shape of the resonator only, n or the position of the note in the musical scale depends on the nature of the gas with which the resonator is filled. And it is important to notice that it is on the nature of the gas in and near the opening that the note depends, and *not* on the gas in the interior of the reservoir, whose inertia does not come into play during vibrations corresponding to the fundamental note. In fact we may say that the mass to be moved is the air in the neighbourhood of the opening, and that the air in the interior acts merely as a spring in virtue of its resistance to compression. Of course this is only true under the limitation specified, that the diameter of the reservoir is small compared to the quarter wave-length. Whether this condition is fulfilled in the case of any particular resonator is easily seen, *à posteriori*, by calculating the value of λ from (6), or by determining it experimentally.

Several Openings.

When there are two or more passages connecting the interior of the resonator with the external air, we may proceed in much the same way, except that the equation of energy by itself is no longer sufficient. For simplicity of expression the case of two passages will be convenient, but the same method is applicable to any number. Let X_1, X_2 be the total flow through the two necks, c_1, c_2 constants depending on the form of the necks corresponding to the constant c in formula (6); then T, the *vis viva*, is given by

$$T = \frac{h_0}{2}\left(\frac{\dot{X_1}^2}{c_1} + \frac{\dot{X_2}^2}{c_2}\right),$$

the necks being supposed to be sufficiently far removed from one another not to *interfere* (in a sense that will be obvious). Further,

$$V = \text{Potential Energy} = \tfrac{1}{2}h_0 a^2 \frac{(X_1 + X_2)^2}{S}.$$

Applying Lagrange's general dynamical equation,

$$\frac{d}{dt}\left(\frac{dT}{d\dot{\psi}}\right) - \frac{dT}{d\psi} = -\frac{dV}{d\psi},$$

we obtain

$$\frac{\ddot{X}_1}{c_1} + \frac{a^2}{S}(X_1 + X_2) = 0, \qquad \frac{\ddot{X}_2}{c_2} + \frac{a^2}{S}(X_1 + X_2) = 0 \ldots\ldots\ldots(7)$$

as the equations to the motion.

By subtraction,

$$\ddot{X}_1/c_1 - \ddot{X}_2/c_2 = 0,$$

or, on integration,

$$\frac{X_1}{c_1} = \frac{X_2}{c_2} \ldots\ldots\ldots\ldots\ldots\ldots\ldots\ldots\ldots(8)$$

Equation (8) shows that the motions of the air in the two necks have the same period and are at any moment in the same phase of vibration. Indeed there is no essential distinction between the case of one neck and that of several, as the passage from one to the other may be made continuously without the failure of the investigation. When, however, the separate passages are sufficiently far apart, the constant c for the system, considered as a single communication between the interior of the resonator and the external air, is the simple sum of the values belonging to them when taken separately, which would not otherwise be the case. This is a point to which we shall return later, but in the mean time, by addition of equations (7), we find

$$\ddot{X}_1 + \ddot{X}_2 + \frac{a^2}{S}(c_1 + c_2)(X_1 + X_2) = 0,$$

so that

$$n = \frac{a}{2\pi}\sqrt{\left(\frac{c_1 + c_2}{S}\right)}. \quad \ldots\ldots\ldots\ldots\ldots\ldots(9)$$

If there be any number of necks for which the values of c are c_1, c_2, c_3 ..., and no two of which are near enough to interfere, the same method is applicable, and gives

$$n = \frac{a}{2\pi}\sqrt{\frac{c_1 + c_2 + c_3 + \ldots}{S}}. \quad \ldots\ldots\ldots\ldots(9')$$

When there are two similar necks $c_2 = c_1$, and

$$n = \sqrt{2} \times \frac{a}{2\pi} \sqrt{\left(\frac{c}{S}\right)}.$$

The note is accordingly higher than if there were only one neck in the ratio of $\sqrt{2} : 1$, a fact observed by Sondhauss and proved theoretically by Helmholtz for the case of openings which are mere holes in the sides of the reservoir.

Double Resonance.

Suppose that there are two reservoirs, S, S', communicating with each other and with the external air by narrow passages or necks. If we were to consider SS' as a single reservoir and to apply equation (9), we should be led to an erroneous result ; for the reasoning on which (9) is founded proceeds on the assumption that, within the reservoir, the inertia of the air may be left out of account, whereas it is evident that the *vis viva* of the motion through the connecting passage may be as great as through the two others. However, an investigation on the same general plan as before meets the case perfectly. Denoting by X_1, X_2, X_3 the total flows through the three necks, we have for the *vis viva* the expression

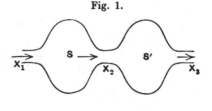

Fig. 1.

$$T = \tfrac{1}{2} h_0 \left\{ \frac{\dot{X}_1^2}{c_1} + \frac{\dot{X}_2^2}{c_2} + \frac{\dot{X}_3^2}{c_3} \right\},$$

and for the potential energy

$$V = \tfrac{1}{2} h_0 a^2 \left\{ \frac{(X_2 - X_1)^2}{S} + \frac{(X_3 - X_2)^2}{S'} \right\}.$$

An application of Lagrange's method gives as the differential equations to the motion,

$$\left. \begin{aligned} &\frac{\ddot{X}_1}{c_1} + a^2 \frac{X_1 - X_2}{S} = 0, \\ &\frac{\ddot{X}_2}{c_2} + a^2 \left\{ \frac{X_2 - X_1}{S} + \frac{X_2 - X_3}{S'} \right\} = 0, \\ &\frac{\ddot{X}_3}{c_3} + a^2 \frac{X_3 - X_2}{S'} = 0. \end{aligned} \right\} \quad \ldots\ldots\ldots\ldots(10)$$

By addition and integration $X_1/c_1 + X_2/c_2 + X_3/c_3 = 0$. Hence, on elimination of X_2,

$$\left. \begin{aligned} &\ddot{X}_1 + \frac{a^2}{S} \left\{ (c_1 + c_2) X_1 + \frac{c_1 c_2}{c_3} X_3 \right\} = 0, \\ &\ddot{X}_3 + \frac{a^2}{S'} \left\{ (c_3 + c_2) X_3 + \frac{c_3 c_2}{c_1} X_1 \right\} = 0. \end{aligned} \right\}$$

Assuming $X_1 = A\epsilon^{pt}$, $X_3 = B\epsilon^{pt}$, we obtain, on substitution and elimination of $A : B$,

$$p^4 + p^2 a^2 \left\{ \frac{c_1 + c_2}{S} + \frac{c_3 + c_2}{S'} \right\} + \frac{a^4}{SS'} \{ c_1 c_3 + c_2 (c_1 + c_3) \} = 0 \quad \ldots\ldots(11)$$

as the equation to determine the resonant notes. If n be the number of vibrations per second, $n^2 = -p^2/4\pi^2$, the values of p^2 given by (11) being of course both real and negative. The formula simplifies considerably if $c_3 = c_1$, $S' = S$; but it will be more instructive to work this case from the beginning. Let $c_1 = c_3 = mc_2 = mc$.

The differential equations take the form

$$\left. \begin{aligned} \ddot{X}_1 + \frac{a^2 c}{S} \{ (1 + m) X_1 + X_3 \} = 0, \\ \ddot{X}_3 + \frac{a^2 c}{S} \{ (1 + m) X_3 + X_1 \} = 0, \end{aligned} \right\} \quad \text{while } X_2 = -\frac{X_1 + X_3}{m}.$$

Hence

$$\left. \begin{aligned} (X_1 + X_3)^{\cdot\cdot} + \frac{a^2 c}{S} (m + 2) (X_1 + X_3) = 0, \\ (X_1 - X_3)^{\cdot\cdot} + \frac{a^2 c}{S} m (X_1 - X_3) = 0. \end{aligned} \right\}$$

The whole motion may be regarded as made up of two parts, for the first

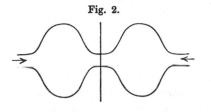

Fig. 2.

of which $X_1 + X_3 = 0$; which requires $X_2 = 0$. This motion is therefore the same as might take place were the communication between S and S' cut off, and has its period given by

$$n^2 = \frac{a^2 c_1}{4\pi^2 S} = \frac{a^2 mc}{4\pi^2 S}.$$

For the other component part, $X_1 - X_3 = 0$, so that

$$X_2 = -\frac{2X_1}{m}, \qquad n'^2 = \frac{a^2 (m + 2) c}{4\pi^2 S} \quad \ldots\ldots\ldots\ldots(12)$$

Thus $\dfrac{n'^2}{n^2} = \dfrac{m + 2}{m}$, which shows that the second note is the higher. It

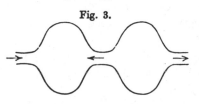

Fig. 3.

consists of vibrations in the two reservoirs opposed in phase and modified by the connecting passage, which acts in part as a second opening to both, and so raises the pitch. If the passage is small, so also is the difference of pitch between the two notes. A particular case worth notice is obtained by putting in the general equation $c_3 = 0$, which amounts to suppressing one of the communications with the external air. We thus obtain

$$p^4 + a^2 p^2 \left(\frac{c_1 + c_2}{S} + \frac{c_2}{S'} \right) + \frac{a^4}{SS'} c_1 c_2 = 0 \; ;$$

or if $S = S'$, $c_1 = mc_2 = mc$,

$$p^4 + a^2 p^2 \frac{c}{S} (m + 2) + \frac{a^4 c^2}{S^2} m = 0,$$

$$n^2 = \frac{a^2 c}{8 \pi^2 S} \{ m + 2 \pm \sqrt{m^2 + 4} \}.$$

If we further suppose $m = 1$ or $c_2 = c_1$,

$$n^2 = \frac{a^2 c}{8 \pi^2 S} (3 \pm \sqrt{5}).$$

If N be the number of vibrations for a simple resonator (S, c),

$$N^2 = \frac{a^2 c}{4 \pi^2 S} \; ;$$

so that

$$n_1^2 \div N^2 = \frac{3 + \sqrt{5}}{2} = 2 \cdot 618, \qquad N^2 \div n_2^2 = \frac{2}{3 - \sqrt{5}} = 2 \cdot 618.$$

It appears therefore that the interval from n_1 to N is the same as from N to n_2, namely, $\sqrt{(2 \cdot 618)} = 1 \cdot 618$, or rather more than a fifth. It will be found that whatever the value of m may be, the interval between the resonant notes cannot be less than $2 \cdot 414$, which is about an octave and a minor third. The corresponding value of m is 2.

A similar method is applicable to any combination of reservoirs and connecting passages, no matter how complicated, under the single restriction as to the comparative magnitudes of the reservoirs and wave-lengths; but the example just given is sufficient to illustrate the theory of multiple resonance. In Part III. a resonator of this sort will be described, which was constructed for the sake of a comparison between the theory and experiment. In applying the formulæ (6) or (12) to an actual measurement, the question will arise whether the volume of the necks, especially when they are rather large, is to be included or not in S. At the moment of rest the air in the neck is compressed or rarefied as well as that inside the reservoir, though not to the same degree; in fact the condensation must vary continuously between the interior of the resonator and the external air. This consideration shows that, at least in the case of necks which are tolerably symmetrical, about half the volume of the neck should be included in S.

[In consequence of a suggestion made by Mr Clerk Maxwell, who reported on this paper, I have been led to examine what kind of effect would be produced by a deficient rigidity in the envelope which contains the alternately compressed and rarefied air. Taking for simplicity the case of a sphere, let us suppose that the radius, instead of remaining constant at its normal value R, assumes the variable magnitude $R + \rho$. We have

$$\text{kinetic energy} = \frac{h_0 \dot{X}^2}{2c} + \frac{m}{2} \dot{\rho}^2,$$

$$\text{potential energy} = \frac{h_0 a^2}{2S} \{X + 4\pi R^2 \rho\}^2 + \tfrac{1}{2}\beta\rho^2,$$

where m and β are constants expressing the inertia and rigidity of the spherical shell. Hence, by Lagrange's method,

$$\left.\begin{array}{l} \ddot{X} + \dfrac{ca^2}{S}(X + 4\pi R^2 \rho) = 0, \\[3mm] m\ddot{\rho} + 4\pi R^2 \dfrac{h_0 a^2}{S}(X + 4\pi R^2 \rho) + \beta\rho = 0, \end{array}\right\}$$

equations determining the periods of the two vibrations of which the system is capable. It might be imagined at first sight that a yielding of the sides of the vessel would necessarily lower the pitch of the resonant note; but this depends on a tacit assumption that the capacity of the vessel is largest when the air inside is most compressed. But it may just as well happen that the opposite is true. Everything depends on the relative magnitudes of the periods of the two vibrations supposed for the moment independent of one another. If the note of the shell be very high compared to that of the air, the inertia of the shell may be neglected, and this part of the question treated statically. Putting in the equations $m = 0$, we see that the phases of X and ρ are opposed, and then X goes through its changes more slowly than before. On the other hand, if it be the note of the air-vibration, which is much the higher, we must put $\beta = 0$, which leads to

$$4\pi R^2 h_0 \ddot{X} - cm\ddot{\rho} = 0,$$

showing that the phases of X and ρ agree. Here the period of X is diminished by the yielding of the sides of the vessel, which indeed acts just in the same way as a second aperture would do. A determination of the actual note in any case of a spherical shell of given dimensions and material would probably be best obtained deductively.

But in order to see what probability there might be that the results of Part III. on glass flasks were sensibly modified by a want of rigidity, I thought it best to make a direct experiment. To the neck of a flask was fitted a glass tube of rather small bore, and the whole filled with water so as make a kind of water-thermometer. On removing by means of an air-pump the pressure of the atmosphere on the outside of the bulb, the liquid fell in the tube, but only to an extent which indicated an increase in the capacity of the flask of about a ten-thousandth part. This corresponds in the ordinary arrangement to a doubled density of the contained air. It is clear that so small a yielding could produce no sensible effect on the pitch of the air-vibration.]

Open Organ-pipes.

Although the problem of open organ-pipes, whose diameter is very small compared to their length and to the wave-length, has been fully considered by Helmholtz, it may not be superfluous to show how the question may be attacked from the point of view of the present paper, more especially as some important results may be obtained by a comparatively simple analysis. The principal difficulty consists in finding the connexion between the spherical waves which diverge from the open end of the tube into free space, and the waves in the tube itself, which at a distance from the mouth, amounting to several diameters, are approximately plane. The transition occupies a space which is large compared to the diameter, and in order that the present treatment may be applicable must be small compared to the wave-length. This condition being fulfilled, the compressibility of the air in the space mentioned may be left out of account and the difficulty is turned. Imagine a piston (of infinitely small thickness) in the tube at the place where the waves cease to be plane. The motion of the air on the free side is entirely determined by the motion of the piston, and the *vis viva* within the space considered may be expressed by $\frac{1}{2}h_0\dot{X}^2/c$, where $\dot{X}$ denotes the rate of total flow at the place of the piston, and c is, as before, a linear quantity depending on the form of the mouth. If Q is the section of the tube and ψ the velocity potential, $\dot{X} = Q\,d\psi/dx$. The most general expression for the velocity-potential of plane waves is

$$\psi = \left(\frac{A}{k}\sin kx + B\cos kx\right)\cos 2\pi nt + \beta\cos kx\sin 2\pi nt, \dots(13)$$

$$\frac{d\psi}{dx} = (A\cos kx - Bk\sin kx)\cos 2\pi nt - \beta k\sin kx\sin 2\pi nt,$$

where $k = 2\pi/\lambda = 2\pi n/a$. When $x = 0$,

$$\psi = B\cos 2\pi nt + \beta\sin 2\pi nt, \qquad \frac{d\psi}{dx} = A\cos 2\pi nt.$$

The variable part of the pressure on the tube side of the piston $= -h_0\,d\psi/dt$. The equation to the motion of the air in the mouth is therefore

$$\frac{Q}{c}\frac{d}{dt}\frac{d\psi}{dx} + \frac{d\psi}{dt} = 0,$$

or, on integration,

$$\frac{Q}{c}\frac{d\psi}{dx} + \psi = 0. \dots\dots\dots\dots\dots\dots\dots(14)$$

This is the condition to be satisfied when $x = 0$.

Substituting the values of ψ and $d\psi/dx$, we obtain

$$\cos 2\pi n t \left(A \frac{Q}{c} + B \right) + \beta \sin 2\pi n t = 0,$$

which requires

$$A \frac{Q}{c} + B = 0, \qquad \beta = 0.$$

If there is a node at $x = -l$, $A \cos kl + Bk \sin kl = 0$; so that

$$k \tan kl = -\frac{A}{B} = -\frac{c}{Q}. \quad\ldots\ldots\ldots\ldots\ldots\ldots(15)$$

This equation gives the fundamental note of the tube closed at $x = -l$; but it must be observed that l is not the length of the tube, because the origin $x = 0$ is not in the mouth. There is, however, nothing indeterminate in the equation, although the origin is to a certain extent arbitrary; for the values of c and l will change together so as to make the result for k approximately constant. This will appear more clearly when we come, in Part II., to calculate the actual value of c for different kinds of mouths. In the formation of (14) the pressure of the air on the positive side at a distance from the origin small against λ has been taken absolutely constant. Across such a loop surface no energy could be transmitted. In reality, of course, the pressure is variable on account of the spherical waves, and energy continually escapes from the tube and its vicinity. Although the pitch of the resonant note is not affected, it may be worth while to see what correction this involves.

We must, as before, consider the space in which the transition from plane to spherical waves is effected as small compared with λ. The potential in free space may be taken

$$\psi = \frac{A'}{r} \cos (kr + g - 2\pi n t), \quad\ldots\ldots\ldots\ldots\ldots(16)$$

expressing spherical waves diverging from the mouth of the pipe, which is the origin of r. The origin of x is still supposed to lie in the region of plane waves.

$*4\pi r^2 \dfrac{d\psi}{dr} =$ rate of total flow across the surface of the sphere whose radius is r

$$= -4\pi A' [\cos 2\pi n t \{\cos (kr + g) + kr \sin (kr + g)\}$$
$$+ \sin 2\pi n t \{\sin (kr + g) - kr \cos (kr + g)\}].$$

* Throughout Helmholtz's paper the mouth of the pipe is supposed to lie in an infinite plane, so that the diverging waves are hemispherical. The calculation of the value of c is thereby simplified. Except for this reason it seems better to consider the diverging waves completely spherical as a nearer approximation to the actual circumstances of organ-pipes, although the sphere could never be quite complete.

If the compression in the neighbourhood of the mouth is neglected, this must be the same as

$$Q \frac{d\psi}{dx=0} = QA \cos 2\pi nt.$$

Accordingly

$$AQ = -4\pi A' \{\cos(kr+g) + kr \sin(kr+g)\},$$
$$0 = \sin(kr+g) - kr(\cos kr+g).$$

These equations express the connexion between the plane and spherical waves. From the second, $\tan(kr+g) = kr$, which shows that g is a small quantity of the order $(kr)^2$. From the first

$$A' = -\frac{AQ}{4\pi},$$

so that

$$\psi_r = -\frac{AQ}{4\pi r} \cos 2\pi nt - \frac{AQk}{4\pi} \sin 2\pi nt,$$

the terms of higher order being omitted.

Now within the space under consideration the air moves according to the same laws as electricity, and so

$$\frac{Q}{c} \frac{d\psi}{dx=0} = -\psi_{x=0} + \psi_r,$$

$$\frac{d\psi}{dx=0} = A \cos 2\pi nt,$$

$$\psi_{x=0} = B \cos 2\pi nt + \beta \sin 2\pi nt.$$

Therefore on substitution and equation of the coefficients of $\sin 2\pi nt$, $\cos 2\pi nt$, we obtain

$$AQ\left(\frac{1}{c} + \frac{1}{4\pi r}\right) = -B, \qquad \beta = -\frac{AQk}{4\pi}.$$

When the mouth is not much contracted c is of the order of the radius of the mouth, and when there is contraction it is smaller still. In all cases therefore the term $1/4\pi r$ is very small compared to $1/c$; and we may put

$$\frac{AQ}{c} = -B, \qquad \beta = -\frac{AQk}{4\pi}, \quad \dots\dots\dots\dots(17)$$

which agree nearly with the results of Helmholtz. In his notation a quantity α is used defined by the equation $-A/Bk = \cot k\alpha$, so that $\cot k\alpha = \tan kl$ by (15), or $k(l+\alpha) = \frac{1}{2}(2m+1)\pi$; α may accordingly be considered as the correction to the length of the tube (measured, however, in our method only on the negative side of the origin), and will be given by $\cot k\alpha = -c/kQ$.

The value of c will be investigated in Part II.

The original theory of open pipes makes the pressure absolutely constant at the mouth, which amounts to neglecting the inertia of the air outside. Thus, if the tube itself were full of air, and the external space of hydrogen, the correction to the length of the pipe might be neglected. The first investigation, in which no escape of energy is admitted, would apply if the pipe and a space round its mouth, large compared to the diameter, but small compared to the wave-length, were occupied by air in an atmosphere otherwise composed of incomparably lighter gas. These remarks are made by way of explanation, but for a complete discussion of the motion as determined by (13) and 17, I must refer to the paper of Helmholtz.

Long Tube in connexion with a Reservoir.

It may sometimes happen that the length of a neck is too large compared to the quarter wave-length to allow the neglect of the compressibility of the air inside. A cylindrical neck may then be treated in the same way as the organ-pipe. The potential of plane waves inside the neck may, by what has been proved, be put into the form

$$\psi = A' \sin k\,(x - \alpha) \cos 2\pi n t,$$

if we neglect the escape of energy, which will not affect the pitch of the resonant note.

$$d\psi/dt = -\,2\pi n A' \sin k\,(x - \alpha) \sin 2\pi n t,$$

$$d\psi/dx = kA' \cos k\,(x - \alpha) \cos 2\pi n t,$$

where α is the correction for the outside end.

The rate of flow out of $S = Q\,d\psi/dx.$

$$\text{Total flow} = Q \int \frac{d\psi}{dx}\,dt = kA'Q \cos kL\,\frac{\sin 2\pi n t}{2\pi n},$$

the reduced length of the tube, including the corrections for both ends, being denoted by L. Thus rarefaction in S

$$= k\,\frac{A'Q \cos kL}{S}\,\frac{\sin 2\pi n t}{2\pi n} = \frac{1}{a^2}\,\frac{d\psi}{dt} = \frac{2\pi n A' \sin kL}{a^2}\,\sin 2\pi n t.$$

This is the condition to be satisfied at the inner end. It gives

$$\tan kL = \frac{a^2}{4\pi^2 n^2}\,\frac{kQ}{S} = \frac{Q}{kS}. \quad\ldots\ldots\ldots\ldots\ldots\ldots\ldots(18)$$

When kL is small,

$$\tan kL = kL + \tfrac{1}{3}\,(kL)^3 = \frac{Q}{kS};$$

so that

$$k^2 = \frac{Q}{LS}\left(1 - \tfrac{1}{3}\frac{LQ}{S}\right),$$

$$n = \frac{a}{2\pi}\sqrt{\frac{Q}{LS}\left(1 - \tfrac{1}{3}\frac{LQ}{S}\right)} = \frac{a}{2\pi}\sqrt{\frac{Q}{L(S + \tfrac{1}{3}LQ)}}. \quad \text{......(19)}$$

In comparing this with (5), it is necessary to introduce the value of c, which is Q/L. (5) will accordingly give the same result as (19) if *one-third* of the contents of the neck be included in S. The first overtone, which is often produced by blowing in preference to the fundamental note, corresponds approximately to the length L of a tube open at both ends, modified to an extent which may be inferred from (18) by the finiteness of S.

The number of vibrations is given by

$$n = \frac{a}{2}\left(\frac{1}{L} + \frac{Q}{\pi^2 S}\right). \quad \text{.....................(20)}$$

[The application of (20) is rather limited, because, in order that the condensation within S may be uniform as has been supposed, the linear dimension of S must be considerably less than the quarter wave-length; while, on the other hand, the method of approximation by which (20) is obtained from (18) requires that S should be large in comparison with QL.

A slight modification of (18) is useful in finding the pitch of pipes which are cylindrical through most of their length, but at the closed end expand into a bulb S of no great capacity. The only change required is to understand by L the length of the pipe down to the place where the enlargement begins, with a correction for the *outer* end. Or if L denote the length of the tube simply, we have

$$\tan k(L + \alpha) = \frac{Q}{kS}, \quad \text{.....................(20 a)}$$

and $\alpha = \tfrac{1}{4}\pi R$ approximately.

If S be very small we may derive from (20 a)

$$n = \frac{a}{4(L + \alpha + S/Q)}. \quad \text{.....................(20 b)}$$

In this form the interpretation is very simple, namely, that at the closed end the shape is of no consequence, and only the volume need be attended to. The air in this part of the pipe acts merely as a spring, its inertia not coming into play. A few measurements of this kind will be given in Part III.

The overtones of resonators which have not long necks are usually very high. Within the body of the reservoir a nodal surface must be formed, and the air on the further side vibrates as if it was contained in a completely closed vessel. We may form an idea of the character of these vibrations from the case of a sphere, which may be easily worked out from the equations given by Professor Stokes in his paper "On the Communication of Motion from a vibrating Sphere to a Gas"*. The most important vibration within a sphere is that which is expressed by the term of the first order in Laplace's series, and consists of a swaying of the air from side to side like that which takes place in a doubly closed pipe. I find that for this vibration

$$\text{radius : wave-length} = \cdot 3313,$$

so that the note is higher than that belonging to a doubly closed (or open) pipe of the length of the diameter of the sphere by about a musical fourth. We might realize this vibration experimentally by attaching to the sphere a neck of such length that it would by itself, when closed at one end, have the same resonant note as the sphere.

Lateral Openings.

In most wind instruments the gradations of pitch are attained by means of lateral openings, which may be closed at pleasure by the fingers or otherwise. The common crude theory supposes that a hole in the side of, say, a flute establishes so complete a communication between the interior and the surrounding atmosphere that a loop or point of no condensation is produced immediately under it. It has long been known that this theory is inadequate, for it stands on the same level as the first approximation to the motion in an open pipe in which the inertia of the air outside the mouth is virtually neglected. Without going at length into this question, I will merely indicate how an improvement in the treatment of it may be made.

Let ψ_1, ψ_2 denote the velocity-potentials of the system of plane waves on the two sides of the aperture, which we may suppose to be situated at the point $x = 0$. Then with our previous notation the conditions evidently are that when $x = 0$,

$$\psi_1 = \psi_2, \qquad \frac{Q}{c}\left(\frac{d\psi_1}{dx} - \frac{d\psi_2}{dx}\right) + \psi = 0, \quad\ldots\ldots\ldots\ldots(20\,c)$$

the escape of energy from the tube being neglected. These equations determine the connexion between the two systems of waves in any case that may arise, and the working out is simple. The results are of no particular interest, unless it be for a comparison with experimental measurements, which, so far as I am aware, have not hitherto been made.]

---

* Professor Stokes informs me that he had himself done this at the request of the Astronomer Royal. [1899. See *Theory of Sound*, §§ 330, 331.]

PART II.

In order to complete the theory of resonators, it is necessary to determine the value of c, which occurs in all the results of Part I., for different forms of mouths. This we now proceed to do. Frequent use will be made of a principle which might be called that of minimum *vis viva*, and which it may be well to state clearly at the outset.

Imagine a portion of incompressible fluid at rest within a closed surface to be suddenly set in motion by an arbitrary normal velocity impressed on the surface, then the actual motion assumed by the fluid will have less *vis viva* than any other motion consistent with continuity and with the boundary conditions*.

If u, v, w be the component velocities, and ρ the density at any point,

$$vis\ viva = \tfrac{1}{2} \iiint \rho\,(u^2 + v^2 + w^2)\,dx\,dy\,dz,$$

the integration extending over the volume considered. The minimum *vis viva* corresponding to prescribed boundary conditions depends of course on ρ; but if in any specified case we conceive the value of ρ in some places diminished and nowhere increased, we may assert that the minimum *vis viva* is *less* than before; for there will be a decrease if u, v, w remain unaltered, and therefore, *à fortiori*, when they have their actual values as determined by the minimum property. Conversely, an increase in ρ will necessarily raise the value of the minimum *vis viva*. The introduction of a rigid obstacle into a stream will always cause an increase of *vis viva*; for the new motion is one that might have existed before consistently with continuity, the fluid displaced by the obstacle remaining at rest. Any kind of obstruction in the air-passages of a musical instrument will therefore be accompanied by a fall of the note in the musical scale.

Long Tubes.

The simplest case that can be considered consists of an opening in the form of a cylindrical tube, so long in proportion to its diameter that the corrections for the ends may be neglected. If the length be L and area of section σ, the electrical resistance is L/σ, and

$$c = \frac{\sigma}{L}. \qquad \dots\dots\dots\dots\dots\dots\dots\dots\dots(21)$$

For a circular cylinder of radius R

$$c = \frac{\pi R^2}{L}. \qquad \dots\dots\dots\dots\dots\dots\dots\dots(22)$$

* Thomson and Tait's *Natural Philosophy*, § 317.

Simple Apertures.

The next in order of simplicity is probably the case treated by Helmholtz, where the opening consists of a simple hole in the side of the reservoir, considered as indefinitely thin and approximately plane in the neighbourhood of the opening. The motion of the fluid in the plane of the opening is by the symmetry normal, and therefore the velocity-potential is constant over the opening itself. Over the remainder of the plane in which the opening lies the normal velocity is of course zero, so that ϕ may be regarded as the potential of matter distributed over the opening only. If the there constant value of the potential be called ϕ_1, the electrical resistance for *one side only* is

$$\phi_1 \div \iint \frac{d\phi}{dn} \, d\sigma,$$

the integration going over the area of the opening.

Now

$$\iint \frac{d\phi}{dn} \, d\sigma = 2\pi \times \text{ the whole quantity of matter;}$$

so that if we call M the quantity necessary to produce the unit potential, the resistance for one side $= 1/2\pi M$.

Accordingly

$$c = \pi M. \quad \ldots\ldots\ldots\ldots\ldots\ldots\ldots\ldots\ldots(23)$$

In electrical language M is the *capacity* of a conducting lamina of the shape of the hole when situated in an open space.

For a circular hole $M = 2R/\pi$, and therefore

$$c = 2R. \quad \ldots\ldots\ldots\ldots\ldots\ldots\ldots\ldots\ldots(24)$$

When the hole is an ellipse of eccentricity e and semimajor axis R,

$$c = \frac{\pi R}{F(e)}, \quad \ldots\ldots\ldots\ldots\ldots\ldots\ldots\ldots(25)$$

where F is the symbol of the complete elliptic function of the first order. Results equivalent to (23), (24), and (25) are given by Helmholtz.

When the eccentricity is but small, the value of c depends sensibly on the area (σ) of the orifice only. As far as the square of e,

$$F(e) = \frac{\pi}{2}(1 + \tfrac{1}{4}e^2),$$

$$\sigma = \pi R^2 \sqrt{1 - e^2} = \pi R^2 (1 - \tfrac{1}{2}e^2), \qquad R = \sqrt{\frac{\sigma}{\pi}}(1 + \tfrac{1}{4}e^2);$$

that

$$c = \pi \sqrt{\frac{\sigma}{\pi}} \div \frac{\pi}{2} = 2 \sqrt{\frac{\sigma}{\pi}} , \quad \dots\dots\dots\dots\dots (26)$$

the fourth power of e being neglected—a formula which may be applied without sensible error to any orifice of an approximately circular form. In fact for a given area the circle is the figure which gives a minimum value to c, and in the neighbourhood of the minimum the variation is slow.

Next, consider the case of two circular orifices. If sufficiently far apart they act independently of each other, and the value of c for the pair is the simple sum of the separate values, as may be seen either from the law of multiple arcs by considering c as the electric *conductivity* between the outside and inside of the reservoir, or from the interpretation of M in (23). The first method applies to any kind of openings with or without necks. As the two circles (which for precision of statement we may suppose equal) approach one another, the value of c diminishes steadily until they touch. The change in the character of the motion may be best followed by considering the plane of symmetry which bisects at right angles the line joining the two centres, and which may be regarded as a rigid plane precluding normal motion. Fixing our attention on half the motion only, we recognize the plane as an obstacle continually advancing, and at each step more and more obstructing the passage of fluid through the circular opening. After the circles come into contact this process cannot be carried further; but we may infer that, as they amalgamate and shape themselves into a single circle (the total area remaining all the while constant), the value of c still continues to diminish till it approaches its minimum value, which is less than at the commencement in the ratio of $\sqrt{2} : 2$ or $1 : \sqrt{2}$. There are very few forms of opening indeed for which the exact calculation of M or c can be effected. We must for the present be content with the formula (26) as applying to nearly circular openings, and with the knowledge that the more elongated or broken up the opening, the greater is c compared to σ. In the case of similar orifices or systems of orifices c varies as the linear dimension.

Cylindrical Necks.

Most resonators used in practice have necks of greater or less length, and even where there is nothing that would be called a neck, the thickness of the side of the reservoir could not always be neglected. For simplicity we shall take the case of circular cylinders whose inner ends lie on an approximately plane part of the side of the vessel, and whose outer ends are also supposed to lie in an infinite plane, or at least a plane whose dimensions are considerable compared to the diameter of the cylinder. Even under this form the problem does not seem capable of exact solution; but we shall be able to fix two slightly differing quantities between which the

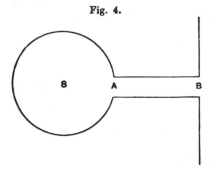

Fig. 4.

true value of c must lie, and which determine it with an accuracy more than sufficient for acoustical purposes. The object is to find the *vis viva* in terms of the rate of flow. Now, according to the principle stated at the beginning of Part II., we shall obtain too small a *vis viva* if at the ends A and B of the tube we imagine infinitely thin laminæ of fluid of infinitely small density. We may be led still more distinctly perhaps to the same result by supposing, in the electrical analogue, thin disks of perfectly conducting matter at the ends of the tube, whereby the effective resistance must plainly be lessened. The action of the disks is to produce uniform potential over the ends, and the solution of the modified problem is obvious. Outside the tube the question is the same as for a simple circular hole in an infinite plane, and inside the tube the same as if the tube were indefinitely long.

Accordingly

$$\text{resistance} = \frac{L}{\pi R^2} + \frac{1}{2R} = \frac{1}{\pi R^2}\left(L + \frac{\pi}{2}R\right). \quad\ldots\ldots\ldots(27)$$

The correction to the length is therefore $\frac{1}{2}\pi R$, that is, $\frac{1}{4}\pi R$ for each end,

and

$$c = \frac{\pi R^2}{L + \frac{1}{2}\pi R}. \quad\ldots\ldots\ldots\ldots\ldots\ldots\ldots\ldots\ldots(28)$$

Helmholtz, in considering the case of an organ-pipe, arrives at a similar conclusion,—that the correction to the length (α) is approximately $\frac{1}{4}\pi R$. His method is very different from the above, and much less simple. He begins by investigating certain forms of mouths for which the exact solution is possible, and then, by assigning suitable values to arbitrary constants, identifies one of them with a true cylinder, the agreement being shown to be everywhere very close. Since the curve substituted for the generating line of the cylinder lies entirely outside it, Helmholtz infers that the correction to the length thus obtained is too small.

If, at the ends of the tube, instead of layers of matter of no density, we imagine rigid pistons of no sensible thickness, we shall obtain a motion whose *vis viva* is necessarily *greater* than that of the real motion; for the motion with the pistons might take place without them consistently with continuity. Inside the tube the character of the motion is the same as before, but for the outside we require the solution of a fresh problem :— To determine the motion of an infinite fluid bounded by an infinite plane, the normal velocity over a circular area of the plane being a given constant, and over the rest of the plane zero. The potential may still be regarded as due to matter confined to the circle, but is no longer constant over its area; but the density of matter at any point, being proportional to $d\phi/dn$ or to the normal velocity, is constant.

The *vis viva* of the motion $= \frac{1}{2} \iint \phi \frac{d\phi}{dn} d\sigma = \frac{1}{2} \frac{d\phi}{dn} \iint \phi d\sigma$, the integration going over the area of the circle.

The rate of total flow through the plane $= \iint \frac{d\phi}{dn} d\sigma = \pi R^2 \frac{d\phi}{dn}$; so that

$$\frac{2 \text{ vis viva}}{(\text{rate of flow})^2} = \frac{\iint \phi d\sigma}{\pi^2 R^4 d\phi/dn}. \quad \dots\dots\dots\dots(29)$$

We proceed to investigate the value of $\iint \phi d\sigma$, which is the *potential on itself* of a circular disk of unit density.

[*Editor's note:* The remainder of this paper is devoted to detailed calculations of the resonances of various types of structures.]

Reprinted from "Scientific Papers of Lord Rayleigh" (Dover Publications, Inc.), Vol. I, 277–279 (1961)

35

40.

OUR PERCEPTION OF THE DIRECTION OF A SOURCE OF SOUND.

[*Nature*, XIV. pp. 32—33, 1876.]

THE practical facility with which we recognize the situation of a sounding body has always been rather a theoretical difficulty. In the case of sight a special optical apparatus is provided whose function it is to modify the uniform excitation of the retina, which a luminous point, wherever situated, would otherwise produce. The mode of action of the crystalline lens of the eye is well understood, and the use of a lens is precisely the device that would at once occur to the mind of an optician ignorant of physiology. The bundle of rays, which would otherwise distribute themselves over the entire retina, and so give no indication of their origin, are made to converge upon a single point, whose excitation is to us the sign of an external object in a certain definite direction. If the luminous object is moved, the fact is at once recognized by the change in the point of excitation.

There is nothing in the ear corresponding to the crystalline lens of the eye, and this not accidentally, so to speak, but by the very nature of the case. The efficient action of a lens depends upon its diameter being at least many times greater than the wave-length of light, and for the purposes of sight there is no difficulty in satisfying this requirement. The wave-length of the rays by which we see is not much more than a ten-thousandth part of the diameter of the pupil of the eye. But when we pass to the case of sound and of the ear, the relative magnitudes of the corresponding quantities are altogether different. The waves of sound issuing from a man's mouth are about eight feet long, whereas the diameter of the passage of the ear is quite small, and could not well have been made a large multiple of eight feet. It is evident therefore that it is useless to look for anything corresponding to the crystalline lens of the eye, and that our power of telling the origin of a sound must be explained in some different way.

It has long been conjectured that the explanation turns upon the combined use of both ears; though but little seems to have been done hitherto in the way of bringing this view to the test. The observations and calculations now brought forward are very incomplete, but may perhaps help to clear the ground, and will have served their purpose if they induce others to pursue the subject.

The first experiments were made with the view of finding out with what degree of accuracy the direction of sound could be determined, and for this it was necessary of course that the observer should have no other material for his judgment than that contemplated.

The observer, stationed with his eyes closed in the middle of a lawn on a still evening, was asked to point with the hand in the direction of voices addressed to him by five or six assistants, who continually shifted their position. It was necessary to have several assistants, since it was found that otherwise their steps could be easily followed. The uniform result was that the direction of a human voice used in anything like a natural manner could be told with certainty from a single word, or even vowel, to within a few degrees.

But with other sounds the result was different. If the source was on the right or the left of the observer, its position could be told approximately, but it was uncertain whether, for example, a low whistle was in front or behind. This result led us to try a simple sound, such as that given by a fork mounted on a resonance-box. It was soon found that whatever might be the case with a truly simple sound, the observer never failed to detect the situation of the fork by the noises accompanying its excitation, whether this was done by striking or by a violin bow. It was therefore necessary to arrange the experiment differently. Two assistants at equal distances and in opposite directions were provided with similar forks and resonators. At a signal given by a fourth, both forks were struck, but only one was held over its resonator, and the observer was asked to say, without moving his head, which he heard. When the observer was so turned that one fork was immediately in front and the other immediately behind, it was impossible for him to tell which fork was sounding, and if asked to say one or the other, felt that he was only guessing. But on turning a quarter round, so as to have one fork on his right and the other on his left, he could tell without fail, and with full confidence in being correct.

The possibility of distinguishing a voice in front from a voice behind would thus appear to depend on the compound character of the sound in a way that it is not easy to understand, and for which the second ear would be of no advantage. But even in the case of a lateral sound the matter is not free from difficulty, for the difference of intensity with which a lateral sound is perceived by the two ears is not great. The experiment may easily be tried roughly by stopping one ear with the hand, and turning round back-wards and forwards while listening to a sound held steadily.   Calculation shows, moreover, that the human head, considered as an obstacle to the waves of sound, is scarcely big enough in relation to the wave-length to give a sensible shadow.   To throw light on this subject I have calculated the intensity of sound due to a distant source at the various points on the surface of a fixed spherical obstacle.   The result depends on the ratio (α) between the circumference of the sphere and the length of the wave.   If we call the point on the spherical surface nearest to the source the anterior pole, and the opposite point (where the shadow might be expected to be most intense) the posterior pole, the results on three suppositions as to the relative magnitudes of the sphere and wave-length are given in the following table :—

	$\alpha = 2$	$\alpha = 1$	$\alpha = \frac{1}{2}$
Anterior Pole	·690	·503	·294
Posterior Pole	·318	·285	·260
Equator	·356	·237	·232

When, for example, the circumference of the sphere is but half the wave-length, the intensity at the posterior pole is only about a tenth part less than at the anterior pole, while the intensity is least of all in a lateral direction. When α is less than $\frac{1}{2}$, the difference of the intensities at the two poles is still less important, amounting to about one per cent., when $\alpha = \frac{1}{4}$.

The value of α depends on the wave-length, which may vary within pretty wide limits, and it might be expected that the facility of distinguishing a lateral sound would diminish when the sound is grave.   Experiments were accordingly tried with forks of a frequency of 128, but no greater difficulty was experienced than with forks of a frequency of 256, except such as might be attributed to the inferior loudness of the former.   According to calculation the difference of intensity would here be too small to account for the power of discrimination.

Reprinted from "Scientific Papers of Lord Rayleigh" (Dover Publications, Inc.), Vol. I, 305–309 (1961)

36

44.

ON THE APPLICATION OF THE PRINCIPLE OF RECIPROCITY TO ACOUSTICS.

[Proceedings of the Royal Society, XXV. pp. 118—122, 1876.]

IN a memoir published some years ago by Helmholtz (*Crelle,* Bd. LVII.) it was proved that if a uniform frictionless gaseous medium be thrown into vibration by a simple source of sound of given period and intensity, the variation of pressure is the same at any point *B* when the source of sound is at *A* as it would have been at *A* had the source of sound been situated at *B,* and that this law is not interfered with by the presence of any number of fixed solid obstacles on which the sound may impinge.

A simple source of sound is a point at which the condition of continuity of the fluid is broken by an alternate introduction and abstraction of fluid, given in amount and periodic according to the harmonic law.

The reciprocal property is capable of generalization so as to apply to all acoustical systems whatever capable of vibrating about a configuration of equilibrium, as I proved in the *Proceedings of the Mathematical Society* for June 1873 [Art. XXI.], and is not lost even when the systems are subject to damping, provided that the frictional forces vary as the first power of the velocity, as must always be the case when the motion is small enough. Thus Helmholtz's theorem may be extended to the case when the medium is not uniform, and when the obstacles are of such a character that they share the vibration.

But although the principle of reciprocity appears to be firmly grounded on the theoretical side, instances are not uncommon in which a sound generated in the open air at a point *A* is heard at a distant point *B,* when an equal or even more powerful sound at *B* fails to make itself heard at *A* ; and some phenomena of this kind are strongly insisted upon by Prof. Henry in opposition to Prof. Tyndall's views as to the importance

of "acoustic clouds" in relation to the audibility of fog-signals. These observations were not, indeed, made with the simple sonorous sources of theory; but there is no reason to suppose that the result would have been different if simple sources could have been used.

In experiments having for their object the comparison of sounds heard under different circumstances there is one necessary precaution to which it may not be superfluous to allude, depending on the fact that the audibility of a particular sound depends not only upon the strength of that sound, but also upon the strength of other sounds which may be heard along with it. For example, a lady seated in a closed carriage and carrying on a conversation through an open window in a crowded thoroughfare will hear what is said to her far more easily than she can make herself heard in return; but this is no failure in the law of reciprocity.

The explanation of his observations given by Henry depends upon the peculiar action of wind, first explained by Prof. Stokes. According to this view a sound is ordinarily heard better with the wind than against it, in consequence of a curvature of the rays. With the wind a ray will generally be bent downwards, since the velocity of the air is generally greater overhead than at the surface, and therefore the upper part of the wave-front tends to gain on the lower. The ray which ultimately reaches the observer is one which started in some degree upwards from the source, and has the advantage of being out of the way of obstacles for the greater part of its course. Against the wind, on the other hand, the curvature of the rays is upwards, so that a would-be observer at a considerable distance is in danger of being left in a sound-shadow.

It is very important to remark that this effect depends, not upon the mere existence of a wind, but upon the velocity of the wind being greater overhead than below. A uniform translation of the entire atmosphere would be almost without effect. In particular cases it may happen that the velocity of the wind diminishes with height, and then sound is best transmitted *against* the wind. Prof. Henry shows that several anomalous phenomena relating to the audibility of signals may be explained by various suppositions as to the velocity of the wind at different heights. When the distances concerned are great, comparatively small curvatures of the ray may produce considerable results.

There is a further possible consequence of the action of wind (or variable temperature), which, so far as I know, has not hitherto been remarked. By making the velocity a suitable function of height it would be possible to secure an actual convergence of rays in a vertical plane upon a particular station. The atmosphere would then act like the lens of a lighthouse, and the intensity of sound might be altogether abnormal. This may perhaps be the explanation of the extraordinary distances at which guns have sometimes been heard.

The difference in the propagation of sound against and with the wind is no exception to the general law referred to at the beginning of this communication, for that law applies only to the vibrations of a system about a configuration of equilibrium. A motion of the medium is thus excluded. But the bending of the sound-ray due to a variable temperature, to which attention has been drawn by Prof. Reynolds, does not interfere with the application of the law.

An experiment has, however, been brought forward by Prof. Tyndall, in which there is an apparent failure of reciprocity not referable to any motion of the medium*. The source of sound is a very high-pitched reed mounted in a short tube and blown from a small bellows with which it is connected by rubber tubing. The variation of pressure at the second point is made apparent by means of the sensitive flame, which has been used by Prof. Tyndall with so much success on other occasions. Although the flame itself, when unexcited, is 18 to 24 inches high, it was proved by a subsidiary experiment that the root of the flame, where it issues from the burner, is the seat of sensitiveness. With this arrangement the effect of a cardboard or glass screen interposed between the reed and the flame was found to be different, according as the screen was close to the flame or close to the reed. In the former case the flame indicated the action of sound, but in the latter remained uninfluenced. Since the motion of the screen is plainly equivalent to an interchange of the reed and flame, there is to all appearance a failure in the law of reciprocity.

At first sight this experiment is difficult to reconcile with theoretical conclusions. It is true that the conditions under which reciprocity is to be expected are not very perfectly realized, since the flame ought not to be moved from one position to the other. Although the seat of sensitiveness may be limited to the root of the flame, the tall column of highly heated gas might not be without effect; and in fact it appeared to me possible that the response of the flame, when close to the screen, might be due to the conduction of sound downwards along it. Not feeling satisfied, however, with this explanation, I determined to repeat the experiment, and wrote to Prof. Tyndall, asking to be allowed to see the apparatus. In reply he very kindly proposed to arrange a repetition of the experiment at the Royal Institution for my benefit, an offer which I gladly accepted.

The effect itself was perfectly distinct, and, as it soon appeared, was not to be explained in the manner just suggested, since the response of the flame when close to the screen continued, even when the upper part

* *Proceedings of the Royal Institution*, January 1875; also Prof. Tyndall's work on *Sound*, 3rd edition.

of the heated column was protected from the direct action of the source by additional screens interposed. I was more than ever puzzled until Mr Cottrell showed me another experiment in which, I believe, the key of the difficulty is to be found.

When the axis of the tube containing the reed is directed towards the flame, situated at a moderate distance, there is a distinct and immediate response; but when the axis is turned away from the flame through a comparatively small angle, the effect ceases, although the distance is the same as before, and there are no obstacles interposed. If now a cardboard screen is held in the prolongation of the axis of the reed, and at such an angle as to reflect the vibrations in the direction of the flame, the effect is again produced with the same apparent force as at first.

These results prove conclusively that the reed does not behave as the simple source of theory, even approximately. When the screen is close (about 2 inches distant) the more powerful vibrations issuing along the axis of the instrument impinge directly upon the screen, are reflected back, and take no further part in the experiment. The only vibrations which have a chance of reaching the flame, after diffraction round the screen, are the comparatively feeble ones which issue nearly at right angles with the axis. On the other hand, when the screen is close to the flame, the efficient vibrations are those which issue at a small angle with the axis, and are therefore much more powerful. Under these circumstances it is not surprising that the flame is affected in the latter case and not in the former.

The concentration of sound in the direction of the axis is greater than would have been anticipated, and is to be explained by the very short wave-length corresponding to the pitch of the reed. If, as is not improbable, the overtones of the note given by the reed are the most efficient part of the sound, the wave-length will be still shorter and the concentration more easy to understand*.

The reciprocal theorem in its generalized form is not restricted to simple sources, from which (in the absence of obstacles) sound would issue alike in all directions; and the statement for *double sources* will throw light on the subject of this note. A double source may be thus defined:—Conceive two equal and opposite simple sources, situated at a short distance apart, to be acting simultaneously. By calling the two sources opposite, it is meant that they are to be at any moment in opposite phases. At a moderate distance the effects of the two sources are antagonistic and may be made to neutralize one another to any extent by diminishing the distance between the sources. If, however, at the same time that we diminish the interval, we augment the

---

* July 13. I have lately observed that the flame in question is extremely sensitive to one of Mr F. Galton's whistles, which gives notes near the limits of ordinary hearing.

intensity of the single sources, the effect may be kept constant. Pushing this idea to its limit, when the intensity becomes infinite and the interval vanishes, we arrive at the conception of a double source having an axis of symmetry coincident with the line joining the single sources of which it is composed. In an open space the effect of a double source is the same as that communicated to the air by the vibration of a solid sphere whose centre is situated at the double point and whose line of vibration coincides with the axis, and the intensity of sound in directions inclined to the axis varies as the square of the cosine of the obliquity.

The statement of the reciprocal theorem with respect to double sources is then as follows:—If there be equal double sources at two points A and B, having axes AP, BQ respectively, then the *velocity* of the medium at B resolved in the direction BQ due to the source at A is the same as the *velocity* at A resolved in the direction AP due to the source at B. If the waves observed at A and B are sensibly plane, and if the axes AP, BQ are equally inclined to the waves received, we may, in the above statement, replace " velocities " by " pressures," but not otherwise.

Suppose, now, that equal double sources face each other, so that the common axis is AB, and let us examine the effect of interposing a screen near to A. By the reciprocal theorem, whether there be a screen or not, the velocity at A in direction AB due to B is equal to the velocity at B in direction AB due to A. The waves received at B are approximately plane and perpendicular to AB, so that the relation between the velocity and pressure at B is that proper to a plane wave; but it is otherwise in the case of the sound received at A. Accordingly the reciprocal theorem does not lead us to expect an equality between the pressures at A and B, on which quantities the behaviour of the sensitive flames depends*. On the contrary, it would appear that the pressure at A corresponding to the given velocity along AB should be much greater than in the case of a plane wave, and then the relative advantage of the position A would be explained.

It will be seen that if the preceding arguments are correct, Prof. Tyndall's experiment does not bear out the conclusions that he has based upon it with respect to the observations of the French Commission at Villejuif and Montlhéry. No acoustic clouds could explain the failure of reciprocity then observed; and the more probable hypothesis that the effect was due to wind is not inconsistent with the observation that the air (at the surface) was moving in the direction against which the sound was best heard.

Further experiments on this subject are very desirable.

* [1899. See however *Phil. Mag.* vol. VII. p. 153, 1879, where it appears that the excitation of a flame is due, not to a variable *pressure*, but to transverse *motion* across the nozzle.]

37

191.

ON THE PHYSICS OF MEDIA THAT ARE COMPOSED OF FREE AND PERFECTLY ELASTIC MOLECULES IN A STATE OF MOTION*.

[*Phil. Trans.* 183 A, pp. 1—5, 1892.]

THE publication of this paper after nearly half a century demands a word of explanation; and the opportunity may be taken to point out in what respects the received theory of gases had been anticipated by Waterston, and to offer some suggestions as to the origin of certain errors and deficiencies in his views.

So far as I am aware, the paper, though always accessible in the Archives of the Royal Society, has remained absolutely unnoticed. Most unfortunately the abstract printed at the time (*Roy. Soc. Proc.* 1846, Vol. v. p. 604; here reprinted as Appendix I.) gave no adequate idea of the scope of the memoir, and still less of the nature of the results arrived at. The deficiency was in some degree supplied by a short account in the *Report of the British Association* for 1851 (here reprinted as Appendix II.), where is distinctly stated the law, which was afterwards to become so famous, of the equality of the kinetic energies of different molecules at the same temperature.

My own attention was attracted in the first instance to Waterston's work upon the connection between molecular forces and the latent heat of evaporation, and thence to a paper in the *Philosophical Magazine* for 1858, "On the Theory of Sound." He there alludes to the theory of gases under consideration as having been started by Herapath in 1821, and he proceeds:—

"Mr Herapath unfortunately assumed heat or temperature to be represented by the simple ratio of the velocity instead of the square of the velocity—being in this apparently led astray by the definition of motion

---

* [From an Introduction to a Memoir, entitled as above, by J. J. Waterston, received Dec. 11, 1845, read March 5, 1846.]

generally received—and thus was baffled in his attempts to reconcile his theory with observation. If we make this change in Mr Herapath's definition of heat or temperature, viz., that it is proportional to the *vis viva*, or square velocity of the moving particle, not to the momentum, or simple ratio of the velocity, we can without much difficulty deduce, not only the primary laws of elastic fluids, but also the other physical properties of gases enumerated above in the third objection to Newton's hypothesis. In the Archives of the Royal Society for 1845—1846, there is paper 'On the Physics of Media that consists of perfectly Elastic Molecules in a State of Motion,' which contains the synthetical reasoning upon which the demonstration of these matters rests. The velocity of sound is therein deduced to be equal to the velocity acquired in falling through three-fourths of a uniform atmosphere. This theory does not take account of the size of the molecules. It assumes that no time is lost at the impact, and that if the impacts produce rotatory motion, the *vis viva* thus invested bears a constant ratio to the rectilineal *vis viva*, so as not to require separate consideration. It also does not take account of the probable internal motion of composite molecules; yet the results so closely accord with observation in every part of the subject as to leave no doubt that Mr Herapath's idea of the physical constitution of gases approximates closely to the truth. M. Krönig appears to have entered upon the subject in an independent manner, and arrives at the same result; M. Clausius, too, as we learn from his paper 'On the Nature of the Motion we call Heat' (*Phil. Mag.* Vol. XIV. 1857, p. 108)."

Impressed with the above passage and with the general ingenuity and soundness of Waterston's views, I took the first opportunity of consulting the Archives, and saw at once that the memoir justified the large claims made for it, and that it marks an immense advance in the direction of the now generally received theory. The omission to publish it at the time was a misfortune, which probably retarded the development of the subject by ten or fifteen years. It is singular that Waterston appears to have advanced no claim for subsequent publication, whether in the Transactions of the Society, or through some other channel. At any time since 1860 reference would naturally have been made to Maxwell, and it cannot be doubted that he would have at once recommended that everything possible should be done to atone for the original failure of appreciation.

It is difficult to put oneself in imagination into the position of the reader of 1845, and one can understand that the substance of the memoir should have appeared speculative and that its mathematical style should have failed to attract. But it is startling to find a referee expressing the opinion that "the paper is nothing but nonsense, unfit even for reading before the Society." Another remarks "that the whole investigation is confessedly founded on a principle entirely hypothetical, from which it is the object to deduce a mathematical representation of the phenomena of elastic media.

It exhibits much skill and many remarkable accordances with the general facts, as well as numerical values furnished by observation.......The original principle itself involves an assumption which seems to me very difficult to admit, and by no means a satisfactory basis for a mathematical theory, viz., that the elasticity of a medium is to be measured by supposing its molecules in vertical motion, and making a succession of impacts against an elastic gravitating plane." These remarks are not here quoted with the idea of reflecting upon the judgment of the referee, who was one of the best qualified authorities of the day, and evidently devoted to a most difficult task his careful attention; but rather with the view of throwing light upon the attitude then assumed by men of science in regard to this question, and in order to point a moral. The history of this paper suggests that highly speculative investigations, especially by an unknown author, are best brought before the world through some other channel than a scientific society, which naturally hesitates to admit into its printed records matter of uncertain value. Perhaps one may go further and say that a young author who believes himself capable of great things would usually do well to secure the favourable recognition of the scientific world by work whose scope is limited, and whose value is easily judged, before embarking upon higher flights.

One circumstance which may have told unfavourably upon the reception of Waterston's paper is that he mentions no predecessors. Had he put forward his investigation as a development of the theory of D. Bernoulli, a referee might have hesitated to call it nonsense. It is probable, however, that Waterston was unacquainted with Bernoulli's work, and doubtful whether at that time he knew that Herapath had to some extent foreshadowed similar views.

At the present time the interest of Waterston's paper can, of course, be little more than historical. What strikes one most is the marvellous courage with which he attacked questions, some of which even now present serious difficulties. To say that he was not always successful is only to deny his claim to rank among the very foremost theorists of all ages. The character of the advance to be dated from this paper will be at once understood when it is realised that Waterston was the first to introduce into the theory the conception that heat and temperature are to be measured by *vis viva*. This enabled him at a stroke to complete Bernoulli's explanation of pressure by showing the accordance of the hypothetical medium with the law of Dalton and Gay-Lussac. In the second section the great feature is the statement (VII.), that "in mixed media the mean square molecular velocity is inversely proportional to the specific weight of the molecules." The proof which Waterston gave is doubtless not satisfactory; but the same may be said of that advanced by Maxwell fifteen years later. The law of Avogadro follows at once, as well as that of Graham relative to diffusion. Since the law of equal energies was actually published in 1851, there can be no hesitation, I think, in attaching Waterston's name to it. The attainment of correct results in the third section, dealing with adiabatic expansion, was only prevented by a slip of calculation.

In a few important respects Waterston stopped short. There is no indication, so far as I can see, that he recognised any other form of motion, or energy, than the translatory motion, though this is sometimes spoken of as vibratory. In this matter the priority in a wider view rests with Clausius. According to Waterston the ratio of specific heats should be (as for mercury vapour) 1·67 in all cases. Again, although he was well aware that the molecular velocity cannot be constant, there is no anticipation of the law of distribution of velocities established by Maxwell.

A large part of the paper deals with chemistry, and shows that his views upon that subject also were much in advance of those generally held at the time..........

With the exception of some corrections relating merely to stops and spelling the paper is here reproduced exactly as it stands in the author's manuscript.—Dec. 1891.

[1901. It may be added that Waterston's memoir contains the first calculation of the molecular velocity, and further that it points out the relation of this velocity to the velocity of sound. The earliest actual *publication* of such a calculation is that of Joule, who gives for the velocity of hydrogen molecules at 0° C. 6055 feet per second (*Manchester Memoirs*, Vol. IX. p. 107, Oct. 1848; *Phil. Mag.* Ser. 4, Vol. XIV. p. 211; Joule's *Scientific Papers*, Vol. I. p. 295), thus anticipating by eight or nine years the first paper of Clausius (*Pogg. Ann.* 1857), to whom priority is often erroneously ascribed.]

Reprinted from "Scientific Papers of Lord Rayleigh" (Dover Publications, Inc.), Vol. IV, 376–381 (1961)

38

244.

ON THE COOLING OF AIR BY RADIATION AND CONDUCTION, AND ON THE PROPAGATION OF SOUND.

[*Philosophical Magazine*, XLVII. pp. 308—314, 1899.]

ACCORDING to Laplace's theory of the propagation of Sound the expansions (and contractions) of the air are supposed to take place without transfer of heat. Many years ago Sir G. Stokes* discussed the question of the influence of radiation from the heated air upon the propagation of sound. He showed that such small radiating power as is admissible would tell rather upon the intensity than upon the velocity. If x be measured in the direction of propagation, the factor expressing the diminution of amplitude is e^{-mx}, where

$$m = \frac{\gamma - 1}{\gamma} \frac{q}{2a}. \qquad \dots\dots\dots\dots\dots\dots\dots\dots\dots\dots\dots(1)$$

In (1) γ represents the ratio of specific heats (1·41), a is the velocity of sound, and q is such that e^{-qt} represents the law of cooling by radiation of a small mass of air maintained at constant volume. If τ denote the time required to traverse the distance x, $\tau = x/a$, and (1) may be taken to assert that the amplitude falls to any fraction, *e.g.* one-half, of its original value in 7 times the interval of time required by a mass of air to cool to the same fraction of its original excess of temperature. "There appear to be no data by which the latter interval can be fixed with any approach to precision; but if we take it at one minute, the conclusion is that sound would be propagated for (seven) minutes, or travel over about (80) miles, without very serious loss from this cause†." We shall presently return to the consideration of the probable value of q.

Besides radiation there is also to be considered the influence of conductivity in causing transfer of heat, and further there are the effects of viscosity.

---

* *Phil. Mag.* [4] I. p. 305, 1851; *Theory of Sound*, § 247.
† *Proc. Roy. Inst.* April 9, 1897. [Vol. IV. p. 258.]

The problems thus suggested have been solved by Stokes and Kirchhoff*. If the law of propagation be

$$u = e^{-m'x} \cos (nt - x/a), \quad \dots\dots\dots\dots\dots(2)$$

then
$$m' = \frac{n^2}{2a^3} \left\{ \tfrac{4}{3}\mu' + \nu \frac{\gamma - 1}{\gamma} \right\}, \quad \dots\dots\dots\dots\dots(3)$$

in which the frequency of vibration is $n/2\pi$, μ' is the kinematic viscosity, and ν the thermometric conductivity. In C.G.S. measure we may take $\mu' = \cdot 14$, $\nu = \cdot 26$, so that

$$\tfrac{4}{3}\mu' + \nu \frac{\gamma - 1}{\gamma} = \cdot 25.$$

To take a particular case, let the frequency be 256; then since $a = 33200$, we find for the time of propagation during which the amplitude diminishes in the ratio of $e : 1$,

$$(m'a)^{-1} = 3560 \text{ seconds.}$$

Accordingly it is only very high sounds whose propagation can be appreciably influenced by viscosity and conductivity.

If we combine the effects of radiation with those of viscosity and conduction, we have as the factor of attenuation

$$e^{-(m+m')x},$$

where
$$m + m' = \cdot 14 (q/a) + \cdot 12 (n^2/a^3). \quad \dots\dots\dots\dots\dots(4)$$

In actual observations of sound we must expect the intensity to fall off in accordance with the law of inverse squares of distances. A very little experience of moderately distant sounds shows that in fact the intensity is in a high degree uncertain. These discrepancies are attributable to atmospheric refraction and reflexion, and they are sometimes very surprising. But the question remains whether in a uniform condition of the atmosphere the attenuation is sensibly more rapid than can be accounted for by the law of inverse squares. Some interesting experiments towards the elucidation of this matter have been published by Mr Wilmer Duff†, who compared the distances of audibility of sounds proceeding respectively from two and from eight similar whistles. On an average the eight whistles were audible only about one-fourth further than a pair of whistles; whereas, if the sphericity of the waves had been the only cause of attenuation, the distances would have been as 2 to 1. Mr Duff considers that in the circumstances of his experiments there was little opportunity for atmospheric irregularities, and he attributes the greater part of the falling off to radiation. Calculating from (1) he deduces a radiating power such that a mass of air at any given excess of temperature above its surroundings will (if its volume remain constant) fall by radiation to one-half of that excess in about one-twelfth of a second.

* *Pogg. Ann.* Vol. cxxxiv. p. 177, 1868 ; *Theory of Sound*, 2nd ed. § 348.
† *Phys. Review*, Vol. vi. p. 129, 1898.

In this paper I propose to discuss further the question of the radiating power of air, and I shall contend that on various grounds it is necessary to restrict it to a value hundreds of times smaller than that above mentioned. On this view Mr Duff's results remain unexplained. For myself I should still be disposed to attribute them to atmospheric refraction. If further experiment should establish a rate of attenuation of the order in question as applicable in uniform air, it will I think be necessary to look for a cause not hitherto taken into account. We might imagine a delay in the equalization of the different sorts of energy in a gas undergoing compression, not wholly insensible in comparison with the time of vibration of the sound. If in the dynamical theory we assimilate the molecules of a gas to hard smooth bodies which are nearly but not absolutely spherical, and trace the effect of a rapid compression, we see that at the first moment the increment of energy is wholly translational and thus produces a maximum effect in opposing the compression. A little later a due proportion of the excess of energy will have passed into rotational forms which do not influence the pressure, and this will accordingly fall off. Any effect of the kind must give rise to dissipation, and the amount of it will increase with the time required for the transformations, *i.e.* in the above mentioned illustration with the degree of approximation to the spherical form. In the case of absolute spheres no transformation of translatory into rotatory energy, or *vice versa*, would occur in a finite time. There appears to be nothing in the behaviour of gases, as revealed to us by experiment, which forbids the supposition of a delay capable of influencing the propagation of sound.

Returning now to the question of the radiating power of air, we may establish a sort of superior limit by an argument based upon the theory of exchanges, itself firmly established by the researches of B. Stewart. Consider a spherical mass of radius r, slightly and uniformly heated. Whatever may be the radiation proceeding from a unit of surface, it must be less than the radiation from an ideal black surface under the same conditions. Let us, however, suppose that the radiation is the same in both cases and inquire what would then be the rate of cooling. According to Bottomley[*] the emissivity of a blackened surface moderately heated is ·0001. This is the amount of heat reckoned in water-gram-degree units emitted in one second from a square centimetre of surface heated 1° C. If the excess of temperature be θ, the whole emission is

$$\theta \times 4\pi r^2 \times ·0001$$

On the other hand, the capacity for heat is

$$\tfrac{4}{3}\pi r^3 \times ·0013 \times ·24,$$

the first factor being the volume, the second the density, and the third the

[*] Everett, *C.G.S. Units*, 1891, p. 134.

413

specific heat of air referred, as usual, to water. Thus for the rate of cooling,

$$\frac{d\theta}{\theta\,dt} = -\frac{\cdot 0003}{\cdot 0013 \times \cdot 24 \times r} = -\frac{1}{r} \text{ very nearly,}$$

whence
$$\theta = \theta_0 e^{-t/r}, \quad \dots\dots\dots\dots\dots\dots\dots\dots\dots(5)$$

θ_0 being the initial value of θ. The time in seconds of cooling in the ratio of $e : 1$ is thus represented numerically by r expressed in centims.

When r is very great, the suppositions on which (5) is calculated will be approximately correct, and that equation will then represent the actual law of cooling of the sphere of air, supposed to be maintained uniform by mixing if necessary. But ordinary experience, and more especially the observations of Tyndall upon the diathermancy of air, would lead us to suppose that this condition of things would not be approached until r reached 1000 or perhaps 10,000 centims. For values of r comparable with the half wave-length of ordinary sounds, e.g. 30 centim., it would seem that the real time of cooling must be a large multiple of that given by (5). At this rate the time of cooling of a mass of air must exceed, and probably largely exceed, 60 seconds. To suppose that this time is one-twelfth of a second would require a sphere of air 2 millim. in diameter to radiate as much heat as if it were of blackened copper at the same temperature.

Although, if the above argument is correct, there seems little likelihood of the cooling of moderate masses of air being sensibly influenced by radiation, I thought it would be of interest to inquire whether the observed cooling (or heating) in an experiment on the lines of Clement and Desormes could be adequately explained by the conduction of heat from the walls of the vessel in accordance with the known conductivity of air. A nearly spherical vessel of glass of about 35 centim. diameter, well encased, was fitted, air-tight, with two tubes. One of these led to a manometer charged with water or sulphuric acid; the other was provided with a stopcock and connected with an air-pump. In making an experiment the stopcock was closed and a vacuum established in a limited volume upon the further side. A rapid opening and reclosing of the cock allowed a certain quantity of air to escape suddenly, and thus gave rise to a nearly uniform cooling of that remaining behind in the vessel. At the same moment the liquid rose in the manometer, and the observation consisted in noting the times (given by a metronome beating seconds) at which the liquid in its descent passed the divisions of a scale, as the air recovered the temperature of the containing vessel. The first record would usually be at the third or fourth second from the turning of the cock, and the last after perhaps 120 seconds. In this way data are obtained for a plot of the curve of pressure; and the part actually observed has to be supplemented by extrapolation, so as to go back to the zero of time (the moment of turning the tap) and to allow for the drop which might occur subsequent to the last observation. An estimate, which cannot be much in error, is thus obtained of the whole rise in pressure during the recovery of temperature, and for the time, reckoned from the commencement, at which the rise is equal to one-half of the total.

In some of the earlier experiments the whole rise of pressure (fall in the manometer) during the recovery of temperature was about 20 millim. of water, and the time of half recovery was 15 seconds. I was desirous of working with the minimum range, since only in this way could it be hoped to eliminate the effect of gravity, whereby the interior and still cool parts of the included air would be made to fall and so come into closer proximity to the walls, and thus accelerate the mean cooling. In order to diminish the disturbance due to capillarity, the bore of the manometer-tube, which stood in a large open cistern, was increased to about 18 millim.*, and suitable optical arrangements were introduced to render small movements easily visible. By degrees the range was diminished, with a prolongation of the time of half recovery to 18, 22, 24, and finally to about 26 seconds. The minimum range attained was represented by 3 or 4 millim. of water, and at this stage there did not appear to be much further prolongation of cooling in progress. There seemed to be no appreciable difference whether the air was artificially dried or not, but in no case was the moisture sufficient to develop fog under the very small expansions employed. The result of the experiments may be taken to be that when the influence of gravity was, as far as practicable, eliminated, the time of half recovery of temperature was about 26 seconds.

It may perhaps be well to give an example of an actual experiment. Thus in one trial on Nov. 1, the recorded times of passage across the divisions of the scale were 3, 6, 11, 18, 26, 35, 47, 67, 114 seconds. The divisions themselves were millimetres, but the actual movements of the meniscus were less in the proportion of about $2\frac{1}{2} : 1$. A plot of these numbers shows that one division must be added to represent the movement between 0^s and 3^s, and about as much for the movement to be expected between 114^s and ∞. The whole range is thus 10 divisions (corresponding to 4 millim. at the meniscus), and the mid-point occurs at 26^s. On each occasion 3 or 4 sets of readings were taken under given conditions with fairly accordant results.

It now remains to compare with the time of heating derived from theory. The calculation is complicated by the consideration that when during the process any part becomes heated, it expands and compresses all the other parts, thereby developing heat in them. From the investigation which

---

* It must not be forgotten that too large a diameter is objectionable, as leading to an augmentation of volume during an experiment, as the liquid falls.

follows *, we see that the time of half recovery t is given by the formula

$$t = \frac{\cdot184\gamma a^2}{\pi^2 \nu}, \quad \dots\dots\dots\dots\dots\dots\dots\dots\dots\dots(6)$$

in which a is the radius of the sphere, γ the ratio of specific heats (1·41), and ν is the thermometric conductivity, found by dividing the ordinary or calorimetric conductivity by the thermal capacity of unit volume. This thermal capacity is to be taken with volume constant, and it will be less than the thermal capacity with pressure constant in the ratio of $\gamma : 1$. Accordingly ν/γ in (6) represents the latter thermal capacity, of which the experimental value is ·00128 × ·239, the first factor representing the density of air referred to water. Thus, if we take the calorimetric conductivity at ·000056, we have in C.G.S. measure

$$\nu = \cdot258, \qquad \nu/\gamma = \cdot183;$$

and thence

$$t = \cdot102a^2.$$

In the present apparatus a, determined by the contents, is 16·4 centim., whence

$$t = 27\cdot4 \text{ seconds.}$$

The agreement of the observed and calculated values is quite as close as could have been expected, and confirms the view that the transfer of heat is due to conduction, and that the part played by radiation is insensible. From a comparison of the experimental and calculated curves, however, it seems probable that the effect of gravity was not wholly eliminated, and that the later stages of the phenomenon, at any rate, may still have been a little influenced by a downward movement of the central parts.

* See next paper.

39

Reverberation

WALLACE CLEMENT SABINE

Wallace Clement Sabine (1868–1919), American physicist, was a member of the Physics Department of Harvard University from 1889 until his death. In the early part of his tenure his interests were confined largely to teaching, the responsibilities of which he took very seriously. He wrote and published a small laboratory manual for the use of college students. His early research was confined to collaboration with older professors at Harvard, mainly on electrical oscillation phenomena. The spark which excited his celebrated career in acoustics came when he was requested by the Harvard Corporation to propose changes to remedy the unsatisfactory acoustical properties of the Fogg Art Museum at Harvard. He at once became interested in the physical problems connected with architectural acoustics. The rest of his career was devoted largely to this field and in the early stages he gained a precise understanding of the influence of reverberation on the reception of sound in a closed space, and developed his famous law connecting the reverberation time with the room volume and the amount of absorbing material. The unit of sound absorption is named after him.

The following extract is taken from the first part of his paper called "Reverberation," published in *The American Architect and the Engineering Record* in 1900. It constitutes the beginning of modern room acoustics.

Reprinted from "Collected Papers on Acoustics," Wallace Clement Sabine (Dover Publications, Inc.), 3–42 (1964)

1

REVERBERATION[1]

INTRODUCTION

THE following investigation was not undertaken at first by choice, but devolved on the writer in 1895 through instructions from the Corporation of Harvard University to propose changes for remedying the acoustical difficulties in the lecture-room of the Fogg Art Museum, a building that had just been completed. About two years were spent in experimenting on this room, and permanent changes were then made. Almost immediately afterward it became certain that a new Boston Music Hall would be erected, and the questions arising in the consideration of its plans forced a not unwelcome continuance of the general investigation.

No one can appreciate the condition of architectural acoustics — the science of sound as applied to buildings — who has not with a pressing case in hand sought through the scattered literature for some safe guidance. Responsibility in a large and irretrievable expenditure of money compels a careful consideration, and emphasizes the meagerness and inconsistency of the current suggestions. Thus the most definite and often repeated statements are such as the following, that the dimensions of a room should be in the ratio 2 : 3 : 5, or according to some writers, 1 : 1 : 2, and others, 2 : 3 : 4; it is probable that the basis of these suggestions is the ratios of the harmonic intervals in music, but the connection is untraced and remote. Moreover, such advice is rather difficult to apply; should one measure the length to the back or to the front of the galleries, to the back or the front of the stage recess? Few rooms have a flat roof, where should the height be measured? One writer, who had seen the Mormon Temple, recommended that all auditoriums be elliptical. Sanders Theatre is by far the best auditorium in Cambridge and is semicircular in general shape, but with a recess that makes it almost anything; and, on the other hand, the lecture-room in the Fogg Art

[1] The American Architect and The Engineering Record, 1900.

3

Museum is also semicircular, indeed was modeled after Sanders Theatre, and it was the worst. But Sanders Theatre is in wood and the Fogg lecture-room is plaster on tile; one seizes on this only to be immediately reminded that Sayles Hall in Providence is largely lined with wood and is bad. Curiously enough, each suggestion is advanced as if it alone were sufficient. As examples of remedies, may be cited the placing of vases about the room for the sake of resonance, wrongly supposed to have been the object of the vases in Greek theatres, and the stretching of wires, even now a frequent though useless device.

The problem is necessarily complex, and each room presents many conditions, each of which contributes to the result in a greater or less degree according to circumstances. To take justly into account these varied conditions, the solution of the problem should be quantitative, not merely qualitative; and to reach its highest usefulness it should be such that its application can precede, not follow, the construction of the building.

In order that hearing may be good in any auditorium, it is necessary that the sound should be sufficiently loud; that the simultaneous components of a complex sound should maintain their proper relative intensities; and that the successive sounds in rapidly moving articulation, either of speech or music, should be clear and distinct, free from each other and from extraneous noises. These three are the necessary, as they are the entirely sufficient, conditions for good hearing. The architectural problem is, correspondingly, threefold, and in this introductory paper an attempt will be made to sketch and define briefly the subject on this basis of classification. Within the three fields thus defined is comprised without exception the whole of architectural acoustics.

1. *Loudness.* — Starting with the simplest conceivable auditorium — a level and open plain, with the ground bare and hard, a single person for an audience — it is clear that the sound spreads in a hemispherical wave diminishing in intensity as it increases in size, proportionally. If, instead of being bare, the ground is occupied by a large audience, the sound diminishes in intensity even more rapidly, being now absorbed. The upper part of the sound-wave escapes unaffected, but the lower edge — the only part that is of service to an

audience on a plain — is rapidly lost. The first and most obvious improvement is to raise the speaker above the level of the audience; the second is to raise the seats at the rear; and the third is to place a wall behind the speaker. The result is most attractively illustrated in the Greek theatre. These changes being made, still all the sound rising at any considerable angle is lost through the opening above, and only part of the speaker's efforts serve the audience. When to this auditorium a roof is added the average intensity of sound throughout the room is greatly increased, especially that of sustained tones; and the intensity of sound at the front and the rear is more nearly equalized. If, in addition, galleries be constructed in order to elevate the distant part of the audience and bring it nearer to the front, we have the general form of the modern auditorium. The problem of calculating the loudness at different parts of such an auditorium is, obviously, complex, but it is perfectly determinate, and as soon as the reflecting and absorbing power of the audience and of the various wall-surfaces are known it can be solved approximately. Under this head will be considered the effect of sounding-boards, the relative merits of different materials used as reflectors, the refraction of sound, and the influence of the variable temperature of the air through the heating and ventilating of the room, and similar subjects.

2. *Distortion of Complex Sounds: Interference and Resonance.* — In discussing the subject of loudness the direct and reflected sounds have been spoken of as if always reënforcing each other when they come together. A moment's consideration of the nature of sound will show that, as a matter of fact, it is entirely possible for them to oppose each other. The sounding body in its forward motion sends off a wave of condensation, which is immediately followed through the air by a wave of rarefaction produced by the vibrating body as it moves back. These two waves of opposite character taken together constitute a sound-wave. The source continuing to vibrate, these waves follow each other in a train. Bearing in mind this alternating character of sound, it is evident that should the sound traveling by different paths — by reflection from different walls — come together again, the paths being equal in length, condensation will arrive at the same time as condensation, and reënforce it, and rare-

faction will, similarly, reënforce rarefaction. But should one path be a little shorter than the other, rarefaction by one and condensation by the other may arrive at the same time, and at this point there will be comparative silence. The whole room may be mapped out into regions in which the sound is loud and regions in which it is feeble. When there are many reflecting surfaces the interference is much more complex. When the note changes in pitch the interference system is entirely altered in character. A single incident will serve to illustrate this point. There is a room in the Jefferson Physical Laboratory, known as the constant-temperature room, that has been of the utmost service throughout these experiments. It is in the center of one wing of the building, is entirely under ground, even below the level of the basement of the building, has separate foundations and double walls, each wall being very thick and of brick in cement. It was originally designed for investigations in heat requiring constant temperature, and its peculiar location and construction were for this purpose. As it was not so in use, however, it was turned over to these experiments in sound, and a room more suitable could not be designed. From its location and construction it is extremely quiet. Without windows, its walls, floor, and ceiling — all of solid masonry — are smooth and unbroken. The single door to the room is plain and flush with the wall. The dimensions of the room are, on the floor, 4.27 × 6.10 meters; its height at the walls is 2.54 meters, but the ceiling is slightly arched, giving a height at the center of 3.17 meters. This room is here described at length because it will be frequently referred to, particularly in this matter of interference of sound. While working in this room with a treble c gemshorn organ pipe blown by a steady wind-pressure, it was observed that the pitch of the pipe apparently changed an octave when the observer straightened up in his chair from a position in which he was leaning forward. The explanation is this: The organ pipe did not give a single pure note, but gave a fundamental treble c accompanied by several overtones, of which the strongest was in this case the octave above. Each note in the whole complex sound had its own interference system, which, as long as the sound remained constant, remained fixed in position. It so happened that at these two points the region of silence for one note coincided with the region of reënforcement in the other, and *vice versa*. Thus the observer in one position heard the fundamental note, and in the other, the first overtone. The change was exceedingly striking, and as the notes remained constant, the experiment could be tried again and again. With a little search it was possible to find other points in the room at which the same phenomenon appeared, but generally in less perfection. The distortion of the relative intensities of the components of a chord that may thus be produced is evident. Practically almost every sound of the voice in speech and song, and of instrumental music, even single-part music so-called, is more or less complex, and, therefore, subject to this distortion. It will be necessary, later, to show under what circumstances this phenomenon is a formidable danger, and how it may be guarded against, and under what circumstances it is negligible. It is evident from the above occurrence that it may be a most serious matter, for in this room two persons side by side can talk together with but little comfort, most of the difficulty being caused by the interference of sound.

There is another phenomenon, in its occurrence allied to interference, but in nature distinct — the phenomenon of resonance. Both, however, occasion the same evil — the distortion of that nice adjustment of the relative intensities of the components of the complex sounds that constitute speech and music. The phenomenon of interference just discussed merely alters the distribution of sound in the room, causing the intensity of any one pure sustained note to be above or below the average intensity at near points. Resonance, on the other hand, alters the total amount of sound in the whole room and always increases it. This phenomenon is noticeable at times in using the voice in a small room, or even in particular locations in a large room. Perhaps its occurrence is most easily observed in setting up a large church organ, where the pipes must be readjusted for the particular space in which the organ is to stand, no matter with how much care the organ may have been assembled and adjusted before leaving the factory. The general phenomenon of resonance is of very wide occurrence, not merely in acoustics but in more gross mechanics as well, as the vibration of a bridge to a properly timed tread, or the excessive rolling of a boat in certain seas. The principle is the same in all cases. The following conception is an easy one to grasp, and is closely analogous to acoustical resonance: If the palm of the hand be placed on the center of the surface of water in a large basin or tank and quickly depressed and raised once it will cause a wave to spread, which, reflected at the edge of the water, will return, in part at least, to the hand. If, just as the wave reaches the hand, the hand repeats its motion with the same force, it will reënforce the wave traveling over the water. Thus reënforced, the wave goes out stronger than before and returns again. By continued repetition of the motion of the hand so timed as to reënforce the wave as it returns, the wave gets to be very strong. Instead of restraining the hand each time until the wave traveling to and fro returns to it, one may so time the motion of the hand as to have several equal waves following each other over the water, and the hand each time reënforcing the wave that is passing. This, obviously, can be done by dividing the interval of time between the successive motions of the hand by any whole number whatever, and moving the hand with the frequency thus defined. The result will be a strong reënforcement of the waves. If, however, the motions of the hand be not so timed, it is obvious that the reënforcement will not be perfect, and, in fact, it is possible to so time it as exactly to oppose the returning waves. The application of this reasoning to the phenomenon of sound, where the air takes the place of the water and the sounding body that of the hand, needs little additional explanation. Some notes of a complex sound are reënforced, some are not, and thus the quality is altered. This phenomenon enters in two forms in the architectural problem: there may be either resonance of the air in the room or resonance of the walls, and the two cases must receive separate discussion; their effects are totally different.

The word "resonance" has been used loosely as synonymous with "reverberation," and even with "echo," and is so given in some of the more voluminous but less exact popular dictionaries. In scientific literature the term has received a very definite and precise application to the phenomenon, wherever it may occur, of the growth of a vibratory motion of an elastic body under periodic forces timed to its natural rates of vibration. A word having this significance is necessary; and it is very desirable that the term should not, even popularly, by meaning many things, cease to mean anything exactly.

3. *Confusion: Reverberation, Echo and Extraneous Sounds.* — Sound, being energy, once produced in a confined space, will continue until it is either transmitted by the boundary walls, or is transformed into some other kind of energy, generally heat. This process of decay is called absorption. Thus, in the lecture-room of Harvard University, in which, and in behalf of which, this investigation was begun, the rate of absorption was so small that a word spoken in an ordinary tone of voice was audible for five and a half seconds afterwards. During this time even a very deliberate speaker would have uttered the twelve or fifteen succeeding syllables. Thus the successive enunciations blended into a loud sound, through which and above which it was necessary to hear and distinguish the orderly progression of the speech. Across the room this could not be done; even near the speaker it could be done only with an effort wearisome in the extreme if long maintained. With an audience filling the room the conditions were not so bad, but still not tolerable. This may be regarded, if one so chooses, as a process of multiple reflection from walls, from ceiling and from floor, first from one and then another, losing a little at each reflection until ultimately inaudible. This phenomenon will be called reverberation, including as a special case the echo. It must be observed, however, that, in general, reverberation results in a mass of sound filling the whole room and incapable of analysis into its distinct reflections. It is thus more difficult to recognize and impossible to locate. The term echo will be reserved for that particular case in which a short, sharp sound is distinctly repeated by reflection, either once from a single surface, or several times from two or more surfaces. In the general case of reverberation we are only concerned with the rate of decay of the sound. In the special case of the echo we are concerned not merely with its intensity, but with the interval of time elapsing between the initial sound and the moment it reaches the observer. In the room mentioned as the occasion of this investigation, no discrete echo was distinctly perceptible, and the case will serve excellently as an illustration of the more general

type of reverberation.   After preliminary gropings,[1] first in the
literature and then with several optical devices for measuring the
intensity of sound, both were abandoned, the latter for reasons that
will be explained later.   Instead, the rate of decay was measured by
measuring what was inversely proportional to it — the duration of
audibility of the reverberation, or, as it will be called here, the duration of audibility of the residual sound.   These experiments may be
explained to advantage even in this introductory paper, for they
will give more clearly than would abstract discussion an idea of the
nature of reverberation.   Broadly considered, there are two, and
only two, variables in a room — shape including size, and materials
including furnishings.   In designing an auditorium an architect can
give consideration to both; in repair work for bad acoustical conditions it is generally impracticable to change the shape, and only
variations in materials and furnishings are allowable.   This was,
therefore, the line of work in this case.   It was evident that, other
things being equal, the rate at which the reverberation would disappear was proportional to the rate at which the sound was absorbed.   The first work, therefore, was to determine the relative
absorbing power of various substances.   With an organ pipe as a
constant source of sound, and a suitable chronograph for recording,
the duration of audibility of a sound after the source had ceased in
this room when empty was found to be 5.62 seconds.   All the cushions from the seats in Sanders Theatre were then brought over and
stored in the lobby.   On bringing into the lecture-room a number
of cushions having a total length of 8.2 meters, the duration of
audibility fell to 5.33 seconds.   On bringing in 17 meters the sound
in the room after the organ pipe ceased was audible for but 4.94

[1] The first method for determining the rate of decay of the sound, and therefore the amount
of absorption, was by means of a sensitive manometric gas flame measured by a micrometer
telescope.   Later, photographing the flame was tried; but both methods were abandoned, for
they both showed, what the unaided ear could perceive, that the sound as observed at any
point in the room died away in a fluctuating manner, passing through maxima and minima.
Moreover, they showed what the unaided ear had not detected, but immediately afterward
did recognize, that the sound was often more intense immediately after the source ceased than
before.   All this was interesting, but it rendered impossible any accurate interpretation of the
results obtained by these or similar methods.   It was then found that the ear itself aided by
a suitable electrical chronograph for recording the duration or audibility of the residual sound
gave a surprisingly sensitive and accurate method of measurement.   Proc. American Institute
of Architects, p. 35, 1898.

seconds. Evidently, the cushions were strong absorbents and rapidly improving the room, at least to the extent of diminishing the reverberation. The result was interesting and the process was continued. Little by little the cushions were brought into the room, and each time the duration of audibility was measured. When all the seats (436 in number) were covered, the sound was audible for 2.03 seconds. Then the aisles were covered, and then the platform. Still there were more cushions — almost half as many more. These were brought into the room, a few at a time, as before, and draped on a scaffolding that had been erected around the room, the duration of the sound being recorded each time. Finally, when all the cushions from a theatre seating nearly fifteen hundred persons were placed in the room — covering the seats, the aisles, the platform, the rear wall to the ceiling — the duration of audibility of the residual sound was 1.14 seconds. This experiment, requiring, of course, several nights' work, having been completed, all the cushions were removed and the room was in readiness for the test of other absorbents. It was evident that a standard of comparison had been established. Curtains of chenille, 1.1 meters wide and 17 meters in total length, were draped in the room. The duration of audibility was then 4.51 seconds. Turning to the data that had just been collected it appeared that this amount of chenille was equivalent to 30 meters of Sanders Theatre cushions. Oriental rugs, Herez, Demirjik, and Hindoostanee, were tested in a similar manner; as were also cretonne cloth, canvas, and hair felt. Similar experiments, but in a smaller room, determined the absorbing power of a man and of a woman, always by determining the number of running meters of Sanders Theatre cushions that would produce the same effect. This process of comparing two absorbents by actually substituting one for the other is laborious, and it is given here only to show the first steps in the development of a method that will be expanded in the following papers.

In this lecture-room felt was finally placed permanently on particular walls, and the room was rendered not excellent, but entirely serviceable, and it has been used for the past three years without serious complaint. It is not intended to discuss this particular case in the introductory paper, because such discussion would be premature and logically incomplete. It is mentioned here merely to illustrate concretely the subject of reverberation, and its dependence on absorption. It would be a mistake to suppose that an absorbent is always desirable, or even when desirable that its position is a matter of no consequence.[1]

While the logical order of considering the conditions contributing to or interfering with distinct hearing would be that employed above, it so happens that exactly the reverse order is preferable from an experimental standpoint. By taking up the subject of reverberation first it is possible to determine the coefficients of absorption and reflection of various kinds of wall surface, of furniture and draperies, and of an audience. The investigation of reverberation is now, after five years of experimental work, completed, and an account will be rendered in the following papers. Some data have also been secured on the other topics and will be published as soon as rounded into definite form.

This paper may be regarded as introductory to the general subject of architectural acoustics, and immediately introductory to a series of articles dealing with the subject of reverberation, in which the general line of procedure will be, briefly, as follows: The absorbing power of wall-surfaces will be determined, and the law according to which the reverberation of a room depends on its volume will be demonstrated. The absolute rate of decay of the residual sound in a number of rooms, and in the same room under different conditions, will then be determined. In the fifth paper a more exact analysis

---

[1] There is no simple treatment that can cure all cases. There may be inadequate absorption and prolonged residual sound; in this case absorbing material should be added in the proper places. On the other hand, there may be excessive absorption by the nearer parts of the hall and by the nearer audience and the sound may not penetrate to the greater distances. Obviously the treatment should not be the same. There is such a room belonging to the University, known locally as Sever 35. It is low and long. Across its ceiling are now stretched hundreds of wires and many yards of cloth. The former has the merit of being harmless, the latter is like bleeding a patient suffering from a chill. In general, should the sound seem smothered or too faint, it is because the sound is either imperfectly distributed to the audience. or is lost in waste places. The first may occur in a very low and long room, the second in one with a very high ceiling. The first can be remedied only slightly at best, the latter can be improved by the use of reflectors behind and above the speaker. On the other hand, should the sound be loud but confused, due to a perceptible prolongation, the difficulty arises from th re being reflecting surfaces either too far distant or improperly inclined. Proc. American Institute of Architects, p. 39, 1898.

will be given, and it will be shown that, by very different lines of attack, starting from different data, the same numerical results are secured. Tables will be given of the absorbing power of various wall-surfaces, of furniture, of an audience, and of all the materials ordinarily found in any quantity in an auditorium. Finally, in illustration of the calculation of reverberation in advance of construction, will be cited the new Boston Music Hall, the most interesting case that has arisen.

ABSORBING POWER OF WALL–SURFACES

In the introductory article the problem was divided into considerations of loudness, of distortion, and of confusion of sounds. Confusion may arise from extraneous disturbing sounds — street noises and the noise of ventilating fans — or from the prolongation of the otherwise discrete sounds of music or the voice into the succeeding sounds. The latter phenomenon, known as reverberation, results in what may be called, with accuracy and suggestiveness, residual sound. The duration of this residual sound was shown to depend on the amount of absorbing material inside the room, and also, of course, on the absorbing and transmitting power of the walls; and a method was outlined for determining the absorbing power of the former in terms of the absorbing power of some material chosen as a standard and used in a preliminary calibration. A moment's consideration demonstrates that this method, which is of the general type known as a "substitution method," while effective in the determination of the absorbing power of furniture and corrective material, and, in general, of anything that can be brought into or removed from a room, is insufficient for determinating the absorbing power of wall-surfaces. This, the absorbing power of wall-surfaces, is the subject of the present paper; and as the method of determination is an extension of the above work, and finds its justification in the striking consistency of the results of the observations, a more elaborate description of the experimental method is desirable. A proof of the accuracy of every step taken is especially necessary in a subject concerning which theory has been so largely uncontrolled speculation.

Early in the investigation it was found that measurements of the length of time during which a sound was audible after the source had ceased gave promising results whose larger inconsistencies could be traced directly to the distraction of outside noises. On repeating the work during the most quiet part of the night, between half-past twelve and five, and using refined recording apparatus, the minor irregularities, due to relaxed attention or other personal variations, were surprisingly small. To secure accuracy, however, it was necessary to suspend work on the approach of a street car within two blocks, or on the passing of a train a mile distant. In Cambridge these interruptions were not serious; in Boston and in New York it was necessary to snatch observations in very brief intervals of quiet. In every case a single determination of the duration of the residual sound was based on the average of a large number of observations.

An organ pipe, of the gemshorn stop, an octave above middle *c* (512 vibration frequency) was used as the source of sound in some preliminary experiments, and has been retained in subsequent work in the absence of any good reason for changing. The wind supply from a double tank, water-sealed and noiseless, was turned on and off the organ pipe by an electro-pneumatic valve, designed by Mr. George S. Hutchings, and similar to that used in his large church organs. The electric current controlling the valve also controlled the chronograph, and was made and broken by a key in the hands of the observer from any part of the room. The chronograph employed in the later experiments, after the more usual patterns had been tried and discarded, was of special design, and answered well the requirements of the work — perfect noiselessness, portability, and capacity to measure intervals of time from a half second to ten seconds with considerable accuracy. It is shown in the adjacent diagram. The current whose cessation stopped the sounding of the organ pipe also gave the initial record on the chronograph, and the only duty of the observer was to make the record when the sound ceased to be audible.

While the supreme test of the investigation lies in the consistency and simplicity of the whole solution as outlined later, three preliminary criteria are found in (1) the agreement of the observations

obtained at one sitting, (2) the agreement of the results obtained
on different nights and after the lapse of months, or even years, by
the same observer under similar conditions, and (3) the agreement
of independent determinations by different observers.   The first
can best be discussed, of course, by the recognized physical methods
for examining the accuracy of an extended series of observations;

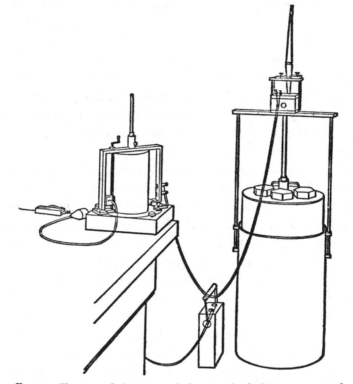

FIG. 1.   Chronograph, battery, and air reservoir, the latter surmounted
by the electro-pneumatic valve and organ pipe.

and the result of such examination is as follows: Each determination
being the mean of about twenty observations under conditions such
that the audible duration of the residual sound was 4 seconds, the
average deviation of the single observations from the mean was .11
seconds, and the maximum deviation was .31.   The computed
"probable error" of a single determination was about .02 seconds;
as a matter of fact, the average deviation of ten determinations
from the mean of the ten was .03 seconds, and the maximum devi-

ation was .05. The reason for this accuracy will be discussed in a subsequent paper. The probable error of the mean, thus calculated from the deviations of the single observations, covers only those variable errors as likely to increase as to decrease the final result. Fixed instrumental errors, and the constant errors commonly referred to by the term "personal factors" are not in this way exposed. They were, however, repeatedly tested for by comparison with a clock beating seconds, and were very satisfactorily shown not to amount to more than .02 seconds in their cumulative effect. Three types of chronographs, and three kinds of valves between the organ pipe and the wind chest were used in the gradual development of the experiment, and all gave for the same room very nearly the same final results. The later instruments were, of course, better and more accurate.

The second criterion mentioned above is abundantly satisfied by the experiments. Observations taken every second or third night for two months in the lecture-room of the Fogg Art Museum gave practically the same results, varying from 5.45 to 5.62 with a mean value of 5.57 seconds, a result, moreover, that was again obtained after the lapse of one and then of three years. Equally satisfactory agreement was obtained at the beginning and at the end of three years in Sanders Theatre, and in the constant-temperature room of the Physical Laboratory.

Two gentlemen, who were already somewhat skilled in physical observation, Mr. Gifford LeClear and Mr. E. D. Densmore, gave the necessary time to test the third point. After several nights' practice their results differed but slightly, being .08 seconds and .10 seconds longer than those obtained by the writer, the total duration of the sound being 4 seconds. This agreement, showing that the results are probably very nearly those that would be obtained by any auditor of normal hearing, gives to them additional interest. It should be stated, however, that the final development of the subject will adapt it with perfect generality to either normal or abnormal acuteness of hearing.

Almost the first step in the investigation was to establish the following three fundamentally important facts. Later work has proved these fundamental facts far more accurately, but the original experiments are here given as being those upon which the conclusions were based.

The duration of audibility of the residual sound is nearly the same in all parts of an auditorium. — Early in the investigation an experiment to test this point was made in Steinert Hall, in Boston. The source of sound remaining on the platform at the point marked

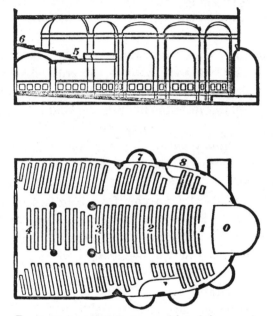

FIG. 2. Steinert Hall, Boston: position of air reservoir
and organ pipe at *0*; positions of observer 1-8.

0 in the diagram, observations were made in succession at the points marked 1 to 8, with the results shown in the table:

Station	Duration	Station	Duration
1	2.12	5	2.23
2	2.17	6	2.27
3	2.23	7	2.20
4	2.20	8	2.26

On first inspection these results seem to indicate that the duration of audibility is very slightly greater at a distance from the source, and it would be easy to explain this on the theory that at a distance the ear is less exhausted by the rather loud noise while the pipe is sounding; but, as a matter of fact, this is not the case, and the variations there shown are within the limits of accuracy of the apparatus employed and the skill attained thus. early in the investigation. Numerous later experiments, more accurate, but not especially directed to this point, have verified the above general statement quite conclusively.

The duration of audibility is nearly independent of the position of the source. — The observer remaining at the point marked O in the diagram of the large lecture-room of the Jefferson Physical Laboratory, the organ pipe and wind chest were moved from station to station, as indicated by the numbers 1 to 6, with the results shown in the table:

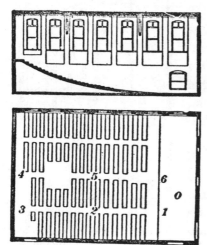

Station	Duration
1	3.90
2	4.00
3	3.90
4	3.98
5	3.95
6	3.96

FIG. 3. Lecture-room, Jefferson Physical Laboratory: position of observer at O; positions of air reservoir and organ pipe 1-6.

The efficiency of an absorbent in reducing the duration of the residual sound is, under ordinary circumstances, nearly independent of its position. — Fifty meters of cretonne cloth draped on a scaffolding under the rather low ceiling at the back of the lecture-room of the Fogg Museum, as shown in the next diagram, reduced the audible duration of the residual sound by very nearly the same amount, regardless of the section in which it hung, as shown in the following table, the initial duration being 5.57 seconds:

Section	Duration
1	4.88
2	4.83
3	4.92
4	4.85

In some later experiments five and a half times as much cretonne draped on the scaffolding reduced the audible duration of the residual sound to 3.25 seconds; and when hung fully exposed in the high dome-like ceiling, gave 3.29 seconds, confirming the above statement.

These facts, simple when proved, were by no means self-evident so long as the problem was one of reverberation, that is, of succes-sive reflection of sound from wall to wall. They indicated that, at least with reference to auditoriums of not too great dimensions, another point of view would be more suggestive, that of re-garding the whole as an energy problem in which the source is at the organ pipe and the decay at the walls and at the contained absorbing material. The above results, then, all point to the evident, but perhaps not appreci-ated, fact that the dispersion of sound between all parts of a hall is very rapid in comparison with the total time re-quired for its complete absorption, and that in a very short time after the source has ceased the intensity of the residual sound, except for the phenom-enon of interference to be considered later, is very nearly the same every-where in the room.

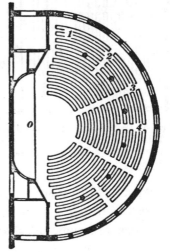

Fig. 4. Lecture-room, Fogg Art Museum: position of observer at *O*; positions of absorbent at 1-4, and in the dome.

This much being determined, the investigation was continued in the fol-lowing manner: Cushions from San-ders Theatre were transferred to the lobby of the lecture-room of the Fogg Museum; a very few were brought into the room and spread along the front row of seats; the duration of audibility of the residual sound, diminished by the introduction of this additional absorbent, was determined, and the total length of cushion was measured. The next row of seats was then covered in the same manner and the two observations made — length of cushion and duration of residual sound. This was repeated till cushions covered all the seats. This work was at first undertaken solely with the intention of determining the relative merits of different absorbing materials that might be placed in the room as a corrective for excessive residual sound, and the account of this application is given in the introductory paper. A subsequent study of these and similar results obtained in many other rooms has shown their applicability to the accurate determination of the absorbing power of wall-surfaces. This application may be shown in a purely analytical manner, but the exposition is greatly helped by a graphical representation. The manner in which the duration of the residual sound in the Fogg lecture-room is dependent on the amount of absorbing material present is shown in the following table:

Length of Cushion in Meters	Duration of Residual Sound in Seconds
0	5.61
8	5.33
17	4.94
28	4.56
44	4.21
63	3.94
83	3.49
104	3.33
123	3.00
145	2.85
162	2.64
189	2.36
213	2.33
242	2.22

This table, represented graphically in the conventional manner — length of cushion plotted horizontally and duration of sound vertically — gives points through which the curve may be drawn in the accompanying diagram. To discover the law from this curve we represent the lengths of cushion by x, and the corresponding durations of sound, the vertical distances to the curve, by t. If we now seek the formula connecting x and t that most nearly expresses the relationship represented by the above curve, we find it to be $(a + x)t = k$, which is the familiar formula of a rectangular hyperbola with its origin displaced along the axis of x, one of its asymptotes, by an amount a. To make this formula most closely fit our curve we must, in this case, give to the constant, a, the numerical value, 146, and to k the value, 813. The accuracy with which the formula represents the curve may be seen by comparing the durations calculated by the formula with those determined from the curve; they nowhere differ by more than .04 of a second, and have, on an average, a difference of only .02 of a second. This is entirely satisfactory, for the calculated points fall off from the curve by scarcely the breadth of the pen point with which it was drawn.

The determination of the absorbing power of the wall-surface depends on the interpretation of the constant, a. In the formula,

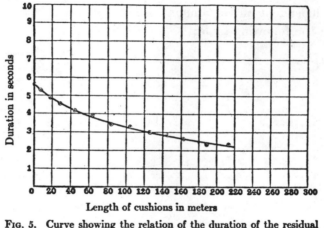

Length of cushions in meters

FIG. 5. Curve showing the relation of the duration of the residual sound to the added absorbing material.

the position of a, indicating that x is to be added to it, suggests that x and a are of a like nature, and that a is a measure of the absorbing power of the bare room; in order to determine the curve this was increased by the introduction of the cushions. This is even better shown by the diagram in which the portion of the curve experimentally determined is fitted into the curve as a whole, and a and x are indicated. Thus, the absorbing power of the room — the walls, partly plaster on stone, partly plaster on wire lath, the windows, the skylight, the floor — was equivalent to 146 running meters of Sanders Theatre cushions.

The last statement shows the necessity for two subsidiary investigations. The first, to express the results in some more permanent, more universally available, and, if possible, more absolute unit than the cushions; the other, to apportion the total absorbing power among the various components of the structure.

The transformation of results from one system of units to another necessitates a careful study of both systems. Some early experiments in which the cushions were placed with one edge pushed against the backs of the settees gave results whose anomalous character suggested that, perhaps, their absorbing power depended not merely on the amount present but also on the area of the surface exposed. It was then recalled that about two years before, at the beginning of an evening's work, the first lot of cushions

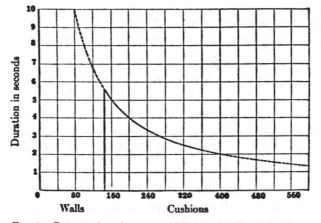

FIG. 6. Curve 5 plotted as part of its corresponding rectangular hyperbola. The solid part was determined experimentally; the displacement of this to the right measures the absorbing power of the walls of the room.

brought into the room were placed on the floor, side by side, with edges touching, but that after a few observations had been taken the cushions were scattered about the room, and the work was repeated. This was done not at all to uncover the edges, but in the primitive uncertainty as to whether near cushions would draw from each other's supply of sound, as it were, and thus diminish each other's efficiency. No further thought was then given to these discarded observations until recalled by the above-mentioned discrepancy. They were sought out from the notes of that period, and it was found that, as suspected, the absorbing power of the cushions when touching edges was less than when separated. Eight cushions had been used, and, therefore, fourteen edges had been touching. A record was found of the length and the breadth of the cushions used, and, assuming that the absorbing power was proportional to the area exposed, it was possible to calculate their thickness by comparing the audible duration of the residual sound in the two sets of observations; it was thus calculated to be 7.4 centimeters. On stacking up the same cushions and measuring their total thickness, the average thickness was found to be 7.2 centimeters, in very close agreement with the thickness estimated from their absorption of sound. Therefore, the measurements of the cushions should be, not in running meters of cushion, but in square meters of exposed surface.

For the purposes of the present investigation, it is wholly unnecessary to distinguish between the transformation of the energy of the sound into heat and its transmission into outside space. Both shall be called absorption. The former is the special accomplishment of cushions, the latter of open windows. It is obvious, however, that if both cushions and windows are to be classed as absorbents, the open window, because the more universally accessible and the more permanent, is the better unit. The cushions, on the other hand, are by far the more convenient in practice, for it is possible only on very rare occasions to work accurately with the windows open, not at all in summer on account of night noises — the noise of crickets and other insects — and in the winter only when there is but the slightest wind; and further, but few rooms have sufficient window surface to produce the desired absorption. It is necessary, therefore, to work with cushions, but to express the results in open-window units.

Turning now to the unit into which the results are to be transformed, an especially quiet winter night was taken to determine whether the absorbing power of open windows is proportional to the area. A test of the absorbing power of seven windows, each 1.10 meters wide, when opened .20, .40, and .80 meter, gave results that are plotted in the diagram. The points, by falling in a straight line, show that, at least for moderate breadths, the absorbing power of open windows, as of cushions, is accurately proportional to the area. Experiments in several rooms especially convenient for the purpose determined the absorbing power of the cushions to

be .80 of that of an equal area of open windows. These cushions
were of hair, covered with canvas and light damask. "Elastic
Felt" cushions having been used during an investigation in a New
York church, it was necessary on returning to Cambridge to determine their absorbing power. This was accomplished through the
courtesy of the manufacturers, Messrs. Sperry & Beale, of New
York, and the absorbing power was found to be .73 of open-window

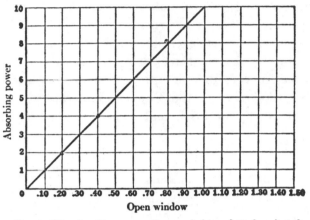

Fig. 7. The absorbing power of open windows plotted against the
areas of the openings, showing them to be proportional.

units — an interesting figure, since these cushions are of frequent
use and of standard character.

Hereafter all results, though ordinarily obtained by means of
cushions, will be expressed in terms of the absorbing power of open
windows — a unit as permanent, universally accessible, and as
nearly absolute as possible. In these units the total absorbing
power of the walls, ceiling, floor, windows and chairs in the lecture-
room of the Fogg Museum is 75.5.

Next in order is the apportionment of the total absorbing power
among the various components of the structure. Let s_1 be the area
of the plaster on tile, and a_1 its absorbing power per square meter;
s_2 and a_2 the corresponding values for the plaster on wire lath; s_3
and a_3 for window surface, etc. Then

$$a_1 s_1 + a_2 s_2 + a_3 s_3 + a_4 s_4, \text{ etc.} = 75.5,$$

s_1, s_2, s_3, etc., are known, and a_1, a_2, a_3, etc. — the coefficients of
absorption — are unknown, and are being sought. Similar equa-

tions may be obtained for other rooms in which the proportions of wall-surface of the various kinds are greatly different, until there are as many equations as there are unknown quantities. It is then possible by elimination to determine the absorbing power of the various materials used in construction.

Through the kindness of Professor Goodale, an excellent opportunity for securing some fundamentally interesting data was afforded by the new Botanical Laboratory and Greenhouse recently given to the University. These rooms — the office, the laboratory and the greenhouse — were exclusively finished in hard-pine sheathing, glass, and cement; the three rooms, fortunately, combined the three materials in very different proportions. They and the constant-temperature room in the Physical Laboratory — the latter being almost wholly of brick and cement — gave the following data:

	Area of Hard Pine Sheathing	Area of Glass	Area of Brick and Cement	Combined Absorbing Power
Office.........................	127.0	7	0	8.37
Laboratory....................	84.8	6	30	5.14
Greenhouse...................	12.7	80	85	4.64
Constant-temperature room.....	2.1	0	124	3.08

This table gives for the three components the following coefficients of absorption: hard pine sheathing .058, glass .024, brick set in cement .023.

APPROXIMATE SOLUTION

IN the preceding paper it was shown that the duration of the residual sound in a particular room was proportional inversely to the absorbing power of the bounding walls and the contained material, the law being expressed closely by the formula $(a + x)t = k$, the formula of a displaced rectangular hyperbola. In the present paper it is proposed to show that this formula is general, and applicable to any room; that in adapting it to different rooms it is only necessary to change the value of the constant of inverse proportionality k; that k is in turn proportional to the volume of the room, being equal to about .171V in the present experiments, but dependent on the initial intensity of the sound; and finally, that by substituting the value of k thus determined, and also the

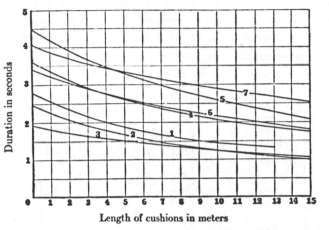

Fig. 8. Curves showing the relation of the duration of the residual sound to the added absorbing material, — rooms 1 to 7.

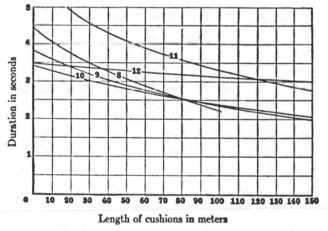

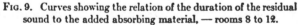

Fig. 9. Curves showing the relation of the duration of the residual sound to the added absorbing material, — rooms 8 to 12.

value of a, the absorbing power of the walls, and of x, the absorbing power of the furniture and audience, it is possible to calculate in advance of construction the duration of audibility of the residual sound.

The truth of the first proposition — the general applicability of the hyperbolic law of inverse proportionality — can be satisfactorily shown by a condensed statement of the results obtained from data collected early in the investigation. These observations were made in rooms varying extremely in size and shape, from a small committee-room to a theatre having a seating capacity for nearly fifteen hundred. Figures 8 and 9 give the curves experimentally determined, the duration of audibility of the residual

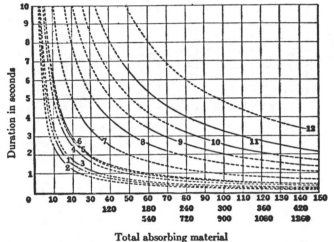

Total absorbing material

FIG. 10. The curves of Figs. 8 and 9 entered as parts of their corresponding rectangular hyperbolas. Three scales are employed for the volumes, by groups 1-7, 8-11, and 12.

sound being plotted against running meters of cushions. Two diagrams are given in order to employ a smaller scale for the larger rooms, this scale being one-tenth the other; and even in this way there is shown but one-quarter of the curve actually obtained in rooms numbered 11 and 12, the Fogg Art Museum lecture-room and Sanders Theatre. In Fig. 10 each curve is entered as a part of its corresponding hyperbola referred to its asymptotes as axes. In this case three scales are employed in order to show the details more clearly, the results obtained in rooms 1 to 7 on one scale, 8 to 11 on another, and 12 on a third, the three scales being proportional to one, three and nine. The continuous portions of the curves show the parts determined experimentally. Even with the scale

thus changed only a very small portion of the experimentally determined parts of curves 11 and 12 are shown. Figures 11 to 16,
inclusive, all drawn to the same scale, show the great variation in
size and shape of the rooms tested; and the accompanying notes
give for each the maximum departure and average departure of the
curve, experimentally determined, from the nearest true hyperbola.

1. Committee-room, University Hall; plaster on wood lath,
wood dado; volume, 65 cubic meters; original duration of residual
sound before the introduction of any cushions, 2.82 seconds; maxi-

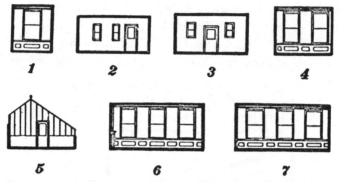

Fig. 11. 1. Committee-room. 2. Laboratory, Botanic Gardens. 3. Office,
Botanic Gardens. 4. Recorder's Office. 5. Greenhouse. 6. Dean's
Room. 7. Clerk's Room.

mum departure of experimentally determined curve from the nearest
hyperbola, .09 second; average departure, .03 second.

2. Laboratory, Botanic Gardens of Harvard University; hard
pine walls and ceiling, cement floor; volume, 82 cubic meters;
original duration of the residual sound, 2.39 seconds; maximum
departure from hyperbola, .09 second; average departure, .02
second.

3. Office, Botanic Gardens; hard pine walls, ceiling and floor;
volume, 99 cubic meters; original duration of residual sound, 1.91
seconds; maximum departure from hyperbola, .01 second; average
departure, .00 second.

4. Recorder's Office, University Hall; plaster on wood lath,
wood dado; volume, 102 cubic meters; original duration of residual
sound, 3.68 seconds; maximum departure from hyperbola, .10
second; average departure, .04 second.

5. Greenhouse, Botanic Gardens; glass roof and sides, cement floor; volume, 134 cubic meters; original duration of residual

FIG. 12. Faculty-room.

sound, 4.40 seconds; maximum departure from hyperbola, .08 second; average departure, .03 second.

6. Dean's Room, University Hall; plaster on wood lath, wood dado; volume, 166 cubic meters; original duration of residual

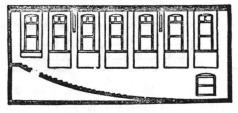

FIG. 13. Lecture-room.

sound, 3.38 seconds; maximum departure from hyperbola, .06 second; average departure, .01 second.

7. Clerk's Room, University Hall; plaster on wood lath, wood dado; volume, 221 cubic meters; original duration of residual

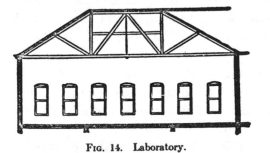

FIG. 14. Laboratory.

sound, 4.10 seconds; maximum departure from hyperbola, .10 second; average departure, .02 second.

8. Faculty-room, University Hall; plaster on wood lath, wood dado; volume, 1,480 cubic meters; original duration of residual sound, 7.04 seconds; maximum departure from hyperbola, .18 second; average departure, .08 second.

9. Lecture-room, Room 1, Jefferson Physical Laboratory; brick walls, plaster on wood lath ceiling; furnished; volume, 1,630 cubic meters; original duration of residual sound, 3.91

Fig. 15. Lecture-room.

seconds; maximum departure from hyperbola, .10 second; average departure, .04 second.

10. Large Laboratory, Room 41, Jefferson Physical Laboratory; brick walls, plaster on wood lath ceiling; furnished; volume, 1,960 cubic meters; original duration of residual sound, 3.40 seconds; maximum departure from hyperbola, .03 second; average departure, .01 second.

11. Lecture-room, Fogg Art Museum; plaster on tile walls, plaster on wire-lath ceiling; volume, 2,740 cubic meters; original duration of residual sound, 5.61 seconds; maximum departure from hyperbola, .04 second; average departure, .02 second. The experiments in this room were carried so far that the original duration of residual sound of 5.61 seconds was reduced to .75 second.

12. Sanders Theatre; plaster on wood lath, but with a great deal of hard-wood sheathing used in the interior finish; volume, 9,300 cubic meters; original duration of residual sound, 3.42

seconds; maximum departure from hyperbola, .07 second; average departure, .02 second.

It thus appears that the hyperbolic law of inverse proportionality holds under extremely diverse conditions in regard to the size, shape and material of the room. And as the cushions used in the calibration were placed about quite at random, it also appears that in rooms small or large, with high or low ceiling, with flat or curved

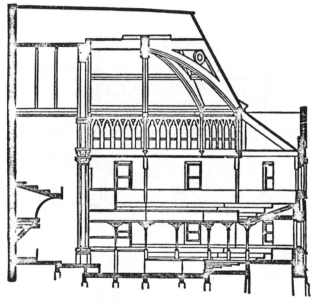

FIG. 16. Sanders Theatre.

walls or ceiling, even in rooms with galleries, the cushions, wherever placed — out from under the gallery, under, or in the gallery — are nearly equally efficacious as absorbents. This merely means, however, that the efficacy of an absorbent is independent of its position when the problem under consideration is that of reverberation, and that the sound, dispersed by regular and irregular reflection and by diffraction, is of nearly the same intensity at all parts of the room soon after the source has ceased; and it will be the object of a subsequent paper to show that in respect to the initial distribution of the sound, and also in respect to discrete echoes, the position of the absorbent is a matter of prime importance.

Having shown that the hyperbolic law is a general one, interest centers in the parameter, k, the constant for any one room, but varying from room to room, as the following table shows:

Room	Volume	Absorbing Power of Walls, etc., $= a$	Parameter k
1. Committee-room, University Hall...	65	4.76	13.6
2. Laboratory, Botanic Gardens........	82	4.65	11.1
3. Office, Botanic Gardens...........	99	8.08	15.4
4. Recorder's Office.................	102	5.91	21.8
5. Greenhouse, Botanic Gardens......	134	5.87	25.8
6. Dean's Room....................	166	7.50	25.4
7. Clerk's Room....................	221	10.6	43.5
8. Faculty-room...................	1,480	34.5	243.0
9. Lecture-room, Jefferson Physical Laboratory, 1.....................	1,630	69.0	270.0
10. Laboratory, Jefferson Physical Laboratory, 41....................	1,960	101.0	345.0
11. Fogg Lecture-room..............	2,740	75.0	425.0
12. Sanders Theatre.................	9,300	465.0	1,590.0

The values of the absorbing power, a, and the parameter, k, are here expressed, not in terms of the cushions actually used in the experiments, but in terms of the open-window units, shown to be preferable in the preceding article.

In the diagram, Figure 17, the values of k are plotted against the corresponding volumes of the rooms; here again three different scales are employed in order to magnify the results obtained in the smaller rooms. The resulting straight line shows that the value of k is proportional to the volume of the room, and it is to be observed that the largest room was nearly one hundred and fifty times larger than the smallest. By measurements of the coördinates of the line, or by averaging the results found in calculating $\dfrac{k}{V}$ for all the rooms it appears that $k = .171V$. The physical significance of this numerical magnitude .171 will be explained later.

This simple relationship between the value of k and the volume of the room — the rooms tested varying so greatly in size and shape — affords additional proof, by a rather delicate test, of the accuracy of the method of experimenting, for it shows that the experimentally determined curves approximate not merely to hyperbolas but to a systematic family of hyperbolas. It also furnishes a more pleasing prospect, for the laborious handling of cushions will be unnecessary. A single experiment in a room and a knowledge of the volume of the room will furnish sufficient data for the calculation of the absorbing power of its components. Conversely, a knowledge of the volume of a room and of the coefficients of absorption of its various components, including the audience for which it is designed, will enable one to calculate in advance of construction the duration of audibility of the residual sound, which measures

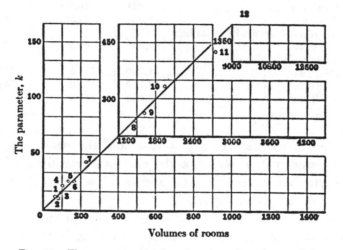

FIG. 17.   The parameter, k, plotted against the volumes of the
rooms, showing the two proportional.

that acoustical property of a room commonly called reverberation. Therefore, this phase of the problem is solved to a first approximation.

The explanation of the fact that k is proportional to V is found in the following reasoning. Consider two rooms, constructed of exactly the same materials, similar in relative proportions, but one larger than the other. The rooms being empty, x, the absorbing power of the contained material, is zero, and we have $a' \, t' = k'$ and $a'' \, t'' = k''$. Since the rooms are constructed of the same materials the coefficients of absorption are the same, so that a' and a'' are proportional to the surfaces of the rooms, that is, to the squares of the linear dimensions. Also, the residual sound is diminished a certain percentage at each reflection, and the more frequent these reflections are the shorter is the duration of its audibility; whence t' and t'' are inversely proportional to the frequency of the reflections, and hence directly proportional to the linear dimensions. Therefore, k' and k'', which are equal to $a' t'$ and $a'' t''$, are proportional to the cubes of the linear dimensions, and hence to the volumes of the rooms.

Further, when the shape of the room varies, the volume remaining the same, the number of reflections per second will vary. Therefore, k is a function not merely of the volume, but also of the shape of the room. But that it is only a slightly varying function, comparatively, of the shape of the room for practical cases, is shown by the fact that the points fall so near the straight line that averages the values of the ratio $\dfrac{k}{V}$.

The value of k is also a function of the initial intensity of the sound; but the consideration of this element will be taken up in a following paper.

RATE OF DECAY OF RESIDUAL SOUND

IN a subsequent discussion of the interference of sound it will be shown by photographs that the residual sound at any one point in the room as it dies away passes through maxima and minima, in many cases beginning to rise in intensity immediately after the source has ceased; and that these maxima and minima succeed each other in a far from simple manner as the interference system shifts. On this account it is quite impossible to use any of the numerous direct methods of measuring sound in experiments on reverberation. Or, rather, if such methods were used the results would be a mass of data extremely difficult to interpret. It was for this reason that attempts in this direction were abandoned early in the investigation, and the method already described adopted. In addition to the fact that this method only is feasible, it has the advantage of making the measurements directly in terms of those units with which one is here concerned — the minimum audible intensity. It is now proposed to extend this method to the determination of the rate of decay of the average intensity of sound in the room, and to the determination of the intensity of the initial sound, and thence to the determination of the mean free path between reflections, — all in preparation for the more exact solution of the problem.

The first careful experiment on the absolute rate of decay was in the lecture-room of the Boston Public Library, a large room,

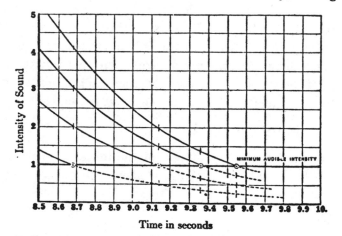

FIG. 18.   Decay of sound in the lecture-room of the Boston Public Library from the initial sound of one, two, three, and four organ pipes, showing only the last second.

finished, with the exception of the platform, in material of very slight absorbing power — tile ceiling, plaster on tile walls, and polished cement floor.[1]   The reverberation was very great, 8.69 seconds.   On the platform were placed four organ pipes, all of the same pitch, each on its own tank or wind supply, and each having its own electro-pneumatic valve.   All these valves, however, were connected to one chronograph, key, and battery, so that one, two, three, or all the pipes, might be started and stopped at once, and when less than four were in use any desired combination could be made.   One pipe was sounded and the duration of audibility of the residual sound determined, of course, as always in these experiments, by repeated observations.   The experiment was then made

[1] Terrazzo cement floor.

450

with two organ pipes instead of one; then with three pipes; and, finally, with four. The whole series was then repeated, but beginning with a different pipe and combining different pipes for the two and three pipe sets. In this way the series was repeated four times, the combinations being so made that each pipe was given an equal weight in the determination of the duration of audibility of the residual sound under the four different conditions. It is safe to assume that with experiments conducted in this manner the average initial intensities of the sound with one, two, three, and four pipes were to each other as one, two, three and four. The corresponding durations of audibility shall be called t_1, t_2, t_3 and t_4. The following results were obtained:

$t_1 = 8.69$ seconds $\qquad$ $t_2 - t_1 = .45$ second

$t_2 = 9.14$ " $\qquad$ $t_3 - t_1 = .67$ "

$t_3 = 9.36$ " $\qquad$ $t_4 - t_1 = .86$ "

$t_4 = 9.55$ "

It is first to be observed that the difference for one and two organ pipes, .45, is, within two-hundredths of a second, half that for one and four organ pipes, .86. This suggests that the difference is proportional to the logarithm of the initial intensity; and further inspection shows that the intermediate result with three organ pipes, .67, is even more nearly, in fact well within a hundredth of a second, proportional to the logarithm of three. This reënforces the very natural conception that however much the residual sound at any one point in the room may fluctuate, passing through maxima and minima, the average intensity of sound in the room dies away logarithmically. Thus, if one plots the last part of the residual sound — that which remains after eight seconds have elapsed — on the assumption that the intensity of the sound at any instant is proportional to the initial intensity, the result will be as shown in the diagram, Fig. 18. The point at which the diminishing sound crosses the line of minimum audibility in each of the four cases is known, the corresponding ordinates of the other curves being multiples or submultiples in proportion to the initial intensity. The results are obviously logarithmic.

Let I_1 be the average intensity of the steady sound in the room when the single organ pipe is sounding, i the intensity at any instant during the decay, say t seconds after the pipe has ceased, then $-\dfrac{di}{dt}$ will be the rate of decay of the sound, and since the absorption of sound is proportional to the intensity

$$-\frac{di}{dt} = Ai, \text{ where } A \text{ is the constant of proportionality,}$$

the ratio of the rate of decay of the residual sound to the intensity at the instant.

$$- \log_e i + C = At,$$

a result that is in accord with the above experiments. The constant of integration C may be determined by the fact that when t is zero i is equal to I_1; whence

$$C = \log_e I_1, \text{ and the above equation becomes}$$

$$\log_e \frac{I_1}{i} = At.$$

At the instant of minimum audibility t is equal to t_1, the whole duration of the residual sound, and i is equal to i', — as the intensity of the least audible sound will hereafter be denoted. Therefore

$$\log_e \frac{I_1}{i'} = At_1.$$

This applied to the experiment with two, three and four pipes gives similar equations of the form

$$\log_e \frac{nI_1}{i'} = At_n,$$

where n is the number of organ pipes in use. By the elimination of $\dfrac{I_1}{i'}$ from these equations by pairing the first with each of the others,

$$A = \frac{\log_e 2}{t_2 - t_1} = 1.54,$$

$$A = \frac{\log_e 3}{t_3 - t_1} = 1.62,$$

$$A = \frac{\log_e 4}{t_4 - t_1} = 1.61,$$

$$A(\text{average}) = \overline{1.59},$$

where A is the ratio between the rate of decay and the average intensity at any instant.

It is possible also to determine the initial intensity, I_1, in terms of the minimum audible intensity, i'.

$$log_e \frac{I_1}{i'} = At_1,$$

$$I_1 = i' \ log_e^{-1} \ At_1 = i' \ log_e^{-1} \ (1.59 \times 8.69) = 1,000,000 \ i'.$$

With this value of the initial intensity it is possible to calculate the intensity i of the residual sound at any instant during the decay, by the formula

$$log_e \ I_1 - log_e \ i = At,$$

and the result when plotted is shown in **Figure 19**, the unit of intensity being minimum audibility.

A practical trial early in the year had shown that it would be impossible to use this lecture-room as an auditorium, and the ex-

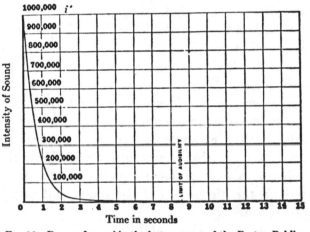

Fig. 19. Decay of sound in the lecture-room of the Boston Public Library beginning immediately after the cessation of one organ pipe.

periments described above, with others, were in anticipation of changes designed to remedy the difficulty. Hair felt, in considerable quantities, was placed on the rear wall. The experiments with the four organ pipes were then repeated and the following results were obtained:

$$t_1 = 3.65 \qquad t_2 - t_1 = .20 \qquad \therefore A = 3.41$$
$$t_2 = 3.85 \qquad t_3 - t_1 = .31 \qquad \therefore A = 3.54$$
$$t_3 = 3.96 \qquad t_4 - t_1 = .42 \qquad \therefore A = \overline{3.29}$$
$$t_4 = 4.07 \qquad\qquad\qquad\qquad\quad A = 3.41 \text{ (average)}$$
$$I_1 = 250,000 \ i'$$

A few nights later the apparatus was moved down to the attendant's reception-room near the main entrance — a small room but similar in proportions to the lecture-room. Here a careful experiment extending over several nights was carried on, and it gave the following results:

$$t_1 = 4.01 \qquad t_2 - t_1 = .19 \qquad \therefore A = 3.65$$
$$t_2 = 4.20 \qquad t_3 - t_1 = .28 \qquad \therefore A = 3.90$$
$$t_3 = 4.29 \qquad t_4 - t_1 = .37 \qquad \therefore A = \overline{3.75}$$
$$t_4 = 4.38 \qquad\qquad\qquad\qquad A = 3.76 \text{ (average)}$$
$$I_1 = 3,800,000 \; i'$$

The first interest lies in an attempt to connect the rate of decay, obtained by means of the four organ pipe experiments, with the absolute coefficient of absorption of the walls, obtained by the experiments with the open and closed windows; and to this end recourse will be had to what shall here be called "the mean free path between reflections." The residual sound in its process of decay travels across the room from wall to wall, or ceiling, or floor, in all conceivable directions; some paths are the whole length of the room, some even longer, from one corner to the opposite, but in the main the free path between reflections is less, becoming even infinitesimally small at an angle or a corner. Between the two or three hundred reflections that occur during its audibility the residual sound establishes an average distance between reflections that depends merely on the dimensions of the room, and may be called "its mean free path."

$$a = \frac{.171 \, V}{t_1}$$

is the absorbing power of the room, measured in open-window units.
Let

s = surface.
V = volume.
A = rate of decay of the sound.
v = velocity of sound, 342 m. per second at 20 degrees C.
p = length of the mean free path between reflections.

Whence $\frac{v}{p}$ = the average number of reflections per second, and

$\frac{a}{s}$ is the fraction absorbed at each reflection, $\frac{av}{sp} = A$, and $p = \dfrac{av}{As} = \dfrac{v.171\ V}{As\ t_1}$, whence may be calculated the mean free path, p.

	V	A	s	t	p
Boston Public Library Lecture-room, bare	2,140.0	1.59	1,160	8.69	7.8
" " " " with felt ..	2,140.0	3.41	1,160	3.65	8.8
" " " Attendant's Room......	63.8	3.76	108	4.01	2.27

The length of the mean free path in the lecture-room, bare or draped, ought to be the same, for the felt was placed out from the wall at a distance imperceptibly small in comparison with the dimensions of the room; but 7.8 and 8.8 differ more than the experimental errors justify. Again, the attendant's room had very nearly the same relative proportions as the lecture-room (about 2 : 3 : 6), but each linear dimension reduced in the ratio 3.22 : 1. The mean free path, obviously, should be in the same ratio; but when the mean free path in the attendant's room, 2.27, is multiplied by 3.22 it gives 7.35, departing again from the other values, 7.8 and 8.8, more than experimental errors justify. The explanation of this is to be found in the fact that the initial intensity of the sound in the rooms for the determination of t_1 was not the same but had the values respectively, 1,000,000 i', 250,000 i' and 3,800,000 i'. Since t_1 has been shown proportional to the logarithms of the initial intensities, these three numbers, 7.8, 8.8 and 7.35, may be corrected in an obvious manner, and reduced to the comparable values they would have had if the initial intensity had been the same in all three cases. The results of this reduction are 7.8, 8.0 and 8.0, a satisfactory agreement.

The length of the mean free path is, therefore, as was to be expected, proportional to the linear dimensions of the room, and such a comparison is interesting. There is no more reason, however, for comparing it with one dimension than another. Moreover, most rooms in regard to which the inquiry might be made are too irregular in shape to admit of any one actual distance being taken as standard. Thus, in a semicircular room, still more in a horseshoe-shaped room such as the common theatre, it is indeterminable what should be called the breadth or what the length. On account, therefore, of the complicated nature of practical conditions one is forced to the adoption of an ideal dimension, the cube root of the volume, $V^{1/3}$, the length of one side of a cubical room of the same capacity. The above data give as the ratio of $\frac{p}{V^{1/3}}$ the value, .62.

It now becomes possible to present the subject by exact analysis, and free from approximations; but before doing so it will be well to review from this new standpoint that which has already been done.

It was obvious from the beginning, even in deducing the hyperbolic law, that some account should be taken of the reduction in the initial intensity of the sound as more and more absorbing material was brought into the room, even when the source of sound remained unchanged. Thus each succeeding value of the duration of the residual sound was less as more and more absorbing material was brought into the room, not merely because the rate of decay was greater, but also because the initial intensity was less. Had the initial intensity in some way been kept up to the same value throughout the series, the resulting curve would have been an exact hyperbola. As it was, however, the curve sloped a little more rapidly on account of the additional reduction in the duration arising from the reduction in initial intensity of the sound. At the time, there was no way to make allowance for this. That it was a very small error, however, is shown by the fact that the departures from the true hyperbola that were tabulated are so small.

Turning now to the parameter, k, it is evident that this also was an approximation, though a close one. In the first place, as just explained, the experimental curve of calibration sloped a little more rapidly than the true hyperbola. It follows that the nearest hyperbola fitting the actual experimental results was always of a little too small parameter. Further, k depended not merely on the uniformity of the initial intensity during the calibration of the room, but also on the absolute value of this intensity. Thus, $k = at_1$, and t is in turn proportional to the logarithm of the initial intensity. Therefore in order to fully define k we must adopt some standard of initial intensity. For this purpose we shall hereafter take as the standard condition in initial intensity, $I = 1,000,000 \, i'$, $(I = 10^6 \, i')$, where i' is the minimum audible intensity, as this is the nearest round number to the average intensity prevailing during these experiments. If, therefore, during the preceding experiments the initial intensity was above the standard, the value deduced for k would be a little high, if below standard, a little low. This variation of the parameter, k, would be slight ordinarily, for k is proportional to the logarithm, not directly to the value of the initial intensity. Slight ordinarily, but not always. Attention was first directed to its practical importance early in the whole investigation by an experiment in the dining-room of Memorial Hall — a very large room of 17,000 cubic meters capacity. During some experiments in Sanders Theatre the organ pipe was moved across to this dining-room, and an experiment begun. The reverberation was of very short duration, although it would have been long had the initial intensity been standard, for in rooms constructed of similar materials the reverberation is approximately proportional to the cube roots of the volumes. There was no opportunity to carry the experiment farther than to observe the fact that the duration was surprisingly short, for the frightened appearance of the women from the sleeping-rooms at the top of the hall put an end to the experiment. Finally, k is a function not merely of the volume but also of the shape of the room; that is to say, of the mean free path, as has already been explained.

It was early recognized that with a constant source the average intensity of the sound in different rooms varies with variations in size and construction, and that proper allowance should be made therefor. The above results call renewed attention to this, and point the way. In the following paper the more exact analysis will be given and applied.

Author Citation Index

Subject Index